TIMES, SPACES, AND PLACES

For Olga

and our children,
Mandy, Jane, and Sally

For Lynda and Victoria

Times, spaces, and places

A Chronogeographic Perspective

Don Parkes
University of Newcastle
New South Wales, Australia

Nigel Thrift
Australian National University
Canberra, Australia

JOHN WILEY & SONS
Chichester · New York · Brisbane · Toronto

British Library Cataloguing in Publication Data:

Parkes, Don
   Times, spaces, and places: A chronogeographic
   perspective.
   1. Anthropo-geography
   2. Time
   I. Title   II. Thrift, Nigel

909	GF50	79-40523

   ISBN 0 471 27616 2 (Cloth)
   ISBN 0 471 27617 0 (Paper)

Typeset by Photo-Graphics, Yarcombe, Honiton, Devon.
Printed by the Pitman Press, Bath, Avon.

Contents

viii

Acknowledgements

Preparation of this book began in 1975 and for three years, until January 1978, the postal services of Australia and the United Kingdom were recipients of substantial payments to deliver our arguments and the countless letters which began, 'What do you think of this idea?'. Of course by the time the letter was delivered what seemed earth-shaking had transformed into just another piece of paper to be screwed up or at best had become a notion that someone else had already thought through, usually much more clearly than we had. To all these people from disciplines as diverse as chronobiology, psychology, sociology, physics, and philosophy we acknowledge a debt. Many of these workers are referred to by name through our references and their permission for us to reproduce maps, diagrams, tables, and other material. To all human geographers we also owe a debt of gratitude, for theirs' is the discipline to which we belong and but for it there would be no book, no ideas; good, bad, or indifferent.

To John Wiley (U.K.) Ltd., and especially Michael Coombs, Celia J.C. Bird and Eileen Wimshurst, our thanks. Without their courage in contracting this book, with its neologistic title and unproven marketability, we would not have had the same confidence to persist. Thank you all very much indeed.

To staff at the Universities of Cambridge (Martin Centre for Architectural and Urban Studies) and Leeds (School of Geography), Nigel Thrift extends his thanks. At Leeds, especially during 1978, Gordon Bryant, John Dickson, and Pamela Talbot gave a lot of time and effort to preparation of material.

To the staff at the University of Newcastle-upon-Tyne where Don Parkes was formerly an undergraduate and where he was a guest of the Department for 12 months in 1974 and for 15 months in 1977-1979, while preparing the final stages of this book, thank you. Especial thanks go to Professor J.W. House for his hospitality, way back in 1974 and in the recent period, to Professor

E.S. Simpson for making so many facilities available; also to Professor John Goddard and his colleagues at the Centre for Urban and Regional Development Studies. John Knipe, Eric Quenet, and Doreen Shanks prepared diagrams and photographs and Norris Riley was a most 'timely', interested and interesting, glaciologist; thank you Norris.

The typing of the final manuscript was undertaken by Mrs. Olga Parkes. She was not always a willing worker; but she was always a very good one! We both thank you, very much.

A final note of thanks must go to the undergraduate students at Newcastle (Australia) who sat through trial runs of the book as a third year elective course in 1977 as well as Ken Lee (Senior Tutor in Geography) and Paul Tranter (Commonwealth Postgraduate Scholar at Newcastle, Australia). All of them have contributed much more than they realize. For typing many early drafts and bibliographic filing cards, thank you Jan Taylor and Val Wiggins at Newcastle, N.S.W.

Preface

Our hope is that this book will help to place time firmly in the minds of human geographers. We sense that, at present, there is something of an impasse in theory development and in the generation of new ideas in human geography. Are human geographers simply running out of new worlds to read and write about? We think not, because new worlds can be discovered when time is given an explicit treatment and when it is *always* considered commensurately with space, or territory.

Solutions to the problems which beset individuals and small groups, as well as the 'larger' issues at aggregate and global scales will continue to elude us unless due consideration is given to the role that social uses of time have had in their creation. While time has not always been an integral part of human geography, it has always played some part in human use of space. It is a principal component in place.

Of course, situations do exist in which a particular interest is only marginally enhanced by a search for the place of time in the issue at hand. But a necessary step in the procedural process of human geographic enquiry should be to establish this. Our perspectives of the world are narrowed unnecessarily if time, in its many facets, is not a part of our vision. We run the risk of conceptualizing a world which, in some mysterious and ideological way, operates *'over time'*. How often we read or hear the phrase that 'such and such has changed over time'. Yet we might as well say, 'It has changed over change'. For such a superficial statement we would substitute an alternative vision of 'a world of times' in which change takes place at many time scales and in many times. Changes are times.

This book adopts a *chronogeographical* perspective. Such a perspective acknowledges that time is many-faceted. It is predicate on the view that it is useful to reserve the more obvious term *time-geography* for the human geographic paradigm recently developed at Lund University, Sweden, which only concerns itself with the familiar time of the clock and the calendar. The term *chronosophy* has already been suggested by the physicist J. T. Fraser (1968) for the interdisciplinary study of time *per se*. Other facets of time, which we are going to call *paratimes,* have however been espoused in the social and biological science literature. They are all a part of the phenomenological, structural and dialectical discourse (Bourdieu, 1976) of human geography. An approach can, therefore, be usefully described as 'chronogeographical' if there is an attempt to determine the temporal and paratemporal constituents of a particular human geographical situation.

However, it should be made clear that there are a number of objectives which this book does not set out to achieve. Firstly we do *not* claim that this book embodies any single new paradigm. Rather it is intended to give food for thought to the supporters of existing paradigms as to the role of time in their interpretation of the human world. At best we hope that it can extend these interpretations. Secondly the book is *not* an all-embracing prospectus of the temporalities which might be involved in every possible human geographical subject area; historical geography is an obvious example that is omitted (Prince, 1978). The burgeoning volume of published material on time in economics and anthropology also receives only brief mention.

In order to allow some level of continuity we have tended to use urban and especially social geographic source material. This has allowed us to develop an argument which might otherwise have been dissipated by a more catholic range of examples. But, on the other hand, *this book is not an introduction to urban and social geography* using a new visiting card. It is an introduction to ideas about time for human geographers.

Thirdly the book is *not* an introduction to human geography, in the sense that it is **elementary** and fully integrated with existing paradigms. We feel confident, however, that it can be of use at undergraduate level, particularly in second or third year stages. Indeed, in its earliest manuscript form the book was used in this way with considerable success. As an undergraduate course book it should service existing systematic courses but would also be of sufficient substance to find a niche of its own as a short elective or optional lecture sequence on Time in Human Geography, or Chronogeography. There are other texts on the market now which we feel will complement this book rather well and they are referenced in the chapters that follow. But apart from the interest which it may have for undergraduates we feel that it will also have a substantial and perhaps more important role as a postgraduate reference volume. For the postgraduate student, we hope that there are enough ideas in the book to stimulate research programmes, either theoretical or empirical. Perhaps the stimulation of empirical work is most important because some of the concepts and principles which are discussed, especially in Part I, need more exposure to the harsh light of empirical verification. The fairly extensive bibliography should also prove useful in this task.

How is the book organized and what is the content of each of the chapters? The book has two major parts. The first three chapters which form Part I introduce the framework of *concepts and principles* which we have found to be useful in arranging our thoughts about time and its possible positions in human geography. Many of them are taken up from time to time in Part II. *Chapter One* therefore introduces ideas about *locations* in space which will all be very familiar but also introduces the notion of *locations* in time. This is followed by a very brief overview of the different philosophic conceptualisations of time. Some of the basic concepts and terms which have been used to describe and explain the characteristics of temporal phenomena are then outlined and the chapter concludes with a somewhat phenomenological perspective on the way in which spaces and times combine to evoke the reality of *place*.

Chapter Two introduces the concept of many-faceted time, first as *universe time,* the familiar 'voice of time' recorded by calendars and clocks. The more difficult concept of *paratimes* is then proposed. Paratimes are the times *beyond* the normal *everyday time* of the clock and calendar. We begin with the internalized paratimes of human life which go under the general category of *life-times*. They fall into two subcategories for our purposes, biological (or somatic) time and psychological time. Both are relevant if we are to improve our understanding of human behaviour and spatial organisation. The second major category of paratimes contains *social times* — the different social uses of time at different scales and in different social formations. The conceptual separation of times and paratimes and their subcategories does however seem to enable the construction of useful principles. This long chapter is particularly important within the context of the book because by bringing together the various times which scholars from diverse disciplines have been studying and by distinguishing them in a rather straightforward way, we hope to show the breadth of the time perspectives available to human geography. On the basis of the concepts and principles contained in this chapter alone, there lies a whole new set of worlds for human geographers to explore. *Chapter Three* considers ways in which notions of space and time are interrelated. How do they operate on one another in the making of human *societies* so that an acceptable level of spatial and temporal coherence (or agreement) is achieved among the system elements?

Part Two has six chapters in it. This part of the book can be thought of as being about *practice*. To the degree that it is possible, with a rather new set of concepts and principles (Part II), we have tried to organize these chapters around examples from published work in geography and other social sciences which seem to us to have (1) an inherently temporal perspective and which thereby contribute to the better understanding of spatial isues and (2) which allow at least some of the concepts and principles of Part One to be identified in real world circumstances.

Chapters Four and *Five* deal with human activity in the framework of time budget analysis. First of all in *Chapter Four* we discuss some aspects of the difficult notion of time allocation and this brings us, albeit briefly, into contact with economics and the problems of scarcity and its allocation mechanism. Time budget accounting methods are then introduced with a number of examples. *Chapter Five* tackles the more difficult topic of trying to explain human activity. We approach this task on the shoulders of the members of three important 'schools' of human activity theory, and of their published empirical results. As with *Chapter Four* we conclude with some examples from recent studies from Britain, the U.S.A., and Australia.

Chapter Six is called *Time-geography: The Lund approach.* Our view is that the contributions which have been made to human geography by the research team under the direction of Torstein Hägerstrand stand as perhaps the only inherently geographic paradigm to have been developed within the discipline. At one and the same time, time-geography is disarmingly simple and inspiringly powerful. The key notions and their dynamic map representation of the possibilities for human action are introduced and some examples of the time-geographic paradigm 'in practice' are also given. These examples vary in their scale and complexity.

Chapters Seven, Eight and *Nine* have a rather different flavour to the first three chapters in this part of the book. They differ not only in their content but especially in the larger time and space scales to which they are addressed. *Chapter Seven* is about movement and begins with a brief and no doubt familiar discussion of diffusion. To have omitted a reference to diffusion processes would have been to have appeared to overlook a topic which has been of paramount interest amongst human geographers. Two other movement categories to which we then draw attention are 'convergence-divergence' and 'periodic market' systems. Each of these, in its way, can be associated with concepts introduced in *Chapters One, Two,* and *Three,* and this is especially true with regard to the ideas of *time spacing* and *space timing* discussed in the third chapter.

Chapter Eight focuses on certain behaviours in and of the city, in a 24 hour period. The emphasis is essentially ecological. Examples from recent empirical work are taken from the United States and Australia. From the *small time* of *Chapter Eight* where concern is very much with the behaviour of the city 'as people' we turn, in *Chapter Nine,* to the *big time* of the built fabric of the city. The concepts of place first discussed in *Chapter One* are reintroduced in the first section. Now, however, it is the impressions of the more distant, historical past and the persistence of the present built environment, into the future which are considered. This is followed by a demonstration of how the built fabric of the city is laid down by successive waves or 'cycles' of economic prosperity and depression and attention is drawn to the growing extent to which international economic buoyancy is a factor in the nature of the townscapes we occupy. Building cycles themselves are then considered and their relation to the spatial characteristics of selected cities based on recent work by human geographers in the United States.

An appendix concludes the book. This is intended as a guide to analytical methods. However it is *not* intended to be a sufficient introduction to the methods discussed. Rather it should be viewed as an expanded glossary which is intended for consultation whenever this is felt to be necessary in reading the chapters. All of the methods included are fully referenced to many excellent texts, which will provide adequate instruction to some of the more difficult techniques which have been used or referred to from time to time.

The time elements of human geography are no new discovery. For instance, Whittlesey (1945) and Ogilvie (1953) both drew attention to time as a key element of the subject, but it is only recently that a flowering of interest has taken place. Our hope is that this book proves equal to the task of capturing this interest and that time flies as it is read!

PART ONE

Concepts and principles

1
Location, time and place

Have you read the Preface? If not, please try and find time to do so before reading on.

1.1 Preliminaries

The study of human geography usually begins with the notion of location in space and the distinction between absolute and relative location is one of the first principles to be learned. Location means position and may refer to the position of items[1] of any sort at all, tangible, or intangible.

Location absolute and relative
Position Items

At an introductory level the word space is usually treated simply as an alternative expression for territory or region, if it is used at all. At more advanced levels of study the notions of social and economic space are encountered. Here again a distinction between absolute and relative location is made. Social and economic spaces differ from territorial or universal space, however, because they may be represented by more than three dimensions or axes. They are sometimes described as *N*-dimensional. We shall describe them as paraspaces, to distinguish them from territorial, physical or universal space and also in order to imply their conceptual equivalence to certain aspects of time which are discussed in Chapter 2.

Universal space

Paraspace

Location in time, however, has not been given the same significance. Possibly this is because we have not really tried to come to grips with what time is and how the items which are of consequence to our daily lives are positioned in it, and in their own changing item relations, to produce specific times; in much the same way as items in relation produce specific spaces. The absolute spatial location of items may be determined with reference to grid coordinate lines based on survey or by reference to some arbitrary

3

grid such as that used on local area street maps. But the absolute location of items may also refer to location in time derived from a clock or calendar. Here clock time acts as a sort of grid and items are put into it. Additionally, however, by considering the position of an item in time in relation to the position of another item in time or relative to some other location in clock time, the concept of relative location is derived.

Distance

Interval

The separation of two items in space may be described by the distance between them and the separation of two items in time by the interval between them. When spatial metrics such as metres or kilometres are used to measure distance we have a measure of absolute distance. If temporal metrics such as hours or days are used to measure interval we have a measure of absolute interval. However, when an aspatial metric is used to indicate distance and an atemporal metric is used to indicate time, for instance money, then distance and interval are being represented in relative terms, as relative distance and relative time.

Space-time

One of the most common relative space measures combines space with time, as distance with interval. Thus in everyday life we consider the time it takes to get somewhere. This notion of distance and interval in combination is now frequently referred to as a space-time metric.[2] Geographic space-time has nothing at all to do with the space-time of philosophy, mathematics, and physics of the 20th century. In geography space-time is akin to the space and time of Newton where, of course, the dimensions of space and the dimension of time may be joined quite legitimately into four dimensional notions, as in motion. The geographer's space-time is not a new physical structure, as is the four dimensional space-time of Minkowski or Einstein, instead it is a technical convenience and a more realistic way of looking at the world.

Motion

Space-time for geography may be illustrated with some simple diagrams; though more complex concepts will be introduced in later chapters. For the sake of simplicity we begin by showing space x as a single axis and space y as the second axis which together delimit two-dimensional space; Figure 1.1(a). Considering an item (or event) in whatever sense you wish, it may be located on a space-space diagram as below.

In Figure 1.1(b) the same items or events may be located on a space-time map. The y axis is now representing changing positions (locations) in time, the x axis provides a basis for locating items in a spatial (usually territorial) direction.

Velocity

Both maps illustrate the fact that movement has occurred. On the space-space map, however, no idea of speed or velocity can be inferred, only the general direction of movement of an item i from a location at i_1 to a location at i_2. But in a space-time map this additional information can be estimated from the interval between two locations displayed by the time axis and the distance between them displayed by the space axis. Because only one space axis is

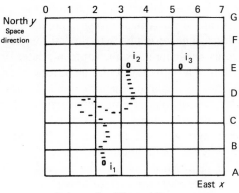

North y
Space direction

East x

Scale : each cell is one kilometre square

(a)

Figure 1.1(a). Location in space-space. i is any item or event, i_1--------i_2 is a route. As relative location i_2 is north and east of i_1. As absolute location i_3 is in cell 5E

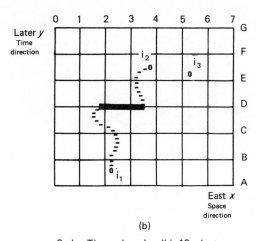

Later y
Time direction

East x
Space direction

(b)

Scale : Time axis each cell is 10 minutes
Space axis each cell is 1 kilometre

Figure 1.1(b). Location in space-time. i is any item or event, i_1--------i_2 is a path which began at i_1 and ended at i_2, but not that unlike in the space-space map we cannot trace a path backward on the y axis. The thick black line at time D represents instantaneous movement from nearly 2 to just east of 3. As relative location i_2 was east and later than i_1. As absolute location i_3 is in space-time cell 5E

6

shown, however, the path which is drawn on the space-time map in 1.1(b) represents a movement along a single line, in other words when the dotted line curves back to the left (or the west) the item is represented as retracing the same route at a later time.

This notion of a path describing the history of an event in space-time is well established in mathematics and physics, for instance to describe the 'contortions of a moving particle' (Davies, 1977, 53). Such a 'physicalist' scheme was taken up by Hägerstrand about fifteen years ago but was little known by human geographers outside of Sweden until about 1970 (cf. Chapter 6 following). Rather than representing space on a single axis, and thereby limiting the representation of movement to that along a line, as illustrated in 1.1(b), the space dimension may be represented by two axes, which is the way it is usually represented on conventional maps. Figure 1.2 illustrates such a space-time scheme, with two dimensions on the space axis and a single vertical time axis.[3]

Path

History

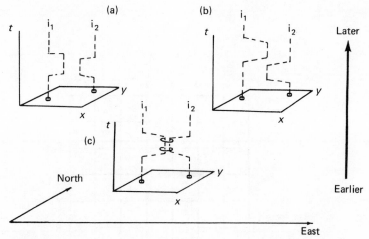

Figure 1.2. Location principles in 3-dimensional 'dynamic maps'
(a) The path of two items that never meet.
(b) The path of two items that occupy the same location at different times.
(c) The paths of two items that meet for an interval. (Based on Carlstein, 1974)

Biography

Geometric and thematic continuity

Succession

The path of an item describes the history or biography of the item. With time constant, i.e. movement is only on the horizontal axis, two items which are located on the same line are instantly located, as by telephone link; they are simultaneous. Of course in the strictest sense we cannot have 'movement' without any passage of time, but for our introductory purposes this should not be misleading to you. The distance between these items provides a thematic spatial continuity, but there is not necessarily any geometric continuity. If two items occur one after the other at the same location, there is succession. Thematic continuity in the time dimension is duration, and usually describes the persistence of an item at a single absolute location in space, though a journey in a car will include geometric and thematic continuity with duration. **Duration**

A simple illustration will clarify these additional features of the space-time map, Figure 1.3 below.

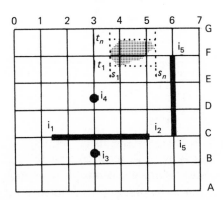

Scale : Time axis each cell is 10 minutes
Space axis each cell is 1 kilometre

 $\cdots t_n$ The temporal properties of events are sequence and duration

t_1 The spatial properties of events are juxtaposition and area

s_1 s_n

Figure 1.3. Constant time and constant space locations.

$i_1 \rightarrow i_2$ are simultaneous: time constant, space distant.

$i_3 \rightarrow i_4$ are sequential: space constant, time interval.

$i_5 \rightarrow i_5$ duration of a persistent item at constant space

In order to maintain orientation we need space and time dimensions (Abler, Adams, and Gould, 1971), and the simple schemes above have illustrated that description by spatial location alone is rather restrictive. Furthermore, whereas the concept of space has been handled in dimensions other than length, width, and height (or depth), by urban and social geographers, the time axis has not been given such attention. Social space is perhaps the most familiar example in urban and social geography and a social space diagram for the most pervasive of these dimensions has been represented as follows in Figure 1.4.

The time dimension, often acknowledged to be an essential component in geography, nonetheless seems to have been treated as a self-evident, axiomatic truth requiring no further intellectual

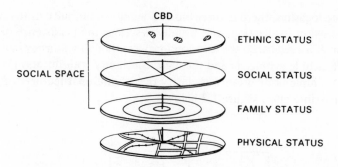

Figure 1.4. Social space-physical space. Represented in this manner there is no linking argument. A view of time which allows for it to be many-faceted provides such an argument (cf. Chapter 8). (After Murdie, 1969)

consideration. This is rather surprising. Perhaps we are satisfied that we know full well what we mean by it: we can tell the time and that is all there is to it. But one cannot help feeling uneasy about such a glib viewpoint. The philosopher Whitehead considered that 'it was impossible to meditate on time... without overwhelming emotion at the limitation of human intelligence' (1920, 73). This suggests that as human geographers we are left with one or other alternative: time does not require our 'meditation', or it does. The latter would seem to be the argument to accept!

Human geography is wrapped in the study of space and we have noted that the first principles of item location have tended to have a spatial emphasis. In some cases this emphasis has excluded location in time completely and yet this seems to be an untenable situation. The meaning of location in space is always bound up with threads of time. To get at the description and explanation of the items that occur in space these threads have to be untied carefully. We cannot simply cut our way through this Gordian knot. Time is not an easy notion to come to grips with. Consider the question that Saint Augustine put to himself, and the only answer he was able to give.

'What is time? If someone asks me I seem to know. But if someone asks me to explain what it is, I cannot.'
(*Confessions* Bk II c.400 A.D.)

A search for the answer to this question would take us into almost every branch of human intellectual endeavour: a daunting task indeed. There must be some limitation to our exploration, at least during these first hesitant steps into the world of time in geography. For instance, it seems reasonable to assume that the time of relativity in contemporary mathematics and physics will take us beyond the bounds of the notion of time required for a study of the space-time organization of human activities and its associated items. All this may change when we travel about at the speed of light or as superluminol tachyons, but for the moment we can handle space-time from a rather less elevated position.

To ask the question, 'What is time?' has been described as 'a subtle way to ask a multitude of questions' (Fraser, 1968, 590), and the particular task ahead of us is to try and gain some knowledge of those aspects of time which are relevant to the study of Man, space, and activity in societal settings. A useful if neologistic term to describe this geographic viewpoint might be chronogeography. **Chronogeography** 'Not another obfuscating term!' we hear you cry. But biologists have found it useful to partition their complex and many-faceted subject into areas of special interest and one of these is now known as chronobiology. *The International Journal of Chronobiology* (1973) defines chronobiology 'as objectively quantifying and investigating mechanisms of biologic time structure, including rhythmic manifestations of life' (p.35). 'But, what is wrong with the simpler term time-geography?' The short answer is that this has become the familiar name given to a particular human geographic approach about which we shall have more to say in Chapter 6, where a single time dimension is used and is represented as shown in the illustrations above (Figures 1.1, 1.2, 1.3). But time can be seen as many-faceted in much the same way as space has been conceived to be. For such reasons it is less confusing to have a more general term such as chronogeography.

To set these ideas in the context of recent developments in human geography we have prepared a simple illustration (Figure 1.5).

Space : locational	Space : experiential	Space and time : locational	Space and time: locational and experiential
Emphasis on space 'outside' Geo-metrics	Emphasis on space 'inside' Psycho-geo-metrics	Emphasis on space and time 'outside'	Emphasis on space and time 'outside' and 'inside'
Locational Analysis c. 1965 cf. Haggett	Mental Maps c. 1968 cf. Gould	Time-Geography c. 1970 cf. Hägerstrand	Chronogeography ? c. 1975 cf. Parkes and Thrift

Figure 1.5. Getting places into focus

The term locational may refer to space or time that is objectively surveyable, typically geographical territory and calendar or clock time of the physical universe. The term experiential may refer to space or time that has been subjectively determined, typically as so-called mental maps and personalized images of time, for instance as 'too soon' or 'too long', 'passing quickly', or 'passing slowly'.

Locational space and time

Experiential space and time

Locational spaces and times are outside the individual, experiential spaces and times are constructed from inside the individual. We will be considering a scheme for their joint role in the realization of spaces into places later in this chapter. For the moment simply note that with these notions of locational and experiential dimensions, the time dimension of the space-time maps in Figures 1.1, 1.2, and 1.3 may be considered in terms of both these time types; i.e. the time outside (locational) and the time inside (experiential). If you are already familiar with the representation of spatial images as mental maps, then the extension of the notion to time should not be too difficult.

The immediate task is to put some of the principal discussions on the development of the intellectual notion of time into perspective. Only a very brief synopsis is given here, following Benjamin (1968). The history of the philosophy of time is covered well by Sherover (1975) and Capek (1976).

1.2 Notions about time

We are aware of time in the most obvious way, simply by virtue of changing events and enduring items. So time has a relational quality that derives from the relationship of our daily lives to these changes and durations. But occasionally time seems to speed up or slow down and this suggests that there is possibly some absolute time against which we make such judgments (Benjamin, 1968, 5).

Puzzling properties of time have interested philosophers at least since Plato (428-347 B.C.) divided the universe into temporal and nontemporal parts. After Heraclitus had claimed that 'all is flux', Plato tried to resolve the opposition which remained after Parmenides and Zeno had to their satisfaction refuted him by showing that nothing changes.[4] The nature of the relation between being and change is the essence or substance which remains when Plato's moral and political philosophy is disregarded (Benjamin, 1968). With Aristotle (384-322 B.C.) however, we can no longer say that time is fast or slow because both are defined by time. Time is now the number of motion in respect of 'before and after'. Time becomes a combination of change and permanence and its measurement is considered.

By the end of the seventeenth century the existence of time had been established beyond the need for further discussion of its ontological status. Thus Locke (1632-1704) was concerned to show how we build up our idea of time. For him it came from (1) sensation and (2) reflection. In other words it is not innate in us. Reflection on several ideas one after another produces the sensation of succession and reflection on the interval between parts of the succession gives the idea of duration. The more complex notions of time measurement and time of past, future, and eternity are built up from simple experiences of succession and duration.

Newton's work on time had been a great influence on Locke. His particular contribution was the introduction of the notion of unobservable, absolute time. Benjamin suggests that this is a somewhat unexpected development in Newton's work because he had always placed great store on the empirical method. His concept of time implies that there is a unique series of moments and that events are distinct from these moments but can occupy some of them (Whitrow, 1972, 102). For Newton there was an absolute and ultimate time in nature. Leibniz (1646-1716) was a contemporary of Newton's and stood somewhere between Locke and Kant 'in his conception of the role which the mind plays in our knowledge of the external world'. He conceived of two kinds of time, one which is potential and ideal and the other actual and real. Thus, when he distinguished space from time such that space was the order of co-existence and time the order of succession, he meant only that these were potential orders which may or may not be exemplified in actual occasions. The ideas which they embody contain more than is involved in any existent situation (Benjamin, 1968, 20).

Leibniz's theory that events are more fundamental than moments is known as the relational theory of time and is a rejection of Newton's idea that moments of absolute time exist in their own right. The impact of the Newtonian point of view, however, was dominant throughout the eighteenth and nineteenth centuries and even by the beginning of the twentieth century it 'had come to be generally assumed that there is but one universal system of time and that it existed in its own right' (Whitrow, 1972, 103).

Kant's views on time (1724-1804) have been described as the most vulnerable concept in his whole system (Smith, 1918, 137 in Benjamin, 1968, 21). To Kant time was one of the forms of our intuition. Coexistence and succession were its essential characteristics but could only be perceived by us with some prior notion of time in our minds. It is inevitable that we see phenomena as temporal because time is merely a form of intuition built into the mind. Set apart from the mind, time is nothing.

Between the time when Kant was writing and the development of Bergson's (1859-1941) work in the later nineteenth and early twentieth century the Darwinian theory of evolution was introduced. Time as the key to the understanding of reality was Bergson's conviction. He argued that time was best exemplified in Man himself and not in the movements of the outer world. Thus it was the overlapping of mental states and their gradual transition into succeeding states that gave us the best impression of time. He proposed that time be conceived as duration (durée), in one of its manifestations and as projection of time-into-space or spatialized time as its other form (Bergson, 1928 , 24; and Cleugh, 1937, 110).

Alexander (1859-1938) argued that he and Bergson were the first philosophers to take time seriously through the method of empirical thought. With no special sense organs for the perception of time and space we become aware of these dimensions in one of two ways:

i. By determination of their character through what they contain, e.g. space by figures (items as we have been calling them) and shapes; time by succession of happenings.
ii. By thinking of them as undifferentiated processes, in which case we end up with Newton's unexperienceable absolute space and time.

Time geometry and geography

Alexander found such a close relation between space and time that he combined them into space-time. 'Space is in its very nature temporal and time spatial' (Alexander, 1920). In movement, motion or mobility we find a perfect correlation between space and time. But 'geometry with its three dimensions is only a chapter in four dimensional physics' (Alexander, 1920) (1966, p.58). By substituting 'human geography' for 'physics' we can quickly suggest that it is only by inclusion of time that the geometric properties of relations among items (or events) become geographic.

From preliminary comments in 1.1 and from this brief overview of the notions of time, one thing seems to be quite clear: space and time are closely, if awkwardly, related. We will now consider some of the more basic terminological and conceptual issues concerning time study.

1.3 Some basic concepts and terminological issues

Cycles

In 1971, Young, a sociologist, and Ziman, a physicist, expressed concern at the lack of systematic enquiry by social scientists (possibly excluding economists) into the nature and behaviour of cycles in social behaviour. To illustrate their concern at this situation, which by the way appears to have changed very little since their paper was published in *Nature,* they discuss a number of time-related concepts and principles and we have freely adapted and adopted from their discussion.

Episodic time

The very simplest form or shape of time is episodic time and this may be illustrated schematically as in Figure 1.6; where a, b, c, d are event episodes. Such a time may be purely subjective having no quantitative measure and being based only on the positioning of remembered or anticipated events. Events are simply recorded as having eventuated or of being eventual. It is possible, however, to produce an episodic time scale by correlating individual sequences of events, so long as some preferred clock is available, though of course this 'clock' need not be finely metricated in any way at all. In this way the different episodic times of distinct social and other abstract spaces could in principle be constructed. Many of the social times which we will be discussing in Chapter 2.4 are of this type. An episodic time scale may be based on a scheme such as that in Figure 1.7.

Social time

Figure 1.6. Episodic time. (Young and Ziman, 1971. Reproduced by permission of *Nature*)

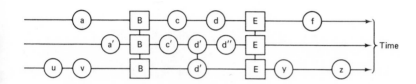

Figure 1.7. Episodic time as the basis for a time scale. (After Young and Ziman, 1971. Reproduced by permission of *Nature*)

The ⃞B and ⃞E chains are acting as simple preferred clocks or markers (Parkes and Thrift, 1975) for fixing other past and future events. A good example of the scaling of an episodic time is found in the estimation of a person's age in societies where there is no precise record of birthday. Episodes or events in the life of the subject are correlated with known or 'preferred' dates or markers, for instance by asking a person to recall mutually remembered events from the past.

Markers

When, however, a string of events known as a sequence, repeats itself, for instance as in Figure 1.8 or represented in a closed loop as in Figure 1.9, the term cycle may be used to describe their form. If there is no closure of a sequence, i.e. identical repetition does not follow immediately, then there is no cycle, but only recurrence, illustrated in Figure 1.10. In human geography it is most likely that we will find recurrence rather than cycles, even though the clock or calendar time which we usually refer to is itself based on cyclic events. Recurrence is most obviously associated with the incidence of social events. However, for most practical purposes we can use the term cycle with sufficient precision for us to be able to adopt certain quite advanced quantitative methods of analysis, some of which are outlined in the Appendix.

Sequence

Cycle

Recurrence

The so-called life-cycle obviously does not repeat itself for any single individual. However, as Young and Ziman point out, it seems admissible to retain the notion of a cycle when we wish to distinguish the recurrent events in the life sequence of an individual

Life-cycle

14

Figure 1.8. A Sequence. (Young and Ziman, 1971. Reproduced by permission of *Nature*)

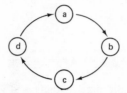

Figure 1.9. A cycle. (Young and Ziman, 1971. Reproduced by permission of *Nature*)

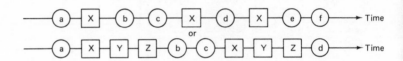

Figure 1.10. Recurrences. (After Young and Ziman, 1971. Reproduced by permission of *Nature*)

from those of all other people alive at the same time. 'A particular human life, as it progresses through various stages in its biological, psychological, and social dimensions, is certainly a sequence of events but the common usage of the term 'life cycle' is justified only if one does not think of it from the point of view of an individual as he sees himself, but rather as he sees others (including himself) as a member of a society or species. Everyone is in some degree repeating the sequence of behaviour of other people who have preceded him doing the same kind of thing at the same age, always doing [more or less] what others have done before.' (Young and Ziman, 1971, 92). This concept of life cycle may be illustrated as in Figure 1.11. In other words, i_3 follows the sequence of both i_2 and i_1 but the phase relations (see below) among the three may

Phase relations

differ; they are not exactly in time. Furthermore, i_3 may move through some stages of the life cycle at a greater rate than i_1 but more slowly than i_2.

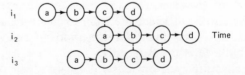

Figure 1.11. Life cycle stage (or phase) relations. (After Young and Ziman, 1971. Reproduced by permission of *Nature*)

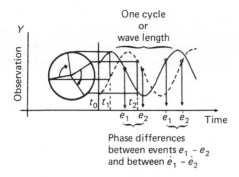

Figure 1.12. Phase difference and phase. (After Sollberger, 1965. Reproduced by permission of Elsevier/North-Holland Biomedical Press)

We use the term phase to describe the position of an event in a cycle relative to some other event in the cycle or relative to some external marker. Any change in these relations is called a phase shift. Phases might possibly be used as an index of urbanness, based on the idea that the greater the number of phases between a single location on one cycle and other markers or cycles the greater the urbanness, the more advanced the scale. At different stages in the life cycle and at different levels in the vertical structure of social classes, phase shifts are experienced, for instance through the process of social mobility. Social mobility puts an individual or group into a new phase relation with the cycles of the antecedent level. Shiftworkers provide an example of a social category who experience phase shifts with rather greater frequency than most. Another familiar example of the phase shift notion is the so-called 'jet-lag' experienced in long-distance intermeridional flight (Blatt and Quinlan, 1972; Strughold, 1971).

Phase shifts
Social mobility

Apart from a few obvious exceptions, such as flashes of lightning or car accidents, events occur with some duration which defines the relative location in time between their beginning and their ending.

The subjective assessment of duration may be different to the objective clock measure, as we will see in 2.3. However, as well as their overall duration, events which recur cyclically may have a measurable average time or a subjectively discernible time for their effect to fade away or relax. This is their relaxation time.

Relaxation time

The length of time before the recurrence of an event is the free time of the event. This interval, averaged over all the recurrences is the mean free time, and its calculation allows us to compare social or geographic variability in terms of event structure. Variance and standard deviation of mean free time may also be calculated. The value τ is simply $1/n (\tau_1 + \tau_2 + \ldots + \tau_n)$. E is the event under consideration, say change of residence or shopping episodes in a month or the incidence of accidents or of other urban social 'pathologies' like those which are discussed in Chapter 8.5.

Mean free time

16

**Periodic
recurrence or pulse**

When the event P recurs with such regularity that the interval between occurrences is more or less equal, such that we have periodic recurrence, the term pulse is preferred. The time behaviour of the event would be described as pulsating and may be illustrated as Figure 1.15.

In studying the city, geographers often have to rely on the number of people involved in an event (e.g. number arriving at work at 0900) or the number associated with a particular situation (e.g. unemployment) as an indicator of the intensity, severity or in general the 'scale' of the issue involved. If there is a variation in this property and if this variability is irregular we describe the behaviour as a fluctuating function of time. On the other hand, regular recurrence (within some bounds of variability, say ± 15 minutes in a day or $\pm$ 'a week' in a year) would indicate a behaviour which is an oscillating function of time. 'A cyclic movement of a measurable quantity up and down would thus be defined as an oscillation' (Young and Ziman, 1971, 93).

**Fluctuating
function**

**Oscillating
Function**

Oscillation

Signals

If we now look at a time plot (Figure 1.17) for the signals of certain pathological events — suicides, self-focussed problems, marriage related problems — recorded by telephone call to a social distress agency over four days in Newcastle, a heavy industrial city on the east coast of Australia, we seem to have something between a fluctuating function of time and an oscillation as shown in Figure 1.16.

Frequence

The number of recurrences of an event or the number of cycles in some longer interval is known as the frequency of the event or the frequency of the cycle. It is the reciprocal of the unit time of the interval chosen. For instance, in urban areas, every 24 hours, almost to the minute, many commuters catch the bus to work. The frequency of this periodic recurrence, for a single individual over an interval of 24 hours, is 1/24. Once every 24 hours the event 'catching the bus to work' occurs. The frequency of the same event over the interval or period of the working week is 5 and at this time scale has distinct pulse characteristics. In Figure 1.18, for example, E is always at 0830, the time that most people leave home for work. You might like to imagine a dispersion around this vertical line, indicating the proportion or number of people leaving for work, earlier or later than 0830. A wave-like pattern can be drawn around the peak, with small perturbations throughout the day and night, representing shift workers and part-time workers. If we consider a number of urban subareas, say neighbourhoods or communities defined in some way, perhaps by factorial ecology (Herbert and Johnston, 1976; Parkes, 1973), and isolate the relaxation time, mean free time, recurrences, periodic cycles and pulsations for similar categories of events in each space, then we can construct the generalized notion of subareal cadence or rhythm.[5]

Cadence

Rhythm

Amplitude

A particularly useful parameter for describing a cycle is its amplitude. The amplitude is the height of the peak of the cycle and

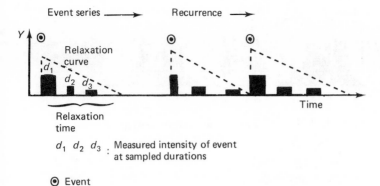

Event series ⟶   Recurrence ⟶

Relaxation
curve

d_1

d_2 d_3

Relaxation
time

Time

d_1 d_2 d_3 : Measured intensity of event
                    at sampled durations

◉ Event

Figure 1.13. Relaxation time. (After Young and Ziman, 1971.
Reproduced by permission of *Nature*)

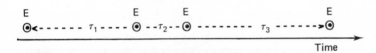

E   E   E   E

τ_1 τ_2 τ_3

Time

Figure 1.14. Free time. (After Young and Ziman, 1971. Reproduced
by permission of *Nature*)

Y

P   P   P   P   P

P   P   P

Time

Figure 1.15. Pulsation. (Young and Ziman, 1971. Reproduced by
permission of *Nature*)

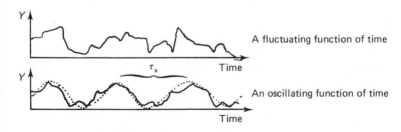

A fluctuating function of time

τ_x Time

An oscillating function of time

Time

Figure 1.16. Cycles, fluctuations, and oscillations. (After Young
and Ziman, 1971. Reproduced by permission of *Nature*)

18

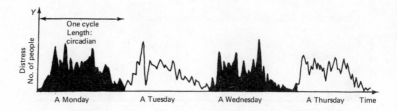

Figure 1.17. Cycles, fluctuations, and oscillations in urban distress

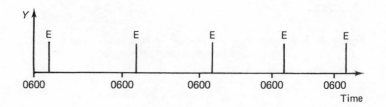

Figure 1.18. Frequency

may be used as an indicator of the scale of the difference between cyclic phenomena, so long as a comparable measure is used. The amplitude of the various cycles which affect human settlements can be quite sensitive indicators and may be used for various comparative purposes in human geography. In an economic geography of East Anglia, Sant (1973) notes that 'though it is only one of the important features in cycle studies, amplitude tends to be given most attention because it is the only one which contributes to the formation of economic policies'. A map of the average amplitude of unemployment cycles in East Anglia will be found in Figure 1.19.

As the family of possible cycles and recurrences which the geographer is likely to face in any particular study will probably be large, the relations among them need to be investigated beyond a simple consideration of their phase relations and independent amplitudes. This is so because individual cycles are often brought into either synchronized or harmonized relation, generally described by the principle of coherence or agreement. The term 'synchronized' is appropriate when cycles have amplitudes which occur at the same time, in other words where the correlation between cycles is very high with little, or ideally no, lag (see Appendix). 'Harmonized' cycles on the other hand show a systematic relationship with each other, but now the individual cycles are some simple integral ratio of each other. In other words, they have a harmonious or harmonic relation, Figure 1.20.

Any functional relation between cycles occurs through the

Synchronization harmony and coherence

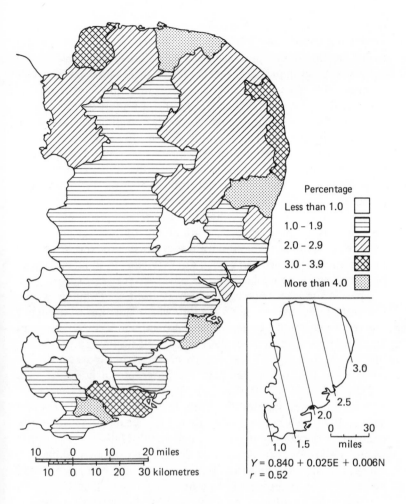

Figure 1.19. Average amplitude of unemployment cycles: East Anglia. (Sant, 1973)

operation of an agent of some sort. We give the name zeitgeber, taken from chronobiology, to that rhythm forcing factor or agent which is itself cyclic. The function of the zeitgeber is to entrain 'weaker' cycles to itself or to impel another cycle to assume synchronization[6] (Halberg and Katinas *et al.,* 1973, 49). Thus entrainment implies that some cycles are susceptible to adjustment by other cyclic items. The entrainment may be in an exact one to one relation (synchronization) or may be some integral multiple such as that illustrated above in discussing harmonized cycles. Many of the shorter day to day, week to week and monthly cycles, (Small Time) and longer, annual (Big Time) cycles which occur in society are either dependent or dominant, antecedent, or

Zeitgeber

Susceptibility

Entrainment

Dependent or dominant cycles

Antecedent or postcedent cycles

System

Pacemakers

postcedent. Dependent cycles get their pace or frequency from exogenous factors which may themselves be cyclic. The impelling cycles are dominant, they are zeitgeber. Antecedent cycles come before other cyclic events, i.e. they lead, while postcedent cycles follow. However, antecedent cycles may be dominant or dependent, as may postcedent cycles. The 'assembly of many interacting cycles' forms a system which may itself have distinctive 'time-like properties' (Young and Ziman, 1971, 95). An urban place or region is just such a system. In this sense a system is a place; a system defines a place. Difficulties arise unless we constrain geographical description and explanation within specified, bounded time intervals; a minute, a day, a year, and this is just as important as the delimitation of territory if geographical argument is to proceed with a minimum of misunderstanding. Only when time scale boundaries associated with the system of interest are established can explanatory statements about aspects of place behaviour be made.

Apart from the zeitgeber, which is itself a dominant cyclic function, there are two other factors which function to influence the relations among cycles. They also influence the phase and amplitude parameters of cycles and the relaxation and mean free times of recurrences. These are pacemakers and markers. These terms were first introduced into the geographical literature in a 'formalized' way by Parkes and Thrift (1975) and have been adopted in somewhat modified form in a recent paper (Parkes and Thrift, 1979). Chronobiologists define a pacemaker as 'an entity controlling or influencing rhythmic activity' (Halberg and Katinas et al., 1973). We can adopt this definition without making any changes. Lynch has suggested that the 'pace of urban places could be retarded by damping the rate of transmission of information: [for example] by substituting the weekly for the daily newspaper, or mail for the telephone' (Lynch, 1972, 78), in other words by adjusting the pacemaker. As another example, consider the advent of microprocessors which now exert increasing influence and control over the structure of production and, therefore, the organization of daily life. Just as the technology of the industrial revolution of the 19th century influenced the daily lives of every man, woman and child, through its new pace and timing, 'The clock and not the steam engine' being the single most important factor of the 'revolution' (Mumford, 1934), so the microprocessor will in turn act as a pacemaker in the determination of life-styles of most economic and social 'item' relations, in other words, with the 'social times' of future generations. Its influence will be as dramatically revealed in the nature of place, region or system morphology and function as were the technologies of yesteryear. To the extent that 'chip technology' imposes certain time

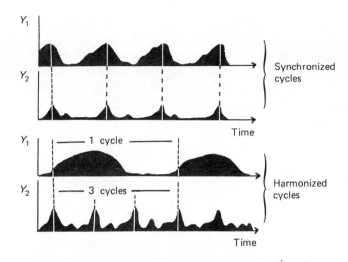

Figure 1.20. Synchronized and harmonized cycles. (After Young and Ziman, 1971. Reproduced by permission of *Nature*)

constraints it acts as pacemaker and in certain circumstances as zeitgeber of future life worlds.

In human geography the source of pacemakers may be 'small' or 'large' spaces (a microprocessor, a building such as the town hall, a city such as London, a region from which raw materials are obtained). But it may also be a set of durations or instants in time (office hours, school terms, bus timetables). Their role is always the same: they pace system behaviour.

We distinguish a marker as a particular type of pacemaker. It is distinguishable from the oscillatory pacemaker because it produces its cycle-controlling or cycle-influencing impact at irregular **Markers** intervals or at such extremely low frequencies that relative to the frequencies of other socially relevant cycles it may be treated as being almost patternless. Markers may also be spaces or times. For instance, so far as the day-to-day rhythms of the London suburb of Wimbledon are concerned, for most of the year the tennis championships which occur annually have little or no controlling or influential significance on the environment. However, for some short period before and after the championships there is a discernible change with billeting of guests, participants and officials, a change in the pace of retail activity, and general invasion of suburban times by new activities. When a time interval of many years is adopted, for instance all the years over which the tournament has been held, the championships in effect behave as a low frequency oscillatory pacemaker, but for any one year they are marker pacemaker in type.

22

Drift time

Fixity

Limits

Constraints
Capability

Coupling

Authority

Persistence

There are three more aspects of time terminology worth discussing at this stage. They relate to the notion of markers mentioned above. When intervals of time are unmarked we use the term drift time. Such a condition is most likely to be experienced in a new environment where neither spatial (landmarks) nor temporal markers have become established. Fixity describes the strength of marking attached to an event (Chapter 5 following) and is the degree of difficulty encountered in altering the time or space location of an event. Finally, and similar to the idea of fixity, is the notion of limits. Limits define the boundaries in space and in time to which an activity or a schedule of activities can be planned as part of a set of sequenced tasks.

Limits related to such 'projects' are also important to the development of the time-geographic approach to 'human system analysis' adopted at the University of Lund in Sweden. Associated with this scheme are three principles, known as constraints to human action, which we have to mention now because they are related to the notion of limits. Their introduction at this stage facilitates certain discussions in the chapters which follow before the time-geographic approach is studied in Chapter 6. The capability constraint limits the activities of the individual because of his biological construction (e.g. the need for sleep) and/or the tools (e.g. telephone, car, door key) which he can command (Hägerstrand, 1970). Tools extend Man's capability to use space-time. Coupling constraints define where, when and for how long an individual has to join other individuals, tools, and materials in order to produce, consume and interact. Finally, there is limitation of action through rules and regulations, defined as the authority constraint. But activities themselves also act as space-time constraints because they limit use of alternative space-time locations: the 'need' to participate in certain activities makes participation in others impossible.

Notions of space, time, and space-time in the context of absolute and relative locations lead us rather straightforwardly into the popular conception of human geography as the study of places. But the relations of space to place (Tuan, 1977) and time to place (Lynch, 1972; Tuan, 1978) are complex. Just what place is seems to be in more doubt amongst students of geography than the population at large. Perhaps this is as it should be! Saint Augustine had difficulty trying to explain what time was, although he felt he knew what it was. Similar difficulties seem to arise over the explanation of place. Where do places come from and how do they persist? The section (1.4) which follows illustrates some of the concepts and principles which might be involved in answering such questions and the key concept is that time, in its locational and experiential context, is a factor in place-making. In Chapter 3 the notions of cycles and persistence are taken up again in an attempt to further understand where and when places come from.

1.4 Time in the notion of place

> 'Time and the way it is handled has a lot to do with the structuring of spaces'.
>
> (Hall, 1966, 163)

The meaning of place has been considered by philosophers, architects and social scientists of many persuasions and by human geographers. There is a rich and varied literature. More specifically **Place** within urban planning, (Lynch, 1972, 1976) and in human geography through the phenomenological emphasis of Tuan (1973, 1977, 1978), Buttimer (1976), and Relph (1976), questions have been asked concerning the essence of place, as Relph (1976, Chapter 3) puts it. However, the most striking feature of much of this work, especially when a phenomenological stance is taken, is the absence of substantial statements as to where place comes from, or what makes places more than a grouping of material artefacts. The contemporary discussion often appears to be somewhat prolix, but in the hands of fluent essayists such as Tuan, Lynch, and Relph, there is usually a richly flavoured, scholarly, and evocative treatment, making for enjoyable reading, in spite of the fact that one is sometimes left without any easily reproducible concepts and principles.

Though geographers have written much about places they 'have generated little theory to account for (their) character' (Hugill, 1975, 215). When geographers theorize about place the importance of their 'meaning' to the individual is always stressed but because definition is often lacking it is not always clear what exactly it is that is supposed to have 'meaning'. Relph suggests that 'the confusion about the meaning of the notion of place results because it is not just a formal concept awaiting precise definition, but is also a naive and variable expression of geographical experiences' (Relph, 1976, 4). For Lynch, more closely attached perhaps to the necessary pragmatism of urban and regional planning, the meaning of place lies in the context of how 'setting symbolizes fundamental feelings' (Lynch, 1976, 172-173). But he also recognizes that data are hard to acquire and personal introspection, memories, and the references to place in novels and memoirs can perhaps be a useful source (Salter and Lloyd, 1976; Thrift, 1978; Seamon, 1978). Again, however, may one ask how reference is to be made to something that is so clearly unclear? Is a place-name all that is needed, or do we get to know a place simply by the company it keeps?

Intuitively we all know places, like or dislike places, and visit or stay away from places. Some people map places and others use maps of places to help them to move about and to avoid becoming lost. Others use maps in order to use time efficiently, such as delivery services and taxis. Places may be small or large. 'Come to my place for the evening' means come to my home (place) which may be a mansion, flat, or room. Such a home (place) may be large or small relative to someone else's home (place). On the other hand consider 'New York is a big place! Where exactly do you live?' This question puts place at another scale level where place is replaced by a location.

Relph (1976) quotes the philosopher Heidegger (1958, 19): '"place" places man in such a way that it reveals the external bonds of his existence and at the same time the depths of his freedom and reality.' Fine sentiments, but what is it that does this placing? Place apparently! But place 'is a profound and complex aspect of man's experience of the WORLD' (Relph, 1971, 1, our emphasis), and so such a comment is not deeply helpful, however eloquent it may be, because it might readily be rephrased in the more obviously tautological form: place is a profound and complex aspect of man's experience of PLACE. Such a substitution of world by place follows Blumer's terminology; world is inclusive of 'the setting, the surroundings, and the texture of things' (Blumer, 1969, 11 cited in Hugill, 1975, 214).

Behaviour or conduct

Some writers consider that 'perceptual unity' is a distinctive characteristic of place (May, 1970) but unity is considered to go beyond the arrangement of the physical items that people move among in their daily social behaviour (conduct); trading off access for comfort, security, commitment, intention, and urgency. Perhaps the sense of simply 'feeling at home' may be the closest one can get to describing a positive attraction to a place, place as a familiar mental and social environment rather than the absolute or relative location of physical items.

Place and placelessness

Our immediate discussion will now centre on certain notions of space and place but this will be followed by a discussion of how it is that time and space 'make' place.

Primitive space

Perceptual space

The first point to establish is that space is many-faceted, ranging from the space of direct experience to the space of abstract geometric and mathematical thought. We can think of many spaces. Thus primitive space is the space of instinctive behaviour devoid of images or concepts. Structuring of such space begins in infancy and space and place are hard to distinguish — 'perhaps space is simply a continuous series of egocentric places' (Relph, 1976, 9). Perceptual space has content and meaning as a space of action centring on immediate needs and practices. Distances and directions are fixed relative to Man. However, perceptual space is not isolated, because we live in a culture-realm of experiences and intentions, an inter-subject(ive) world (our parentheses and hyphenation), (Schutz, 1962). Existential space (Figure 1.21) is lived-space; the inner structure of space as it appears to us in our concrete experiences of the world as members of a culture group socialized according to a common set of experiences, signs, and symbols.[7] Existential space is never entirely passive but is contin-

Existential space

uously remade by human activity including architectural and other planning processes. Norberg-Schulz (1971) has given existential space a vertical and a horizontal structure; the latter based firmly on Lynch's approach (Figure 1.21). Sacred space is continuously differentiated and replete with symbols, sacred centres and meaningful objects, but is now essentially an archaic space replaced in modern society by profane space. A space which is unique and has its own name is a geographic space (Relph, 1976). Such space reflects experiences and intentional links with environment and this is just where time begins to exert its influence on place making because spaces have meaning in terms of human tasks and lived experiences. The structure of geographical space has been examined, for instance in the work of Lynch in Boston (1960), and more recently in Los Angeles (1976), through the images (or mental maps) that people seem to hold. Figure 1.22 illustrates a comparison of objectively surveyed territory in Los Angeles with the 'mental maps' of Los Angeles, held by subjects centred on two geographically distant districts within Los Angeles.

Sacred space

Profane space

Geographic space

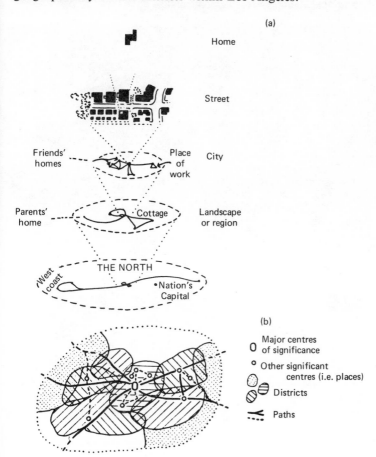

Figure 1.21. Existential space. (After Relph, 1976, 21. Reproduced by permission of Pion Ltd.)

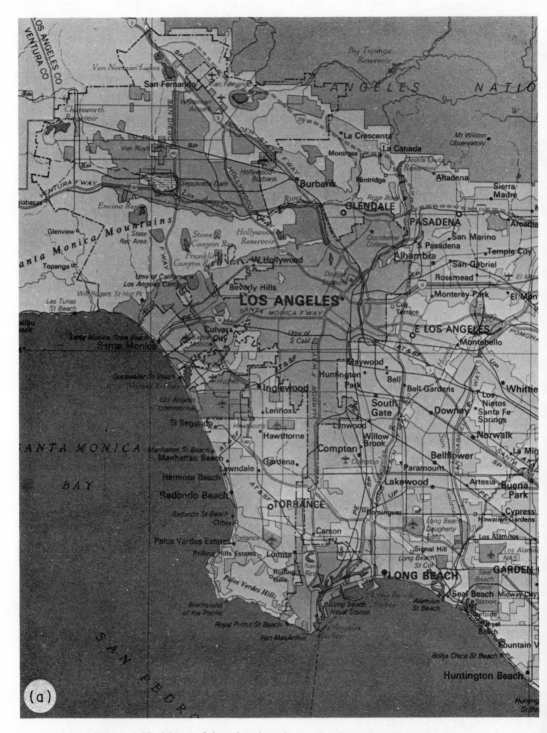

Figure 1.22. Maps of locational and experiential space, Los Angeles.
(a) Locational (reproduced by permission of US Geological Survey),
(b) and (c) are from Lynch, 1976. Reproduced by permission of the MIT
Press. © 1976)

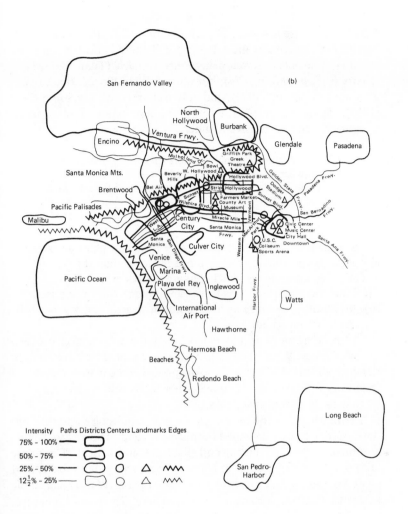

Figure 1.22. continued

A further distinction between cognitive and abstract space is made by Relph. The former involves development of theories about space as an object of reflection. Geometries of one sort or another, for instance as map transformations, depict place as a location which is definable by sets of co-ordinates. This is the space which we have referred to in Figure 1.5 as locational. Abstract space, however, is the space of logical relations and does not need empirical observations. This is essentially the experiential space of Figure 1.5 and of the scheme to be outlined in a moment. But it is now clear that time needs to be seen as an integral part of place, and it is to the incorporation of time into place that we now turn.

Time is an essential ingredient in the realization of place (Parkes and Thrift, 1978). Thus 'repetitive tradition re-establishes place and expresses its stability and continuity' (Relph, 1976, 32).

Cognitive and abstract space

Locational space

Experiential space

Repetitive tradition

Persistence

Repetition (or recurrence) is perhaps the key to the character of persistence that places must have, for if there is only change 'we should not be able to develop any sense of place' (Tuan, 1977, 177). Briefly in *Topophilia* (1974), but more fully in 'Space, time, place: a humanistic frame' (1978), and in two chapters of *Space and Place* (1977), Tuan introduces time into the discussion of place in an imaginative and persuasive manner.

> 'The time dimension matters more (than the spatial) because people appear to be more interested in narrative than in static pictures, in events that unfold in time than in objects deployed in space'
>
> (Tuan, 1974, 216).

Three approaches help to explain how time and place are related, although the problem is rather more intricate than this.

The three approaches which Tuan explores are:
 i) Time as motion or flow and place as a pause;
 ii) Attachment to place as a function of time;
 iii) Place as time made visible.

With 'place as pause in movement' we have one relation between time and place. With 'the city as time made visible we have another relation' (Tuan, 1978, 15). A growing attachment to a place as a function of the passage of time is sufficiently self-evident for us to leave it aside in this discussion.

A path identifies the movement of any item. Space and time are the currency to be exchanged in this movement. The key to Tuan's argument is that movement and place are antithetical. The pause 'allows a location to become a centre of meaning' but 'space, as it gains familiarity and meaning (over time) is barely distinguishable from place'. Some comment is required here. There can, of course, be no pause in time as such unless the basic items which are used for time sensing are clearly identified and defined. But this having been achieved it seems possible that unlike the continuous everyday time of clocks with which we are all familiar, there can be a

Discontinuous time

discontinuous time which has relevance to human activity, one form of which is composed of segments of movement. This being so, pause in time, i.e. intervals between movements and not just pause in space, becomes possible. The so-called stages of the life-cycle provide an example of pauses in time.

This is a rather special notion of time, as the time of social life, which has only a crudely calibrated clock of between five and eight stages, depending on which life-cycle model one accepts.

The other relation of time to place which Tuan recognizes is that of time made visible in the city. 'As an historical document the city is highly selective in regards to the facts (buildings, streets, monuments) that are maintained; as a record of time it is full of gaps' (Tuan, 1977, 196). But for Lynch time takes on a more positive, constructive role as a many-faceted and integral aspect of

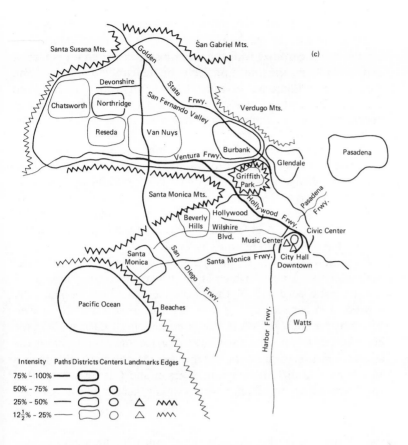

Santa Susana Mts.

San Gabriel Mts.

Golden

(c)

Devonshire

State   Frwy.

Chatsworth   Northridge

San Fernando Valley

Verdugo Mts.

Reseda   Van Nuys

Burbank

Pasadena

Ventura Frwy.

Glendale

Griffith Park

Santa Monica Mts.

Hollywood Frwy.

Pasadena Frwy.

Beverly Hills   Hollywood

Wilshire Blvd.

Civic Center

Music Center

San Diego Frwy.

Santa Monica

Santa Monica Frwy.

City Hall Downtown

Pacific Ocean   Beaches

Harbor Frwy.

Watts

Intensity   Paths Districts Centers Landmarks Edges

75% – 100%

50% – 75%

25% – 50%

$12\frac{1}{2}$% – 25%

Figure 1.22. continued

place (Lynch, 1972). The internalized times of human life (Chapter 2.3), the discontinuous times of social events (Chapter 2.4) as when; 'We are suddenly a year older', the sense of future and past as represented in the present (Chapter 2.3), the appropriate times of socially relevant conduct. All these make time structure 'subject to modification by internal state or external suggestion. The location of self in time feeds on thinner stuff than the sense of place, and that may be a reason for a conscious policy of thickening our external references to time' (Lynch, 1972, 123). With this more expansive notion of time Lynch looks through the lens of a camera to realize places in the city of Boston in photographs. He captures the 'city as time made visible' (Tuan, 1977), but while Tuan's gaps are still there we now see a many-faceted time in action. The juxtaposition of buildings, old and new, public clocks and public buildings, provide a mass of time marks and landmarks for the organisation of daily activity. Together with variable flows of people and vehicles these landmarks and time marks provide a sense of urban 'pace'.

Time marks

Land marks

Urban pace

City parks, for instance, allow an opportunity to decelerate our personal time as the changing item relations in the parkland scene are felt to synchronize with the times inside us, (the same space at two o'clock in the morning may be fearfully dangerous and dyschronic with internal rhythms, the human heart beats fast, and the city park and the bustling main street of twelve hours earlier become a slow and empty time where anticipated events are likely to be injurious).[8]

'The order which people give to the world — a perceived, remembered, and expected order about objects, events and human nature' (Golledge and Moore, 1976, 192) provides the focus on place perspective which Gerson and Gerson adopt (1976). One of the four social contexts which they explore is the temporal context. From day to day it is in fact the temporary (and temporal) context of life that we have to learn to live with.

As sociologists they argue that it is 'the kinds of social order maintained at a place' which is of primary interest. Places have characteristic pace and rhythm, a theme which was emphatically expressed in a paper by the neo-classical urban ecologist, Engel-Frisch as far back as 1948 (Chapter 8 following). Each space in the city has a different pace: the CBD is hurried and conduct along the streets is measured in hours and minutes. Sub-urban places have a slower pace, a different rhythm. Gerson and Gerson suggest that the following appropriate questions might therefore be asked.

> 'What sort of time does this place exhibit? What rates of conduct? What rhythms and what periods are required, permitted, ignored? How do people in the place negotiate differences between their personal temporal characteristics and those of the place? How are the various kinds of social time (cf. Chapter 2.4), work times (which is also a category of social time in the terms which we will be developing later), sleep time and ceremonial time related to one another in the perspectives of the people?'
>
> (Gerson and Gerson, 1976, 198)

These questions lead us rather easily into our final approach to the illustration of the role of time in place, to which the discussion so far in this section has been a prelude. The question we seek an answer to is 'Where does place come from?' We leave until Chapter 3 the question of where its persistence comes from.

Societal level

An individual has a number of positions (locations) within society and is more or less aware of the various physical artefacts of **Information** the containing environment entirely as a result of the information he receives, following the filtering effects that occur at each societal level. Four societal levels are distinguishable, each one acting as a **Filtering** mediator between the individual and the information which the society utilizes.

Space and time are brought into conjunction as place, for the individual, through the receipt of information which is always filtered through a number of social and physical agencies before it is received (including the physical element of the environment, natural and manmade). The four levels which have been disinguished are illustrated in Figure 1.23.

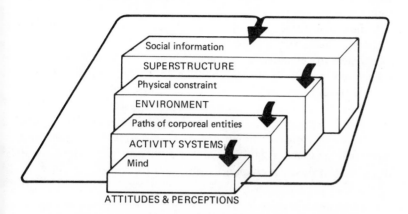

Figure 1.23. Information and levels in the social system. (Parkes and Thrift, in Carlstein, Parkes and Thrift, 1978, 120. Reproduced by permission of Carlstein *et al.* and Edward Arnold (Publishers) Ltd)

Information (as defined by everyday usage) includes the 'semantic value' or contextual meaning attached to items, events and symbols, and we will differentiate between information as hard or soft. Our treatment of information is not to be confused with information as defined in Information Theory (cf. Chapman, 1977; Chapter 8). Hard information promotes obligatory responses. There is a clearcut, almost mandatory, meaning attached to it because of the context it belongs in. It is manifest in particular behaviour or conduct which is often immediate. If delayed beyond the present it is not readily open to postponement, and nor is it likely to be possible to conduct the behaviour in alternative spatial locations, without some loss in perceived suitability or congeniality (Chapter 5.2, the transductive approach). This sort of information may be in the form of appointments or contracts or, more subtly, it is realised as the result of recurrent, perhaps habitual, activities, such as catching the 0732 bus every morning. In Chapter 5.2, the 'deliberated' choice model implies hard information.

Hard information

Soft information

Hard information REALIZES place immediately because absolute and relative location in space and time are fixed (say by a dental appointment). In the everyday life of city dwellers, superstructural authorities 'transmit' hard information by regulations; licencing laws, school attendance, bus timetables. Place is well determined because the space-time location is hard to change; it is inelastic[9]. Soft information, on the other hand, also

32

contributes to place realization but may not enable it.[10] This is because, by definition, there is always a high degree of elasticity in soft information; meaning is unclear, or has too many 'degrees of freedom' associated with it. For instance, the information may relate an activity to either of two locations in three dimensional space, x, y, z or x', y', z', and the time locations may be at t_1 or t_2. In other words the activity or event is relatively elastic and in space-time the mapping of space and time is ill-defined. The essence of soft information is the degree to which it 'can fit around the demands made upon it by hard information and whether the exigencies of experience enable such soft information to be adequately processed into a concrete form' (Parkes and Thrift, 1978, 121-122) upon which activity can be effected.

Apart from illustrating that the levels of the social system act as filters and mediators to both hard and soft information, the scheme for place realization distinguishes space and time dimensions in a particular manner (Figure 1.24) and this is germane to our discussion in the next chapter of time and paratimes.

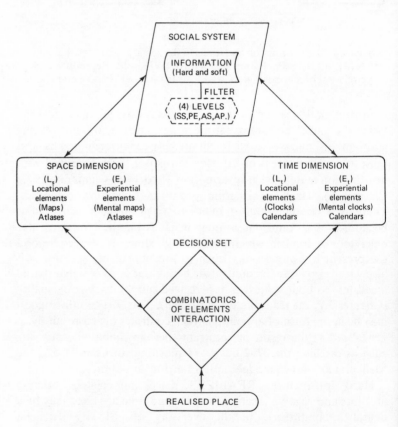

Figure 1.24. The realization of place. (Parkes and Thrift 1978, 121. Reproduced by permission of Carlstein *et al.* and Edward Arnold (Publishers) Ltd)

The space dimension transmits and receives information. As transmitter it may signal that it is 'too extensive', 'too crowded with symbols or items', 'too small', 'too high', or 'too deep'. As receiver it is susceptible to mediation by authorities. Planning regulations are perhaps the most obvious example. The same circumstances apply to time. As transmitter the time may be 'too soon', 'too late', 'insufficient', and as receiver it is also open to system level mediation such as licencing laws or parking time restrictions or marketing conventions for opening and closing times of facilities.

The space dimension is composed of two distinct but related elements, (Figure 1.24). The first is the locational element of space (L_s)[11] and the second the experiential element of space (E_s). Locational elements are those which are capable of precise position on 'universally' accepted co-ordinates; national grids or intersecting lines of latitude and longitude; they may be located by ground survey. Experiential space elements are not precisely positioned in any universally accepted manner. They are based on personal, unique codes derived from information in the social and physical environment and are liable to change their relative locations with learning experiences (cf. Golledge, 1978, for a discussion on learning processes in relation to environmental knowing, and Chapter 8.4 following, on images of space at different times).

Locational elements

Experiential elements

The time dimension (Figure 1.24) must also be considered in terms of locational (L_t) and experiential elements (E_t). Locational time is the time of the calendar and the clock. It is universally co-ordinated to specific reference points such as GMT or is locally specified (British Summer Time) so that events or items in general may be located with objective invariance.[12] The experiential time element is perhaps less easy to grasp because our discussion has not yet considered Man's sense of time; psychological time and other less familiar notions of time's complex. In the course of realization of place, these experiential elements of time refer to our personal sense of time and to certain value-loaded nuances which we attach to the location of events which have occurred, will occur or are occurring in space (locational and experiential) as well as in locational time. The appropriateness of the time at which an event occurs is an example. Thus, the experiential time elements, mediated by one or more of the social system levels, may indicate a right time for play, an appropriate duration for a social event, an acceptable interval in a bus timetable, etc. In a sense there are mental clocks and mental calendars analogous to the mental maps and mental atlases which human geographers have come to be aware of, and like 'mental maps' these 'mental clocks' have a role in the determination of human spatial behaviour.

Mental maps

Mental clocks

The four elements of space and time (L_s, E_s, L_t, E_t) have a variety of possible combinations. In some instances the locational elements

Encounters with place

Realized place

may be dominant and be strictly tied to hard information controlled by law or contract. The work*place* provides an obvious example of a realized place in such terms. In other instances the experiential elements will dominate and soft information will be the generator of spatial behaviour. Much passive recreation is conducted in realized places of this sort.[13] For any individual a combination of the four space and time elements produces a structured space-time which is place. Thus we have a notion of place which seems to accommodate some of the phenomenological concepts which writers such as Tuan and Relph have discussed, but also perhaps the germs of an explanation as to how place can become so highly personalized. Place is rather like an UFO. We may encounter it at various levels. With the realized place scheme (Figures 1.23 and 1.24) we have an encounter of the fourth kind! The first encounter is in locational space, the second is in locational space and locational time, the third is in both of the former plus experiential space whilst the encounter of the fourth kind involves experiential time as well as the first three encounters and is realized place.

The next chapter takes us into a discussion of many-faceted time. An urban or social GEOMETRY may confront absolute and relative locations of items with the rigour of Euclidean and other geometric theorems. Given a proper and concise statement of objectives this is clearly a legitimate and worthwhile paradigm to adopt. An urban or social GEOGRAPHY, however, cannot justify a time-free paradigm. But just as the limitation of a geometric emphasis to Euclidean notions will miss the powers of detection of other geometries, so too a geographic approach, which must involve time, is likely to be less fruitful if time is not given a basis in item relations other than the familiar relations among planets and stars; i.e. physical or locational time. Together, territories, societies and times are the principal components of urban and social geography. The chapter begins with a brief overview of the basic principles of physical time reckoning, which we call UNIVERSE TIME, followed by an introduction to the ideas of paratimes: the times beyond the familiar universe time of the calendar and clock. First of all we discuss LIFE-TIMES as biological and psychological TIMES and then social times. Their separation here is a conceptual device. Its function is to extend the urban and social geographic imagination.

Notes

1. The term item will be used frequently throughout this book. It may be substituted by the terms event or object whenever this seems to clarify the particular context.

2. The terms space-time and time-space seem to be used interchangeably in most of the recent geographical literature, although Carlstein (1978) does make an explicit distinction between the two. We shall stick to 'space-time' following the convention of expressing the x-axis before the y-axis. This will be clear to you from Figure 1.1 following.

3. In Chapter 6 some of the principles of the time-geographic scheme are introduced along with examples of the application of the approach.

4. A good explanation of Zeno's Paradox and other puzzles of time will be found in Capek (1976).

5. In Chapter 8 we will illustrate such notions from empirical work in an Australian city.

6. There are various orders of synchronization but we cannot go into them here (cf. Sollberger 1965, 1.17-1.27).

7. In Chapter 2.4 the role of symbols, signs, and signals is discussed in relation to one of the levels of social time in Gurvitch's theory, *The Spectrum of Social Time* (Gurvitch, 1964).

8. In Chapter 8 when we discuss Time in the City: some empirically derived support for these concepts will be given.

9. The elasticity notion is taken up in some detail in Chapters 4, 5, and 8.

10. Enabling factors are discussed in Chapter 8 within the process of time colonization; the city is becoming an incessant place.

11. Locational space is akin to Relph's notion of cognitive space, but his term seems rather confusing in the light of the indefinite terminology in recent behavioural geography where terms such as cognitive, perceptual, image and imagined are used almost interchangeably. Tuan (1975) discusses some of this terminological and conceptual confusion.

12. Certain culture specific anomalies associated with time reckoning exist and we shall illustrate some of these when we discuss social time in Chapter 2.4, but these anomalies are now strictly found only between small scale social systems and the industrial and post-industrial, large scale cultures. We are concerned essentially with the latter, but reference to small scale society will sometimes help to put the discussion into focus.

13. This is possibly the form that much spontaneous recreation adopts, e.g. as a family picnic or a stroll in a park or an amble down Main Street 'window shopping'. Quality of life may have a lot to do with the opportunity for behaviour which is based on soft information and experiential space-time. We will take up one or two ideas on this track in Chapter 9. You might also care to refer to Parkes and Thrift (1975), 667-668.

2
Time and paratimes

'For man the problem of time is not merely one of theory but one which is supremely and intimately related to the conduct of his life.'

(Kümmel, 1968, 31)

2.1 Preliminaries

In everyday life, time is revealed by experience of the regular recurrence of selected sequences of events, such as the alternation of day and night. Time itself, however, 'is a mental device to give order to [these] events, by identifying them as co-existing or successive' (Lynch, 1972, 120). Events are a particular category of
Events as items *items*. The changing relation among items is the basis for time construction and the items may be material (for instance planets and stars (2.2)) or non-material (for instance periods of rest and periods of activity (2.3 and 2.4)). The particular value and
Material and non-material items interpretation put upon all item relations depends on various preconditioning and predisposing issues in the human social context. These may themselves be fundamentally temporal, involving for instance biological (say, age), psychological (say, sense of urgency) and social (say, sense of appropriateness) elements.

In this chapter three conceptually different categories of time are introduced. Their distinction seems to clarify certain processes which operate in urban society. It is always necessary to try to 'understand the nature of time itself with respect to the phenomena (items) we are trying to explain' (Harvey, 1969, 425).

The three time categories to be discussed are:

(i) Universe time, which is familiar to all of us by its record in calendars and clocks, and is sometimes referred to as standard

36

time. We prefer universe time because this incorporates the item relations upon which it is based.

(ii) Life-times, which are further distinguished as biological and psychological time. Together they operate to sustain life (biological time) and provide a basis for experiential orientation by facilitating the sequential ordering of items (activities) into the past and into the future, through 'sense of time' (psychological time).

(iii) Social times derive from group use and awareness of the frequency, duration and sequence of relations among socially relevant items. There is no single social time which has universal meaning but each social time derives from the time and timing of human organization (Diamant, 1970).

We will use the term paratimes to classify life and social times because they are beyond the normal, everyday notion of time; that is all. Time and paratimes are equivalent to change. The phrase 'change over time' is in fact rather meaningless, rather as 'change over change' would be. The first point to establish is that the changing item relationships that are observed to occur, given certain conditions which are to be outlined below (2.2), in fact define their own time. 'Times' do not exist of themselves, but are constructed. What we call 'time' is a description of the way certain selected items change in relation to one another in a recurrent fashion. Items in changing relation, having been monitored in some way, give us a basis for time, which in turn helps us to order our behaviour and to observe the behaviour of other people and of material systems. Rather importantly it also gives us the ability to identify processes. In the strictest sense processes do not 'occur over time' but *are* time.

2.2 Universe time

This is the locational time discussed in the previous chapter. It is the most pervasive and possibly the most important time for the ordering of human actions in advanced industrial societies not least because 'the definition and measurement of time has always presented greater difficulties than that of the other major dimension of our existence — space' (Diamant, 1970, 93).

(i) The items in relation

'There is no period in the story of life when the cyclic process of day and night and that of seasons did not operate. The unquestioned rank of the rotating heavens and of the changing seasons above all other cyclicities, is probably responsible for the fact that we measure all human activities in terms of suns, moons, and seasons (days, months, and years) and not *vice versa*. The origins of archaic time keeping are indistinguishable from those of interpreting and measuring the motions of the stars, planets, and the moon' (Fraser, 1975, 48). These items and their observed relations are the basis of universe time; they are in continuous relation and in turn give a sense of continuity to our daily lives.

Measurement requirements

Accurate measurement of time must always satisfy the following requirements:

i Accessibility to all who need it.

ii Continuity

iii Invariability (though this is seldom achieved)

However, *'any repetitive phenomenon (item) whatever, the recurrence of which can be counted, is a measure of time'* (Clemence, 1968, 406, our emphasis). This is the most important principle to have in mind in the remaining discussions in this chapter. Whether one set of recurrences or another is selected is a matter of 'judgement about permanent regularity' (Fraser, 1975, 62) and is conditioned by need.

Until 1956 'the rotation of the earth has been the fundamental measure of time for all purposes' (Clemence, 1968, 406). Problems related to the invariability criterion have arisen from time to time and so have certain technical issues associated with what are known as inertial frames of reference, but we need not consider these any further here in detail. By computing experimental error limits and comparing motions among three major systems, the stars in motion, the orbits of the planets, and the system of galaxies, it has been possible to improve on the accuracy of universe time measurements in recent years. In 1956 a new definition of the second was adopted. Until then it had been 1/86,400th of a day, but in ephemeris time, as it is now defined, the second is precisely 1/31556925.9747 of the tropical year. The tropical year is based on the earth's revolution about the sun and is defined 'as the time between two successive passages of the sun through the spring equinox' (Whitrow, 1972, 84): the spring equinox being the day when the periods between sunset and sunrise are equal. There are years and days of differing length in universe time and rather careful adjustments have to be made among them in order that the calendars and clocks which we use in everyday life may be kept consistent. The best known of these adjustments, of course, is the leap year. There are mean solar days and apparent solar days, the former being a calculation describing the motion of a mean sun so as to allow for correction to observations of the apparent sun, which has a variable velocity due to the fact that the earth's orbit around the sun is an ellipse rather than a circle. The difference between the two produces the so-called equation of time shown in Figure 2.1.

Ephemeris time

Mean solar days

Mechanical clocks and sundials do not agree automatically because in fact the clock day is an averaged day. Positive values for

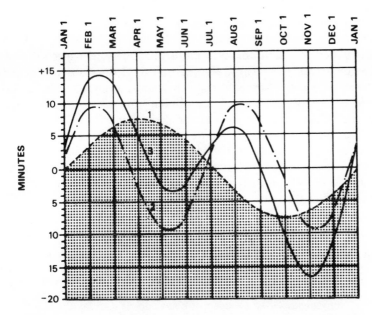

Figure 2.1. The equation of time. (After Whitrow, 1972, 78. Reproduced by permission of Thames and Hudson)

The 'equation of time' curve (3) is the sum of two components. Because the Earth's orbit is an ellipse, and the Earth-Sun distance waxes and wanes through the seasons, a discrepancy is set up (Curve 1). Because the Sun's apparent annual motion through the sky is along the ecliptic, not the equator, it seems to run 'fast' or 'slow' at different seasons of the year (Curve 2). Add these together and we get Curve 3 — the amount by which, at any date, the apparent or sundial day differs from the mean or clock day

the equation of time (see the explanation under Figure 2.1) indicate the times of the year when a sundial is running fast on a mechanical clock, the constant discrepancy being due to location within a time zone.

Solar time is more difficult to accurately determine than sidereal time which is based on the interval between successive transits of the same stars across a meridian. A meridian is the imaginary line (of location) running through the zenith and between the celestial poles. As the star crosses the meridian it is said to be in culmination and the interval between successive culminations of a star located on the celestial equator (Figure 2.2) is at present about 23 hours, 56 minutes, and 4.1 seconds (O'Neill, 1975, 16, to which further reference is recommended for a detailed account of universe or astronomic time reckoning methods). Thus the sidereal day is about 3 minutes and 56 seconds shorter than the solar day. These and other irregularities and anomalies in the item relations upon which universe time is based have made an unfavourable impact in those areas of activity, in advanced scale urban societies, where there is heavy reliance on field and laboratory experimentation

Sidereal time

which requires precise time reckoning. The tracking of satellites has demanded the use of atomic clocks and these demands have stimulated a search for ever more precise, system-specific time reckoning schemes. Thus, although universe time is the principal device by which we order our social systems, it is not necessarily an adequate basis for precise understanding and control of all man's needs. Although it seems most unlikely that we shall replace it, within any conceivable future, there are already signs that advanced societies are becoming aware that for all its needs and purposes, universe clock time is not inviolate. There may well be other systematic recurrent relations which prove to serve man's needs better than does the familiar clock, whose ticking has for so long been both a comfort and an irritant!

(ii) Measurement principles

'There is not one way of measuring time more true than another: that which is generally adopted is only more convenient. Of two watches, we have no right to say the one goes true, the other wrong; we can only say that it is advantageous to conform to the indications of the first'
(Poincaré cited in Whitrow 1972, 97; and Whitrow, 1961, 45).

As we have noted accessibility, continuity and invariability are the principle criteria of time measurement. A clock is not adequately defined merely as a device or as an instrument for measuring time. It is 'generally speaking... simply a very observable phenomenon, the temporal law of which is known' (Gonseth, 1971, 277). That clocks 'measure time' assures us only that they employ predictable regularities' (Fraser, 1975, 64). In fact, any phenomenon which evolves into a series of states of observable magnitude and precise form (such as the planets moving among other heavenly bodies, the position of which can be precisely fixed as co-ordinates) is a clock. But 'in order to set up a clock, one has to know how to utilize the temporal law of a relatively well-determined and clearly observable phenomenon' (Gonseth, 1971, 277). This means that two clocks are always required. The practical problem then involves the transformation of a set of observational tables to a table of expectations (which may be theoretically or empirically based), within acceptable boundaries of experimental measurement error. The 'table of expectations' itself may be used as the other clock. Measurement of anything involves the assignment of numbers to observed items according to rules, and the clock (or the table of expectations of recurrence), is simply a means of doing this.

The idea of time measurement and the precise determination of what a clock is, remains as an area of indecision. In 1932, Percy Bridgeman who was then Hollis Professor of Physics and Natural Philosophy at Harvard University, wrote: 'If anyone should ask me

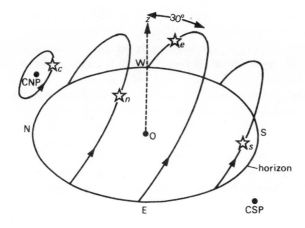

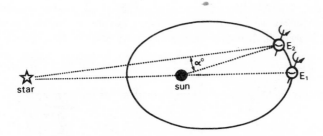

Figure 2.2. Universe item relations as basis for time reckoning. (After O'Neill, 1975. Reproduced by permission of the publishers, Sydney University Press)

(a) The apparent paths of four stars for an observer, 0, at 30 °N in terrestrial latitude. The zenith, z, is directly over his head. A star such as e which rises due east and sets due west traces the Celestial Equator which is tilted 30 °S of the Zenith; its visible path is half a circle. Stars such as n and s also follow a path tilted south by 30 °, n tracing more than half a circle and s less than half above the horizon. A star such as c, a circumpolar, never cuts the horizon; the centre of its circle is the Celestial North Pole (CNP). Below the southern horizon is the Celestial South Pole (CSP)

(b) As the Earth rotates in approximately one day it revolves around the Sun in approximately 365·25 days. Thus on occasion E_1, the Sun and some fixed star will be in conjunction, whereas the occasion E_2, the Sun and that fixed star will be separated by $\alpha°$. On average the Earth's rotation requires about 4 minutes per day to cover this angle

for directions for constructing a clock or for specifications by which he could determine whether an instrument which purported to be a clock was actually one or not, I could give him no satisfactory answer. In fact, here at the very root of one of our most fundamental concepts [time] is a surprising deficiency — no satisfactory definition has yet been given of a clock. The best that we can do is to point to actual mechanical clocks, and you are of course proudly conscious that the clocks par excellence are afforded by the bodies of astronomy' (Bridgeman, 1932, 87-88). In the 45 years since Bridgeman wrote there has been very little clarification of the situation so far as we are aware.

Calendars and clocks are the means which we use most often to tell time, but the milkman, postman or bank clerk who lives across the road may behave with sufficient regularity to serve the same purpose for the timing of certain items.

The accuracy of mechanical clocks has been increasing at a dramatic rate over the past 275 years (Figure 2.3) and this increase is highly correlated with the growth of advanced, technologically sophisticated, high energy consumption urban systems. There has

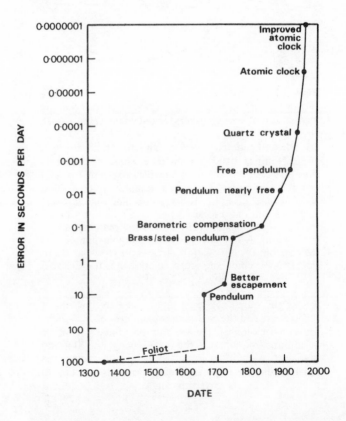

Figure 2.3. The rise in accuracy of the clock. (Data based on Ward, 1961)

been a rapid and penetrating diffusion of timepieces to all corners of the world. In 1945 the geographer Derwent Whittlesey commented that their widening diffusion was a 'symbol of the fact that life moves in a time-dominated world' (Whittlesey, 1945, 23).

The following snippet from the British daily newspaper, *The Guardian,* indicates the heightened public awareness of time in modern society as well as providing rather a good summary of the contemporary state of the art of time reckoning. You will note that 'universal' time is the term used in the international reference standard, and not standard time. Note also that the earth is accelerating. When we discuss social time you will come across the notion of acceleration once again.

Just a second ...... you mean the clock's gone wrong?

By Anthony Tucker,
Science Correspondent

IF THAT new crystal controlled clock or watch of yours is right on the ball, then you might just be able to confirm that December 31 will be precisely one second longer than the 24 hours we usually allot to a single day. The extra second may well vanish in revelry as the clock strikes midnight but it has a serious purpose.

Without it, the clock time set by international reference standard from Paris and known as Co-ordinated Universal Time would drift slowly away from the practical time set by the rotation of the earth and the position of the sun.

Just as an extra day has to be inserted into a leap year to prevent calendar midsummer drifting slowly toward solar midwinter, the leap second prevents clock midday drifting slowly toward solar midnight.

The trouble is that the two time scales based on movements of the earth, which have served the purposes of men since the beginning of recorded history, are disturbingly erratic by comparison with the atomic time which has served as a reference standard since midnight on December 31, 1971.

Not only is the earth slowly accelerating in its orbit round the sun, so that the period from one vernal equinox to the next (a tropical year) is shortening by about five thousandths of a second a year, but the earth's rotation on its axis is slowing down. The year is getting shorter while the day is getting longer.

True, in a million years the mean solar day will be only about twenty seconds longer than now, and by then the orbital year will have reduced from 365.242 mean solar days (1977 length) to about 365.1 days. Earthlings will then have a leap year once every 10 years instead of four, a point of interest, perhaps, only to those born on February 29.

In the old days the mean solar day used to contain 86,400 seconds exactly, because that was one basic way of defining the length of a second. But, armed with the great precision of the atomic clock — which was refined to international time standards at the National Physical Laboratory in England during the late 1950s and early 1960s — the 1967 International Genera! Conference on Weights and Measures in Paris

declared all heavenly bodies and astronomical measurements redundant as a means of determining units of time.

From that year onward the second became a duration of time measured by 9,192,631,770 radiation transitions between two hyperfine levels of the caesium 133 atom. Sadly, however, the new atomic second did not fit exactly into the old solar day. When atomic time began in 1972, a mean solar day contained not 86,400 old seconds, but 86,400.002 new seconds.

This discrepancy, which will get larger as the earth slows down, amounts to almost a second a year, a quarter of an hour in a millennium. Such an appalling discrepancy is not being allowed to accumulate. Every now and then since 1972 Universal Atomic Time has been stopped for a clear second to let the earth catch up. If you had not noticed there is no cause for alarm: it's all under control (more or less).

(Science Correspondent, *The Guardian,* 1977, Reproduced by permission of Anthony Tucker).

44

(iii) Calendars and clocks

A calendar normally uses units of time which are based on the intervals and cycles of natural processes. The year and the day are the most important, but there are 'years' which are not determined in this way and 'days' which are not: for instance: financial years, from June to June in Australia or April to April in Britain, and workDAYS that are only about eight hours long. The natural or universe calendar year is 365.242199 days long, using the Gregorian calendar system of 1582. The subdivision of the year into months of different lengths however, does not satisfy the measurement criterion mentioned above, that units should be of equal length.

Synodic month

Week

There was a time before the introduction of efficient artificial light when the month varied in length according to recurrent phases of the moon and this was called the 'synodic' month. The week, however, is an entirely cultural unit of time. It is not based on any natural cyclic periods. It is the most fundamental unit of social time (Chapter 2.4) and was established in the Roman era, with no changes being made when the Gregorian calendar replaced the Julian system. The twenty-four units used to subdivide the day are also quite arbitrary; as is the sixty minute hour and the sixty second minute. The twelve hour day and the twelve hour night of ancient custom, adjusted to daylight and darkness gave way to a single twenty-four hour day. This may have resulted from the simple arithmetic fact that 12 is the smallest number divisible by 2, 3, 4, and 60 is divisible by 5 and 6 as well; so 2, 3, 4, 5, 6 are convenient factors. But other systems might just as well have been used. As the incidence and complexity of experimental work grows, replacing the traditional 'way of doing things', the need for ever more accurate time devices increases (see also Section 2.4 iv). These needs are most apparent in experimentally based cultures — urban cultures.

The first written record of a man-made subdivision of the year is accorded to the Sumerians in about 3500 B.C. The same people are also often credited with rocking the 'cradle' of urban civilization. Their ideal year was of twelve months, each day was twelve danna long and each danna was subdivided into thirty ges (making about 360 ges in a day). A ges was about four modern minutes. By 2000 B.C. the Egyptians were using a shadow clock which was portable and shaped much like a draughtsman's T-square. The hours which it marked off were not, however, of equal length (Figure 2.4).

By 1470 B.C. the Egyptians had built a 97 foot high obelisk at Karnak which was in fact an enormous sundial (Figure 2.5). There were serious difficulties in standardization of the metric due to variations in the length of seasons and the occasional lack of sun did not allow 'continuous' reference!

If instruments to measure time like the Karnak obelisk were capable of being built, and often at considerable expense, then

45

Figure 2.4. The Egyptian shadow clock.

The clock was pointed towards the sun so that the shadow of the cross-bar fell on the hour scale marked on its handle. The scale included five hour lines plus the noon line. In the morning, the shadow clock was held with the cross-bar towards the east; then it was turned to the west for the afternoon. (See also *Time,* Time-Life Books, 1970)

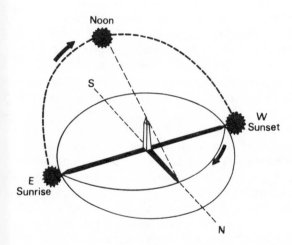

Figure 2.5. The clock obelisk at Karnak.

This was the regional time piece. As the sun travelled from east to west its shadow moved steadily in the opposite direction within the oval area around the obelisk. The three shadows shown here are those thrown by the obelisk at sunrise, noon, and sunset. (See also *Time,* Time-Life Books, 1970)

presumably there was a sense of the need for them. In effect this meant that time had been culturally identified. Cultures would never be the same again, in particular because time measurement increased the authority of the governing groups. Time became a valuable 'resource' of government, a source of social and political power, and it has remained so. Time boundaries set up by super-structural authorities are as effective as real fences in restricting access or exit!

Time as a resource

Mechanical time pieces have become ever more readily available and accurate, and Mumford (1934) has gone so far as to suggest that 'the clock, not the steam engine, is the key machine of the modern industrial age. . . no other machine is so ubiquitous'. With the Industrial Revolution everyday life became a servant to the tyranny of the clock. To the present day the regularity of daily urban life is a direct result of the ubiquity and consistency of clocks and of the stability of calendars. How has this come about? In the Middle Ages the bell was an institutional time piece which reminded and instructed men when to start work, when to finish work and for some it indicated that a 'life time' had come to an end!

Knowledge of the relative position of stars and planets was closely related to the likely success of all kinds of earthly activities. Thus both astrology and astronomy were a part of the development of mechanical clocks, often designed to embody the 'heavenly clockwork'. The great clock at Padua designed and built by Giovanni de Dondi between 1348-1364 was built in the likeness of the universe and thus was 'also started by God'. But early clocks, however elaborate they may have been, were notoriously inaccurate, and until the pendulum was successfully incorporated into clockworks, by Huygens in Holland in the mid-seventeenth century, clocks often only had an hour hand. But there was a growing awareness of the importance of accurate time measurement, in an environment of increasing scientific curiosity and economic activity. Commercial organisation promoted a growing awareness of the value of clocks as aids to navigation in an era of New Worlds, new markets, and new production methods (Le Goff, 1978).

Only from about 1674 were accurate pocket watches available, and then, of course, only to a wealthy minority. They were not generally available to the large majority of the public until about 1850 (Thompson, 1967). Time pieces were first used by the upper classes to usher in industrial capitalism. Although 'the worker' did not have a clock or watch, their effect upon him or her was still great. The factory hooter, timed by the factory clock, became a strict marker and pacemaker for social behaviour and the development of a strict work ethic (Weber, 1958; 1920). In the interval between its signals activities could either be arranged with some discretion or (more to the point) with none. The Protestant work ethic was reinforced by, and in turn reinforced, the authority of the clock. It took many years, perhaps a generation or more, to discipline craftsmen into the factory regime, and this learning process was still diffusing outwards from the cities and large towns in the mid 1880's in some parts of England, Scotland, and Wales. A similar process has occurred in the technologically young countries of today. Working hours took over from working days. As industrial capitalism was transformed into monopoly capitalism, with its burgeoning bureaucracy of decision makers and non-executive administrators, the working minute took over from the working hour and the working day. The 'Marxian evaluation of the clock did not reinstate a relationship between the individual and the clock but only between society and the clock. "The clock is the first automatic machine applied to practical purposes," wrote Marx to Engels in 1863, "the whole theory of production of regular motion

Factory clock as marker and pacemaker

was developed through it.'" (cited in Fraser, 1975, 66). The place of time in the social history and geography of the past two centuries, is firmly rooted in the centre of every major issue related to production, consumption and organization.

Universe time is given to us by astronomers and is used as a co-ordinate for ordering events (Haber, 1972). It is the time which we are all most familiar with, but it is nonetheless a mutilated version of the purest universe time which is available to us. Social constraints of authority, tradition, and ignorance have added their own voices to the rhythms provided by the planets. The voices of time are always a chorus in rehearsal, never to reach perfection!

(iv) Time and regionalization

The regionalization of the world into 24 time zones (Figure 2.6) displays the intervention of socio-cultural 'whim and fiat' (Clemence, 1968, 408) into the otherwise objective, locational symmetry of universe time. The light shaded areas of irregular shape show the national boundaries of political states and the time which they adopt relative to the meridian zone time. The black shaded areas identify states which have not adopted zone time since agreement was first reached at the Washington Meridian Conference in October 1844.

Apart from the differences which are directly related to international zone time, further differences within a single zone result from the practice of Daylight Saving. The usual form which this practice takes is the forward movement of the clock by one hour at some time around the end of March in the northern hemisphere and about the middle of October in the southern hemisphere. The clock is then 'put back' an hour in the autumn months. These variations in local time create social and commercial difficulties when the terrestial relations between two zones is north-south and one space adopts time saving while the other does not. In Australia, for instance, New South Wales adopts daylight saving by putting all clocks forward an hour in October, but Queensland, immediately to the north, does not. At the territorial boundary between the two States distances are transformed (relative locations are adjusted) so that Queensland comes one hour nearer to New South Wales,[1] at least until the end of the 'working day' in New South Wales, say at 1700. This is so because communication from New South Wales to Queensland has to *wait* for an hour (i.e. about 550 miles in terms of jet flight speeds!) whereas from Queensland to New South Wales there is no such delay until the end of the working day. In Indiana some counties in the state do not go onto Daylight Saving, others do. The significance of time zonation becomes most apparent in relation to communications; the best known example of its effects being associated with the so-called jet-lag syndrome in which dysrhythmia of body and psychological

48

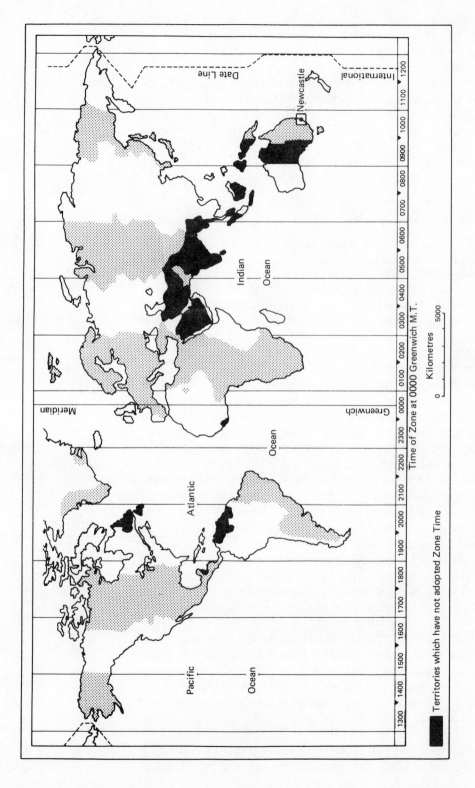

Figure 2.6. Universe time regionalization

Territories which have not adopted Zone Time

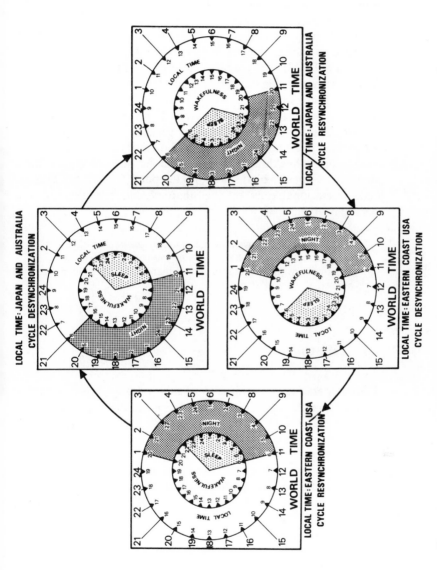

Figure 2.7. Dysrhythmia and long distance flight. Phase shift of the circadian cycle (left) flight from the eastern United States to Europe and back (right) flight from the eastern United States to Japan or Australia and back. The outer area of the squares shows Greenwich Time (universal time). The outer large circle gives the local geographic time and the smaller inner circle the physiological time. The black section of the outer circle indicates night and the shaded section of the inner circle the normal time of sleep. (Reprinted by permission of Charles Scribner's Sons from *Your Body Clock* by Hubertus Strughold. Copyright © 1971 Hubertus Strughold)

functions occurs. An illustration of this effect is shown in Figure 2.7.

Communication by media other than air flight is also affected by regional time zones. The severity of dislocation is related to distance, international zone, and any other social intervention which operates between the origin and the destination. 'A general problem we could face over the next several decades is that of co-ordinating human activity over continents or over the entire earth... a major problem which Ecumenopolis (Doxiadis, 1968) will have to face is that of different times' (Abler, Adams, and Gould, 1970, 564). For commercial linkage within single countries of vast east-west scale such as Australia, for instance, there may be only four hours during the day for commercial interaction between Sydney on the eastern seaboard and Perth on the western seaboard.

At local scales time is also used to zone space. Car parking restrictions, retail hours, school, and university terms, surgery and clinic hours in health care, are all examples of local space timing. All of these schemes bind space into temporal frames. But they are made operational only by the ubiquity and precision of time keeping devices.

> 'Clocks represent ways in which man the toolmaker identifies and puts to use what he regards as the lawful content of his experience of time. It is appropriate therefore to consider man himself.'
>
> (Fraser, 1975, 71)

2.3 Life-times

'Time is particularly significant to man because it is inseparable from the concept of self. We are conscious of our own organic and psychological growth in time' (Mayerhoff, 1960, 1). However, 'we might (either) consider time as diluted poison, administered slowly in increasingly harmful doses' or as Remarque once wrote, 'as a diluted poison administered slowly in harmless doses!' (Fraser, 1968). Whichever view we adopt, and all three are relevant, one thing is clear, time and life are inseparable.

We have been discussing mechanical devices called clocks, which we accept as measuring time in a way that is invariant between users. At least this is apparently so once we pass the age of seven (Piaget, 1968, 203). But we also have a personal, particular sense of time, our bio-psychological times that influence behaviour. Somehow we manage to navigate a path through regions of space and durations of time which may be as spatially extensive as the entire world and as enduring as a life-time or as spatially contained as a room and lasting no longer than a minute. We do this by combining our own sense of time with that of the clock and our own image of space with that of a map. In doing so we use many times and not only the universe time described by the hands of the clock.

(i) The items in relation

'We are so familiar with some of the main rhythms of life — sleeping and waking, menstruation, etc, that we organize our personal activities to fit in with them. It is probably not the case, however, that social organizations take these things into account to any greater extent than having employment organized round an 8-hr working day. If we understood more about the effects of rhythms and how to manipulate them, then there might well be opportunity for increases both in efficiency of work and enjoyment of leisure'.

(Oatley and Goodwin, 1971, 35)

'What constitutes a physiological clock? Is there a master clock? How are clocks coupled to the light-dark cycle?' The biologist Pittendrigh has asked whether biologists should 'abandon the common current view that (their) problem is to isolate and analyse the endogenous rhythm or internal clock. Instead of a single clock controlling or modulating the activities of what would otherwise be steadily maintained processes, we should, rather, expect that the organism comprises a population of quasi-autonomous oscillatory systems'

(Pittendrigh, 1960, 165; cited in Oatley and Goodwin, 1971, 34)

Physiological clock or clocks?

Whatever our spatial or category scale of research interest might be, human geographers are always involved with human individuals and it behoves us to be aware of the temporalities existent in spatial structures and therefore of the temporalities which operate within the individual. Hägerstrand (1973) has stressed that there should not be any sacred lower limit to scale in human geography. This is apposite because it is at the smallest human scale levels that life-times become rather important determinants of behaviour.

Life is a complex system of interlocking items, relating to one another periodically and often in a systematic or rhythmic manner. Though much of the contemporary scientific knowledge about the times of life has been founded on theory and experiment related to plant and simple animal organisms, there is a growing literature about the times of the 'human life sequence' (Halberg, 1969, 1975; Reinberg, 1974; Orme, 1969; Colquhoun, 1971). A useful distinction may be made between somatic and psychological times, but both are in the strictest sense a part of biological time: a part of life time. The items in relation are various — biochemical substances, neural networks and stimuli from the environment beyond the co-ordinates of the individual body.

Human life sequence

Life time

Many years ago Korzybski (1921) argued that the distinguishing factor between Man and other animals was his ability to make conscious decisions that enabled him or her to bind time. This is evident in the ability to make future-related decisions about time use, especially on times which are beyond his/her life expectancy. But there are certain somatic and psychological recurrences over which there is very little control, except in some cases by drug induction. Many are thought to be closely related to the development of personality, mood, and behaviour (Blake, 1971; Orme, 1969) and Orme has summarized some of the experimentally determined relations, between internal clock speeds and typologies of mood, personality, and behaviour. Having studied Table 2.1 you may well wonder what on earth it has to do with human geography. The answer is that it cautions us, as human geographers, to be aware that the study of human individual behaviour cannot be divorced from some appreciation of human biology and psychology, brought together here by internal clocks. It is rather surprising that a detailed understanding of biochemical processes in plants is an assumed component in the study of biogeography and non-human ecology, but for *human* geography it seems to be totally unnecessary to have any appreciation at all about biological Man and how he 'works'. The study of life-times can put this omission right, without too much difficulty, at an introductory level at least.

(ii) Biological time

This is time within the coordinate system of the organism (Fischer, 1968). It may be 'defined as a creature's awareness of its own duration and location within a restricted frequency range, in contra-distinction to traditional concepts of time which attempt to be independent of life. . . biological time is the transformation of perceived sequences into durations' (Fischer, 1968, 358). But what do biological clocks do? They 'enable living organisms to reproduce required responses at appropriate times' (Whitrow, 1972, 68). The bioclock is a name given to the system of regulators within a living organism. These regulators operate with sufficient synchronization of rhythms to give the impression that there is a single bioclock. The notion of 'innate rhythms requires an explanation in terms of some internal, autonomous oscillator' (Oatley and Goodwin, 1971, 9). Two types of biorhythm may be identified: the exogenous biorhythm dependent on external environmental cycles and an endogenous biorhythm having an independent origin as an inherent property of life.

Bioclock as a system of regulation

Exogenous
Endogenous

Biological clocks have been described in terms of three broad categories. First there are continuously consulted clocks, involving systems concerned with time-compensation, for instance when animals are navigating to a celestial system. Second, there are

Time compensation

Table 2.1. Personality, mood, and internal clock speeds (Orme, 1969, 87. Reproduced by permission of J.E. Orme)

See also Blake (1971) and Colquhoun (1971) for a critical discussion.

	Internal clock slow Inner time units long with:	Internal clock fast Inner time units short with:
1.	(ii) Filled time interval	Unfilled time interval
2.	(ii) Difficult tasks	Easy tasks
3.	(ii) Subject interested	Subject bored
4.	(iii) (B) (iv) Stress, pain	Sensory deprivation
5.	(i) Passive attitude of subject	Active attitude of subject
6.	(i) Introversion	Extraversion
7.	(iii) Authoritarian parents	Non-authoritarian parents
8.	(i) High academic achievement	Low academic achievement
9.	(i) Ability to inhibit behaviour for a future reward	Inability to inhibit behaviour for a future reward
10.	(i) Relatively more phantasy	Relatively less phantasy
11.	(i) Smaller 'dissociation' score	Larger 'dissociation' score
12.	(i) Rise in skin temperature under stress	Decrease in skin temperature under stress
13.	(i) More error on a vigilance task	Less error on a vigilance task
14.	(i) Slow, careful	Quick, careless
15.	(i) Longer spiral after-effect	Shorter spiral after-effect
16.	(i) Anxiety states, neurotic depressives	Hysterics, psychopaths, delin- quents
17.	(ii) Melancholics	Manics
18.	(iii) Non-paranoid schizophrenics	Paranoid schizophrenics

interval timers controlling discrete events such as the onset of an activity and third there are clocks or systems of regulators controlling pure rhythms where the amplitude of a continuous process varies with time. Richter, on the other hand, 'postulates three types of internal clock: peripheral, central, and homeostatic' (cf. Cohen, 1967, 8). When the 'fabric of time in biology' (Fischer, 1967) is isolated for study, the difficulties are considerable because it is hard to isolate an organism from its environment. 'The synchronizing of environmental information with the cycle-bound information decoding process of self-organizing, self-referential systems is biological time, that is, the inseparable meaning of systems and their environments' (Fischer, 1967, 479).

Interval timers

Rhythm control

Biological time

The fundamental control over rhythm in living organisms is probably genetic, but other controls are apparent, and they also influence timing. The sex ratio in a group and the number of individuals in a population may influence the timing of behaviour. Changes in rhythm occur in some individuals when they are isolated, but when they are returned to their group or species, rhythm is rather quickly re-established. The inner clock, whatever it is, seems to maintain its rhythm except under the influence of unusual stimuli when it may drop its measurable threshold. A truly biorhythmic condition, once 'set by environmental stimulus will persist in favourable conditions (but)... when environmental variables are altered, the rhythm might alter' (Orme, 1969, 102) with consequential changes in behaviour.

The American biologist, Brown, proposes that the earth's geomagnetic field is in fact an exogenous influence on the maintenance of many internal rhythms. 'Circadian (i.e. approximately one day in period) rhythms appear to be simply one component within a "bio-geo-solunar" rhythm complex' (Brown, 1969, 199, 200). He argues for two types of timing in association

Co-ordinate timing

Durational timing

with biological clocks; co-ordinate timing and durational timing. The former refers to the timing of rhythmically recurrent events through 'responsiveness to the systematically changing environmental fields' and the latter to 'some fully independent, internal timing mechanism capable of reproducing the natural geophysical periods' (Brown, 1969, 199, 200). Some of his views differ from those of many other biologists (Cloudesley-Thompson, 1969, 299; and Ward, 1972, 204) and are possibly most at odds to those of other 'bioclock watchers' when he argues for organismic sensitivity to geo-electromagnetic fields of influence (Brown, 1969, 291-296). The important point to grasp, however, is that Brown is arguing for internal clocks and not against them. It is on the question of the source of their periodicities and the mechanisms by which they are set which he differs.

Two cyclic periods of different length are of particular significance to biological time and the organization of social conduct. They are the circadian and the circannual. The month is of obvious significance for biorhythms in women, but is of less direct significance to the organization of social behaviour than the day and the year, and so we will not say much about it here. The week, on the other hand, is a period of greater social consequence than it is of biological significance, although there have been published findings of the existence of circaseptan (approximately 7 day) rhythms (Halberg, 1969, 1975). Circadian rhythms in man are generally defined as medial frequency rhythms and circannual

Circadian and circannual cycles

Medial and low frequency

rhythms as low frequency. There is nothing 'micro-scaled' about the study of human geography in a day. The 'day' is a clearly observable scale, it is 'macro-scaled', as is the individual.

(iii) Medial and low frequency rhythms

A circadian rhythm is an oscillation with a period of about twenty four hours. Biological circadian rhythms have been defined as 'statistically validated physiologic changes recurrent with reproducible waveform about every twenty-four hours' (Halberg and Katinas, 1973). They often display a remarkably symmetrical sinusoidal wave form as shown in Figure 2.8 below. In the chronobiological literature such periodicities or cycles are often transformed to a circular graph representation as in Figure 2.9, known as a cosinor (*cosine* and vec*tor*) graph. Some discussion of this graph will be found in the Appendix and elementary aspects of the method will be used to analyse some urban distress rhythms which are to be discussed in Chapter 8.

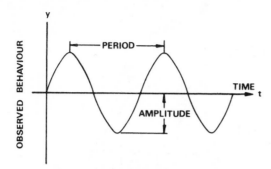

Figure 2.8. A sinusoidal waveform

A circadian rhythm will usually be composed of higher frequency or ultradian rhythms. That is to say that during the course of a twenty four hour period, there will be fluctuations or oscillations occurring, some of which are in themselves cyclic. They may form the basis for rather important behavioural routines, as for instance when associated with stress (Cullen, 1978; and Chapter 5 following). These oscillations may in fact be linked or even synchronized to environmental signals, more clearly than is the medial circadian frequency rhythm itself. Figure 2.10 illustrates these rhythms.

Oscillators are responsible for rhythms. Oscillators may be linear or non-linear. In Figure 2.8 the pure sinusoid would represent a linear oscillator. More precisely a linear oscillator 'is produced when two energy stores transfer energy alternatively from one to the other at rates which depend linearly on the amount stored at each instant of time' (Oatley and Goodwin, 1971, 3). Non-linear oscillators are more typical of biological systems; the most common form is the so-called relaxation oscillation type, Figure 2.11. 'Here a steadily rising [or falling] variable crosses a threshold and then resets to a previous value' (Oatley and Goodwin, 1971, 4). An important characteristic of non-linear oscillators — which we shall be taking up in the next chapter — is that they tend to 'entrain or synchronize with a periodic driving stimulus, or with each other when coupled together even weakly. Coupling in this sense is any influence that one oscillator has on another. An alarm clock is an oscillator which is more or less coupled into the rhythm of sleeping and waking' (Oatley and Goodwin, 1971, 5). **Entrainment**

The principle of entrainment may be a fruitful one in certain aspects of urban social geography, and we shall be pursuing it in the next chapter. For the moment it is simply as a 'characteristic' of life time that we note it.

56

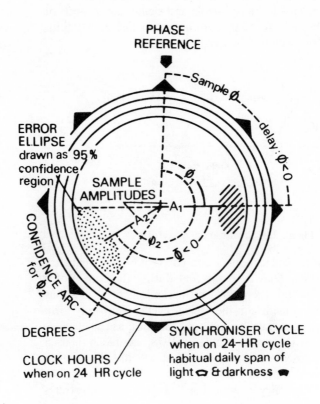

PHASE
REFERENCE

Figure 2.9. A cosinor display for biological rhythm analysis.

Results of the analyses of bioperiodic phenomena are given as polar coordinates. The circumference represents the considered period τ. τ is equal to 360°, irregardless of the length of the period, be it 24 h, 7 days, 30 days, 1 year, etc. The synchronizer cycle can also be represented with its timing.

The error ellipse surface depends essentially upon the dispersion of phase (ϕ) and amplitude (A) computed from time series on the one hand, and from the statistical p value.

When the error ellipse overlaps the pole it means that no rhythm is detected with (a) the considered period, τ, (or frequency $1/\tau$) (b) with similar phases and amplitudes. (The number of samples may be too small; the dispersion of ϕs and As may be too wide or the biologic period of the rhythm does not correspond to that which was taken for the

analysis.) On the contrary a statistically significant rhythm is detectable when the error ellipse does not overlap the pole with 95% as security limits.

Then, the rhythm amplitude (A) is given by the length of the vector. The error ellipse gives the amplitude confidence interval where it intersects the vector and the prolongation of the vector.

The acrophase (ϕ) is given by the location of the prolonged vector on the circumference. It can be then expressed as a phase angle with respect to a phase reference, or expressed in units of time as a fraction of the considered period τ. The confidence arc of the acrophase is given by the tangents of the error ellipse drawn from the pole.

(Reinberg, 1974; and Halberg *et al.*, 1967)

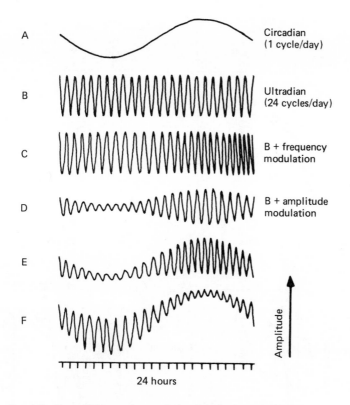

A — Circadian (1 cycle/day)

B — Ultradian (24 cycles/day)

C — B + frequency modulation

D — B + amplitude modulation

E

F

Amplitude

24 hours

E and F : Curve 'B' + superimposed ultradian rhythms combining curves 'C' and 'D'. Frequency modulation phase constant. Amplitude modulation phase varied through 180 degrees

Figure 2.10. Ultradian rhythms with frequency and amplitude modulation. (After Kripke, 1974)

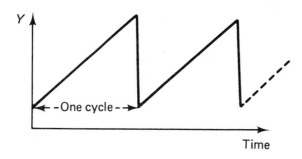

Y

←-One cycle-→

Time

Figure 2.11. A relaxation oscillation

58

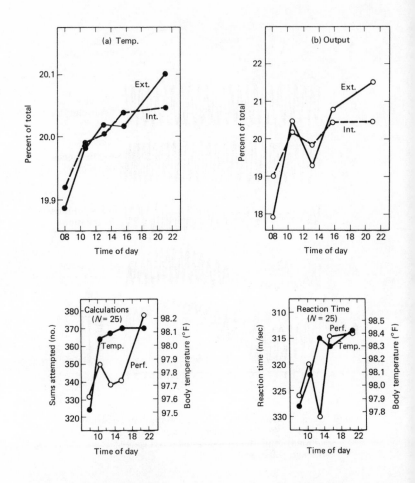

Figure 2.12. Relationship between time of day and body temperature and associated variations in performance with comparison between extroverts and introverts. (Reproduced with permission from Blake, 1971. Copyright by Academic Press Inc. (London) Ltd)

The circadian cycle is usually given a range between 20 hr and 28 hr by chronobiologists. Within this range the rhythms of various functions peak at different times and their degree of synchronization generates different aptitudes for certain tasks, as has been suggested above. For instance, nearly everyone tends to be tired for a short while around 1400. This may or may not be the result of the effects of midday food intake causing a so-called 'post lunch dip' in the curves of vigilance, calculation accuracy, serial reaction rates, etc. (Blake, 1971). Table 2.2 illustrates part of the spectrum of biological rhythms as summarized for chronobiologists by Halberg (1969).

Table 2.2. Illustrative spectrum of rhythms (After Halberg, 1969). Reproduced with permission from the *Annual Review of Physiology*, Volume 31. © 1969 by Annual Reviews Inc

Domain[a]	High frequency $\tau < 0.5$ hours	Medial frequency 0.5 hours $< \tau < 6$	Low frequency $\tau < 6$ days
Regions[a]	$\tau < 1$ second $\tau < 1$ second	Ultradian $(0.5 < \tau < 20$ hours) Circadian $(20 < \tau < 28$ hours) Infradian $(28 < \tau < 6$ days)	Circaseptan $(\tau \cong 7$ days) Circavigintan $(\tau \cong 20$ days) Circatrigintan $(\tau \cong 30$ days) Circannual $(\tau \cong 1$ year)
Type of process in which rhythms are found	Electro- encephalogram Electrocardiogram Respiration	Rest-activity Sleep-wakefulness Response to drugs Blood constituents Urinary variables Metabolic pro- cesses, generally	Mensturation 17-Ketosteroid excretion with special components in all regions indi- cated above and in other domains

[a]Domains and regions (named according to frequency criteria) delineated according to the reciprocal of frequency, i.e. period (τ) of the function approximating rhythm. It should be noted that several variables examined thus far in chronobiology exhibit statistically significant components in *several* spectral domains (After Halberg, 1969)

A foreshortening of time horizons in early afternoon has also been noted by Thor (1962). There are also some observed differences in the length of time perspectives between men and women, Figure 2.13. Body temperature falls to its lowest point between 0100 and 0700; oxygen consumption increases during normal peak hours of the body's activity whether there is an increase in 'normal' social activity or not; heart beat drops to its lowest between 2200 and 0700; adrenal gland secretion dips during sleeping hours but rises before waking to prepare the body for the day's activities; kidney excretion is at its peak at midday. There are many other examples which might be selected. All point firmly to the healthy human being as a rather finely tuned 'creature', and human geographers may find connections because *behaviour* is apparently susceptible to periodic variations in certain of these functions.

We sleep for nearly ⅓ of our life-sequence and this factor alone is crucial to the spatial organization of society, although why we sleep is still not all that clear. The obvious answer is because we need to recuperate, but this reason would not receive a favourable hearing among physiologists any longer. Webb suggests that sleep is a necessary *behaviour controlling* function. It 'is a process in aid of preventing non-adaptive, ineffective, or destructive behaviour from occurring rather than a time period in which recovery from behaviour episodes occurs' (Webb, 1971, 170). The implications

Sleep
Behaviour control

Human ecology

for human ecosystems and their contemporary thrust towards a colonization of night time (Chapter 8) are quite significant. The spatially and temporally mobile society is *not* absolutely free to occupy incessant locations in space and time simply to suit its commercial and hedonistic wants. If it does so, there may well be a very high price to pay. Sleep at regular intervals and at night is a very significant human need.

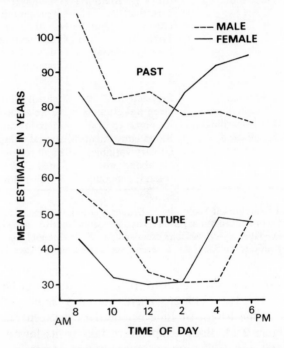

Figure 2.13. Sex differences in past and future time perspectives and time of day. (Reprinted with permission of the author and publisher from: Thor, D.H. Diurnal variability in time estimation, *Perceptual and Motor Skills,* 1962, 15, Figure 3)

Part of the explanation that 'the waking day begins to break up into something more gregarious after mid-day' and that 'the variety, speed and complexity of behaviour must be at its highest from mid-morning to mid-afternoon' (Cullen and Godson, 1972) may lie in certain rhythmic features of circadian neuro-hormonal structure changes. In the work of two American urban and regional planners we read that 'there is ample evidence of normal periodic variation in activities. After some increment of time spent on a single activity the likelihood of a change to a different activity

Human activity

increases rather rapidly... a period of work is likely to be followed by a period of relaxation, a period of active recreation is likely to be followed by one of a passive character' (Chapin and Hightower, 1966, 63-64). It has also been suggested, following empirical study,

Stress

that the stress involved in daily routine correlates with efficiency curves (Cullen and Phelps, 1975, 43-44). Thus much daily variation in activity and in the amount of attention applied to activity is possibly associated with the somatic rhythms in life-times.

(iv) Biological time and regionalization

At frequencies lower than the circadian cycle, there are also important biological rhythms. Death from cerebral vascular lesions has been found to have a circannual rhythm in France (Reinberg, 1974, 426-427) which appears to be unrelated to any exogenous oscillator or zeitgeber in the 'physical' environment. Like the circadian rhythm, the circannual rhythm may also be displayed by chronogram (a) and by a cosine wave as the first stage in a cosinor (b) analysis, a normal two dimensional graph, Figure 2.14.

Circannual cycles

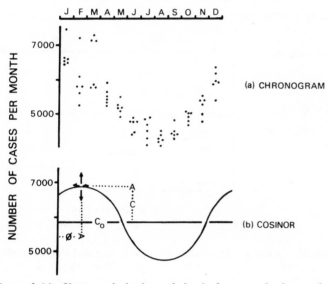

Figure 2.14. Circannual rhythm of death from cerebral vascular lesions in France. Chronogram (Top). For macroscopic inspection, monthly data are plotted as a function of time for each of the considered years. Circannual variation is apparent, yet the task remains to objectively describe the rhythm and to quantify its characteristics. 'Cosinor' (Bottom). The rhythm is approximated by the least squares method with the 365·25-day cosine function best fitting all data. A statistically significant circannual rhythm ($p < 0.005$) is detected; it can be characterized by the point estimation of several of its parameters with their respective 95% confidence limits. 'Acrophase'. ϕ (peak of the best fitting cosine function. ϕ = Feb.20 (Feb.9 to March 2). 'Rhythm — adjusted level', C_0 (or 'Mesor', M) (here monthly mean of death) = 5837 ± 133. 'Amplitude' C (or A) = 1057 (685 to 1430). In other words, at the time of the circannual acrophase the monthly number of deaths from cerebral vascular lesions in France averages 6894 ± 373, whereas six months earlier or later, the corresponding value is 4780 ± 373. (After Reinberg, 1974. Reproduced by permission of A. Reinberg)

Comparison of a range of pathologies for France as a whole and for the Hôpital Ferdinand in Paris are shown in the circannual **acrophase** (or peak of rhythm) chart in Figure 2.15. The 95% confidence interval is shown by a line on each side of the acrophase. A generalization which follows is that the acrophase for death from the pathologies shown here occurs earlier for France as a whole than for deaths recorded in the Hôpital Ferdinand in Paris; except for coronary disease when the acrophase is the same, in mid-February, and for malignant tumours including leukemia when the Paris peak precedes that for the whole of France by 8 months. Data for suicide and acute intoxication were not available for both France as a whole and for the Paris hospital. For suicide there was not enough data for a statistically significant value to be computed for the Hôpital Ferdinand (people tend not to commit suicide in hospital!) and for acute intoxication only figures from the Hôpital Ferdinand were readily available.

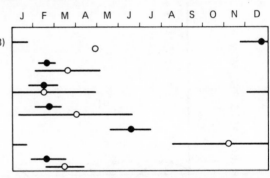

Figure 2.15. Circannual acrophase chart for France and Paris. (Reinberg, 1974, from Reinberg, Gervais *et al.*, 1973. Reproduced by permission of A. Reinberg)

Further comparison of the circanual pathology phenomenon can be illustrated by comparing death from heart disease, suicide, and accidents in Minnesota during the period 1941-1967. The representation is now by cosinor chart (Appendix A) but the similarity of the acrophase for heart disease and suicide is quite striking. From Figure 2.16 note that the suicide data in Minnesota has a circannual acrophase on June 20th for the period 1960-1967 with a 95% confidence arc between early June and July 1st. (An identical pattern was obtained by independent research in France). Each acrophase in the circannual rhythm of pathology identifies the times when social and geographical spaces are most susceptible to pathological disturbances affecting work, family, and in general the social geographic system. Mapped in charts of this sort, the conceptual similarity between the spatial maps of medical geography and the temporal map, is revealed. The reference base is

Susceptibility

still universe time, however. With sufficient and precise observations on the item relations (death from heart disease), using any 'other' precise clock as a reference frame, it is theoretically possible to construct the time of heart disease, but it may not be a 'good clock' because it may not satisfy the criterion of equal spacing or invariability — but it would be more accurate within its own system.

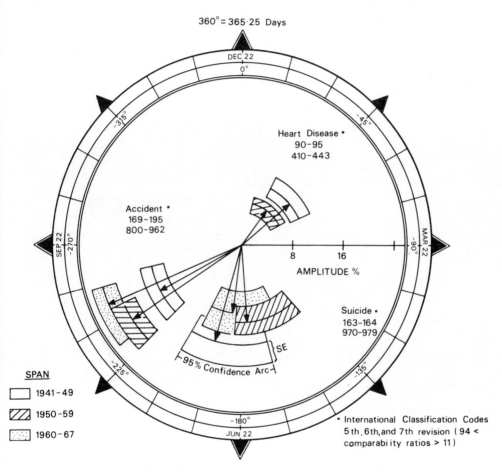

Figure 2.16. Circannual rhythm of mortality in Minnesota, USA, 1941-67 (Halberg, 1973). The reproducibility of findings in consecutive decades is very high

The geographic pattern of circannual rhythm for cardiac death in the USA, recorded by State, has been investigated by Smolensky, Halberg, and Sargent II (1972) and a demonstrable rhythm at the 5% level in 41 of the 48 states studied can be shown (Figure 2.17). The timing of cardiac death is earliest in the west and northwest, about December 15, then between mid-January and mid-February for the northeastern States, and finally between mid-February and mid-March for the southeast and midwest. Thus, while climatic influences (zeitgeber) cannot be ruled out, and nor is it suggested that they should be (Reinberg, 1974), the relatively narrow band of about 71° of arc (i.e. about 70 days) in the year seems rather too narrow compared to the range in the climatic regimen for this to be the main synchronizing agent. Other factors are involved; some are endogenous perhaps, and some are probably related to socio-cultural factors. Reinberg proposes that because circadian and other ultra annual rhythms have been identified, the role of the climatic factor is only an influence on underlying endogenous timing characteristics creating susceptibility peaks or troughs. (Reinberg, 1974).

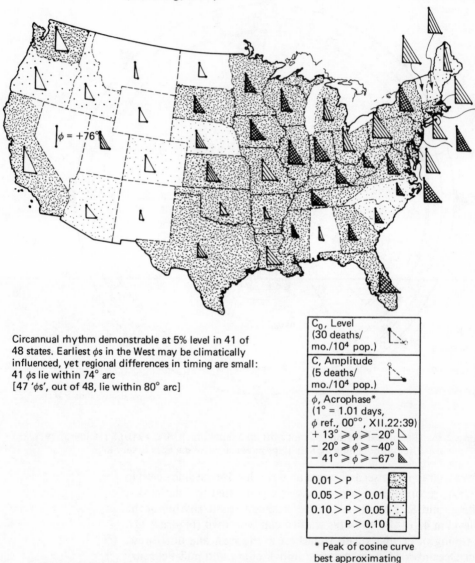

Circannual rhythm demonstrable at 5% level in 41 of 48 states. Earliest ϕs in the West may be climatically influenced, yet regional differences in timing are small: 41 ϕs lie within 74° arc
[47 'ϕs', out of 48, lie within 80° arc]

C_0, Level
(30 deaths/ mo./10^4 pop.)

C, Amplitude
(5 deaths/ mo./10^4 pop.)

ϕ, Acrophase*
(1° = 1.01 days,
ϕ ref., 00°°, XII.22:39)
$+ 13° \geqslant \phi \geqslant -20°$
$- 20° \geqslant \phi \geqslant -40°$
$- 41° \geqslant \phi \geqslant -67°$

$0.01 > P$
$0.05 > P > 0.01$
$0.10 > P > 0.05$
$P > 0.10$

* Peak of cosine curve best approximating a given state's data

Figure 2.17. Month of changing risk and cardiac death in USA (1940). (Smolensky, Halberg, and Sargent II, 1972. Reproduced by permission of the authors)

(v) Psychological time

'Sometimes it takes all the running one can do just to stay in the same place.'

Alice in Wonderland

What about the other aspect of life-time which we call psychological time? First of all note carefully that there is no undisputed evidence that somatic aspects of life-time are independent of psychological time; our mental sense of time. Though as we shall see there are some researchers who argue that at certain time scales there is some evidence for independence.

Why consider man's sense of time in human geography? There is one good reason at least: psychological time is lived time (Piaget, 1968, 211), the time of the actions and activities which make up a life sequence. For instance, it is experiential time (Chapter 1) which is possibly a component in the realization of place, operating through recurrent and innovative actions. Lived time or psychological time is based on the dynamics of relations among all sorts of items, events, or actions and some psychologists have argued (Fraisse, 1964) that it results as a direct experience of the number of changing events, objects, or items noticed by the subject.

Lived time

'Perhaps no area is more in need of exploration for its temporal implications than the field of human conduct (behaviour) and none offers more promise of fruitful reward for imaginative speculation, since all human conduct is conditioned by time perspectives of the individual and of his culture.... As will be realized the various time perspectives of a culture give the dimensions of the values that are operating in the lives of those living in that culture by specifying the conduct that must be observed in response to each situation, wherein that immediate situation is to be seen as instrumental to a more remote (future) or referred (past) situation... A time perspective (may be) imposed upon events which thereby assume an altered meaning or value that is a function of that time perspective, similar to the spatial perspectives which the individual imposes upon the objective world, enlarging the contracting dimensions by reasons of their nearness or remoteness and learning to observe these distortions in his conduct.... Each individual creates a private, personal world by structuralizing his life space' (Lewin, 1935, 1936, cited in Frank 1939).

Time perspective

Frank, who is generally considered as the initiator of the notion of time perspective, continues with the proposition that 'whole social classes may be described by the time perspectives that dominate their lives as revealed in the range of their planning, their prudential calculations, their forethought, their abstinence and so on' (Frank, 1939, 297, and Chapter 8 following). Others, such as Horton (1967) and Le Shan (1952) have expressed similar views and presented some empirical support for their hypotheses. Kendall and Sibley (1970), however, have suggested that the Le Shan study may be producing results which are an artifact of the experimental design. Ruiz, Reivich, and Krauss (1967) have also drawn attention to the possibility that many time perspective studies may not in fact be measuring the same construct. These are technical difficulties associated with time perspective measurement. That time perspectives exist is intuitively accepted.

Kastenbaum (1964) writes that 'in general, time perspective relates to one's orientation beyond the present moment'. The four dimensions of time perspective are extension, coherence, density, and directionality. **Extension** is the length of the time span involved, **coherence** is the degree of organisation of the events in the time span, **density** is the number of significant events within the span and **directionality** is the pace at which the individual feels he is moving, usually, but not always, into the future, of course. Some individuals and groups are thought to occupy preferred directionalities, either a past, present, or future orientation.

A time perspective scheme has also been developed by Gioscia (1972, 73-141). Individuals may be aligned along a vector or dimension of sensitivity to time, which has as its poles the **hyperchronic** (for high sensitivity) and **hypochronic** (for high insensitivity). Their position on this dimension may change, but it will always relate to a position on each of two other dimensions. The first is a catachronic-epichronic dimension. The **epichronic** situation is one in which the individual has power over time and therefore over others. The reciprocal is the **catachronic** position. In this situation the individual 'feels that the processes of events which constitute his situation is too heavy to be altered by his poor strengths' (*sic*) (Gioscia, 1972, 92). The third dimension is the anacronic-metachronic dimension. In the **metachronic** situation 'a goal is achieved before the participants are ready for it' (Gioscia, 1972, 89). One is ahead of universe time, in effect. The **anachronic**, or reciprocal situation is one of 'falling behind'. Gioscia suggests that 'the time lag between biological and sociological maturity which seems to accompany every urbanization of a formerly agrarian culture is thus... an anachronizing process for the young' (Gioscia, 1972, 89). Gioscia's work has been taken up by Rose (1977) in his evaluation of the Swedish time-geographic approach (Chapter 6 following).

Age is perhaps the most discriminatory variable, so far as the directionality of time perspective is concerned. Movement towards the past occurs at an increasing rate beyond the age of about 40. Frank suggested that one might differentiate time perspectives on the basis of different aspects of living. Kastenbaum (1964) later called them content areas. Frank's idea was that time perspectives were applicable to different aspects of living, e.g. economic,

Extension
Coherence
Density
Directionality

Hyperchronic
Hypochronic

Epichronic
Catachronic

Metachronic
Anachronic

Content areas and aspects of living

political, and sexual realms. Kastenbaum develops the idea further. The functional significance of time perspective to everyday living is best illustrated by the individual who has no time perspective due to some pathology. Such an individual is at the mercy of his immediate situation, because it is through a time perspective that social man is able to 'manipulate his environment rather than passively respond to it' (Kastenbaum, 1964, 104). A truncated or limited range of activity alternatives seems likely to narrow time perspectives (as on an anachronic-metachonic dimension), as does isolation.

Time perspective and pathology

Some relations between time perspective, socio-economic group and age have been reported by Freeman (1964). In Chapter 8 empirical results from a study in an Australian city (Parkes, 1974) are presented. The importance of the place of time perspectives in social behaviour cannot be overstressed because for most of us a 'subjective future is proposed in all our activities. Without a tacit belief in tomorrow nearly everything we do today would be pointless' (Cohen, 1968, 262).

Where, then, is man's sense of time located, and what are the items in relation which permit its generation? Providing the answers to such specific questions is one of the tasks of psychologists and chronobiologists. Being aware of the possible significance of time sense in human geography is the geographer's responsibility. Time sense is a component of behaviour which consequently makes its impress on the world. 'Time like space is a primitive form of human experience', (Harvey, 1969, 410).

The source of our experience of time is possibly not independent of various somatic functions, some of which were outlined in 2.3 (ii) above. In 1923, Henri Piéron 'suggested that we could artificially cause our bodily processes to speed up or slow down and thus bring about a corresponding ''over'' or ''underestimation'' of clock time' (Cohen, 1967, 3). Hoagland also reported experiments which indicated that 'human time sense is basically dependent upon the velocity of oxidative metabolism in some cells of the brain' (Hoagland, 1933, 267 and Hoagland, 1968, 312). Cohen has pointed out that the implications of the issues raised by Piéron's supposition and by the experimental work of Hoagland and others 'are wider and more complex than they may appear at first sight. Since man's earliest history, his patterns of social behaviour have reflected the natural periodicities in his environment: the breeding periods of domesticated animals, seed time, and harvest time, seasonal changes in climate, the lunar cycle all became, in one situation or another, social time setters (Section 2.4 following) or zeitgeber' (Cohen, 1967, 6-7).

Social time setters

There seems to be a 90-100 minute cycle in many processes during hours of wakeful activity, especially in relation to concentration. In infants this cycle is closer to 50-60 minutes. Thus

Lynch has suggested that the 60 minute hour upon which so many of our activities are based is an 'inhuman unit' (Lynch, 1972, 118), because very few, if any, 'life-time' item relations have had a place in its determination.

Pulse counter approach

Two major schools of thought exist about the source of man's time sense: the pulse counter approach and the cognitive process approach. We will consider the pulse counter approach first. Physiologically based theories abound as a basis for explanation of time experience. These theories incorporate various pulse mechanisms (e.g. heart beat) to account for the way that time is measured subjectively. There is, in other words, some sort of time base, a cumulative, pulse-dispensing mechanism which delivers interval time signals. The objective of much of this research has been to isolate some organism or complex of organs which is the physiological source of time keeping, and which provides the fundamental temporal interval of about 0.6-0.8 of a second known as the indifference interval or indifference zone. The theory is that this is the time which is equivalent to certain critical body rhythm intervals such as heart beat or the interval between normal walking paces. Below the indifference zone lies the perceptual moment, about 0.1 of a second. For very small intervals of time such as those around the interval of the indifference zone, the possibility that somatic rhythms and psychological rhythms are causally connected seems quite high. As clock time intervals increase, however, the link between the two seems to become more tenuous.

Indifference interval, or zone

Perceptual moment

Opposing the pulse counter approach is the cognitive process mechanism. This approach sees the human organism as an active information processing system (Ornstein, 1969; Michon, 1967, 1972). As a theory of human time sense it contributes to the understanding of *duration* and not of sequence *order*. Michon has suggested that the pulse counter approach has the 'elegance of flexibility' (Michon, 1972, 244). In the cognitive theory approach 'the pulse generating and counting mechanism is replaced by a judgemental process that bases its estimates about duration on the number and nature of the events that occur during the interval' (Michon, 1972, 245). Available evidence at present seems to favour the latter approach where longer durations are involved. For instance, as geographers we might consider long term residential mobility behaviour to be a function of subjective time durations that have their basis in cognitive processing mechanisms rather than in pulse counter processes. However, the intrusion of the biological age factor into this mechanism cannot be overlooked, and here the impact of changing metabolism may indeed influence time sense. Controlling for age in urban social surveys means, in part, imposing a biological time control, but this has not usually been appreciated or, if it has, the anticipated implications of the biological time (age) factor have not been stated in the reported findings.

Cognitive process approach

Duration

Sequence and order

Judgemental process

69

(vi) Patterning time experiences

There is one theory of time experience which seems particularly likely to provide a useful basis for the development of geographic research on human behaviour, and that is Doob's theory of the patterning of time.[2] The focus of his approach is 'the human propensity, shared to some degree with other animals as the capacity to feel, think and behave in a temporal frame of reference, which means... that the past, the present and the future influence each other. Yes, the future influences the past' (Doob, 1971, 3-4, our emphasis). This focus is achieved through the principles of potential and judgement and the circumstances (see also Shackle, 1978) which produce changes in the temporal orientation of individuals. Figure 2.18 summarizes Doob's scheme as a taxonomy of time. (Refer repeatedly to Figure 2.18 as you follow the discussion).

Doob's central notion is the total potential which figures in a complex of feedback loops of temporal and behavioural potentials interacting with stimulus and primary judgements. Primary judgement 'is immediate, spontaneous, phenomenological judgement' which feeds into secondary judgement following consideration. This is unavoidably related to the socialization process but it is also increasingly under the dictate of (universe) clock time because this is the time which usually sets the boundaries to decisions. An assessment of 'appropriateness' is sometimes applied to the temporal information component as one of the considered factors, and as secondary judgement it is fed back into the total potential. It is also fed back into the stimulus as a temporal symbol. The result is that time perception is affected. There is then a further adjustment of temporal knowledge, drives and events. A two way interaction with total potential occurs as well as an influence on primary judgement.

The two major components of total potential are behavioural potential and temporal potential. The behavioural potential reflects at least three dispositions: biochemical, personality, and culture processes.[3] The biochemical processes relate to age, body temperature and in general to the milieu that makes up biorhythmic profile. Personality processes are the actual, unique organization of traits, culturally founded beliefs and attitudes within the individual, including intelligence or skill as well as temporal perspective, i.e. degree of orientation to past, present, or future. The modal beliefs and their media for transmission and reception construct the culture process input to behaviour potential. Certain behaviours are simply 'out of bounds' because of cultural constraints and the temporal potential is required in order to initiate orderly behaviour. It is composed of temporal motive, perspective, and information. The temporal motive is defined by the time of arousal in relation to the interval being judged, whether it is prior, interim, immediate, subsequent, or delayed subsequent and in regard to its importance in connection with the goal or goals

Potentials

Judgements

Total potential

Temporal potential

Behavioural potential
Stimulus

Primary judgement
Secondary judgement
Temporal symbol

Dispositions
Biochemical personality culture

Temporal motive

70

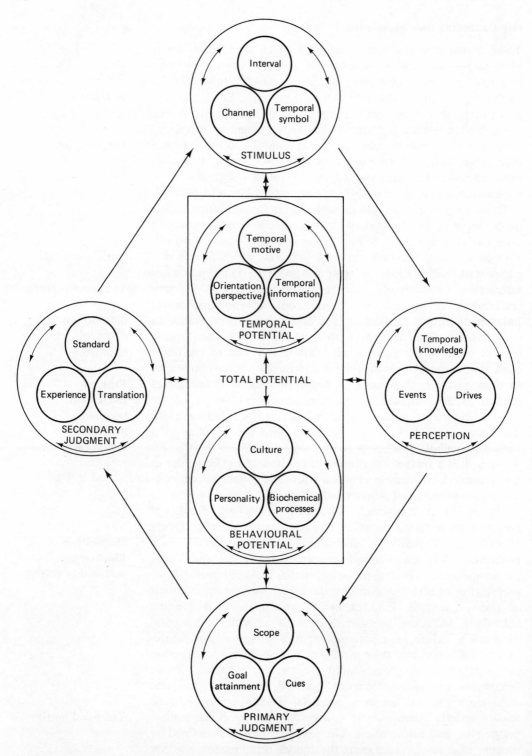

Figure 2.18. Doob's taxonomy of time. (Reproduced by permission of Yale University Press, from Patterning of Time by L.W. Doob. Copyright © 1971)

being sought during the interval. Perspective relates to whether the behaviour is oriented to past, present, or future and coincidentally whether the behaviour is temporary or enduring. Temporal information in the broadest sense includes data about the hour, date, age, or other temporal attribute of an event.

Temporal perspective

Temporal information
Socialization

All processes of individual development which are associated with socialization have a temporal dimension: remembering involves the past, intervention involves the present and anticipation and renunciation the future. It is with aging that the most important and clearly manifest differences in social behaviour appear. They result from the relation of the individual to elapsed intervals (the past) and to the future. Temporal potential and behavioural potential both contribute to total potential. The dispositions which make up behaviour potential are likely to be involved in decisions about space utilization, and so for instance a geography of the aged in cities will find a theoretical footing in this scheme.

Intervention
Anticipation
Renunciation

'It is not my time [lifetime] that is thus arranged; it is time in general, such as it is objectively thought of by everybody in a single civilization... What the category of time expresses is a time common to the group, a SOCIAL TIME, so to speak' (Durkheim, 1965 (1915) 23, cited in Zerubavel, 1976, 87). The subjective sense of time (as we have discussed it above), cannot be allowed to assume absolute dominance because this would negate the possibility of a social time. Schutz and Lukmann (1973) for instance, argued for the 'existence' of intersubjective temporality. Time had an intersubjective character and this enabled it to be shared by at least two people[4] but reference to a 'standard' time was also necessary, acting as a sort of grid system. This standard time is usually the universe time which was discussed above, at least in advanced scale societies. In The Dimensionality of Nations Project, Rummell (1970, 1) remarked 'If anything seems to be assumed by students of "international relations", it is that time has an objective meaning. Time can be calibrated. Events can be tagged as to their day, hour, minute. And nations order their behaviour in correspondence with a fixed, universally relevant standard of time [our universe time] which has linear extension in the past, present, and future. While time thus partitioned, measured, and standardized, satisfies practical need and common sense, does this deeply ingrained orientation represent the only scientific reality — the reality which is constrained and moulded by our scientific frameworks and theories?'... The shift in philosophical orientation which followed Einstein's theory of relativity means that, 'in the contemporary view, scientific theories are our reality and this is the only reality we are privileged to know' (Rummell, 1970, 3). In the earlier 'Newtonian philosophy' there was a reality — out there, so to speak — which could be described by scientific theories (p.3).

Intersubjective temporality and social time

The subjective component in time observation has been accepted into social science theorizing by a number of writers and furthermore, time also varies culturally, in that different cultures perceive and treat it differently (Hall, 1959). Perhaps 'time flows differently for each nation and so time in the international system is multidimensional' (Rummell, 1970, 6). It may also be the case for much smaller social groupings as well.

2.4 Social times

'We are made aware of the error of treating time as a unity, when in fact it is multiple' (Gurvitch, 1964). Social roles, attitudes, values, etc., move in their own characteristic time. They vary in their durations, in their rhythm, in the degree to which they are dominated by the past or projected into the future, etc.' ((Korenbaum, 1964, xxii). Gurvitch's *Spectrum of Social Time* is one of three theoretical approaches which have been published over the past fifteen years which will be introduced in this section. These three approaches and the reference to other work on the nature of social time should serve to illustrate the concept adequately enough.

The section is concluded by considering time in small scale (traditional) and large scale (innovative urban) social systems. In dealing with social Man, time becomes a more diverse phenomenon than is revealed by universe time and life-times alone, although both contribute to and are influenced by Man's social conduct. Indeed it appears that 'sense of time becomes increasingly dependent on place in the social structure and on lived experience' (Horton, 1967). To a degree, as biological Man has come to be influenced by social Man, our 'reliance on watches and other artificial aids has led to an atrophy of our sense of (universe and life) time' (Haldane, 1963, 337). However, we tend to develop a sharpened sense of the relevant times of our own ecological niche within the social system, as a compensatory mechanism.

(i) The items in relation

'Men have always been concerned with *timing,* however little or much they may care about the distant past or future. They are necessarily preoccupied with the practical problems of allocating time, of co-ordinating joint activities. But beyond that they attempt to harmonize their perceptions of inner [life-time] and outer [universe and social time] to feel the fullness of life, and to still the anxiety of death. If this is the aim, then environment ought to support it. Environment is the clock we read to tell REAL time, to tell personal time'
(Lynch, 1972, 66; our emphasis, refer to Chapter 1 and
REALIZED place).

When the items we use to tell this REAL time are drawn out of the social context or social environment, and are used in the co-ordination of activities, they define social time. In Doob's terms the conjuncture of universe and life time is found in social time. The social or cultural environment is a source of behaviour potential. Certain sequences of events (activities) with prescribed or normally acceptable temporal locations and durations set our actions within a temporal framework of laws and conventions. We make secondary judgements about temporal information in the social system by referring to norms, values, and mores and in this way develop predisposing attitudes to certain conduct and strategies which enable us to cope with a range of preconditioning factors. Were it not for such mediation we would only experience *IM*mediate[5] conduct and this might threaten the continuation of social communication and cohesiveness.

The organization of social activities requires the organization of time in order to achieve coherence because of the 'divergent and often contradictory manifestations of social time that social life must always take place in' (Gurvitch, 1964, 13). This organization must include,

(i) Systems of time measurement.
(ii) Allocation and scheduling of time by individuals.
(iii) A set of attitudes toward time past, time present and time future (Goody, 1968, 30).

A visitor from one society often feels a stranger in another. He or she will not feel 'at home', being unable to tell the social time of the system without a learning experience, without socialization into the temporal conventions. Refer back to Doob's comments on memory, intervention, anticipation, and renunciation. For the newcomer social events (items) are put into relation against the 'wrong' set of observation tables of temporal laws (2.2). One is 'out of place' as a result of inadequate temporal information to meet the requirements of temporal potential and the 'wrong' culture disposition in the behavioural potential.

Perhaps the most important point to grasp at the outset is that there is no universally accepted definition of social time. There are too many cultural differences and differences within the social structure of a single social system to allow this. Differences between male and female and young and old, are only the most obvious ones. But the concept of social time is universally applicable.

Social time definition

Doob (1971) argues that there is little reason to accept that males and females are equivalent in their transition from primary to secondary judgement because (to date) their socialization experience has been markedly different. Their biological life-times are also markedly different, affecting temporal perspectives, and a number of cyclic efficiencies.

Time esperanto

Social times arise out of social differentiation, from specialization of role and function and the widening area of social interaction or scope for interaction. They are therefore the result of the need for collaboration (Sorokin and Merton, 1937, 215). Timing is an intrinsic quality of all personal and collective behaviour. If activities have no temporal order they have no order at all (Moore, 1963, 8-9). While the order of activities will stay the same whatever time system is used, there are particular attributes of social time which, in certain circumstances, suggest that unlike the continuous time of astronomic observations, social time is discontinuous (Sorokin and Merton, 1937, 619), and also that it operates on a clock which tends to accelerate, based on observation of the pace of social change (Moore, 1963, 146). It expresses the 'change of social phenomena (items) in terms of other social phenomena (items) taken as points of reference' (Sorokin and Merton, 1937, 618). Reference to events in relation to other social events, rather than in relation to the universe time of the clock or calendar, 'usually establishes an added significant relation between the event and the temporal frame of reference' (Sorokin and Merton, 1937, 618). It must be emphasized, however, that universe time, as a time esperanto, is still a social emergent, and through its worldwide diffusion there is a convergence occurring among the social and cultural spaces of the world. As Doob observed, secondary judgements, upon which behaviour is partially predicated, have been and will continue to be influenced by this process. (Section 2.3, (vi) above).

Sorokin and Merton argued that the need for a framework of temporal reference resulted from the widening need for coordinating social activity and the growing inadequacy of local systems. Urbanization and social differentiation induced a more rapid development of universe time and with the 'extension of multi-dimensional social space[6] the organization of otherwise chaotic, individually varying activities became possible' (Sorokin and Merton, 1937, 627, 628). However, as Zerubavel (1976) notes, Sorokin and Merton's main contribution was the characterisation of social or sociocultural time. It was 'clearly distinct from physico-mathematical, biological, or psychological time. Social time is qualitatively differentiated, and unlike quantitative (universe) time it is discontinuous, not infintely divisible, and does not flow evenly and uniformly. As is evident from the study of calendars and chronological systems, its points of reference [the items in relation in our terms] are SOCIAL phenomena, events which are regarded as ones of peculiar social significance. It is also 'artificial' in that it is based almost entirely on social CONVENTIONS rather than on purely natural laws and relations. [UNIVERSE and some LIFE-TIME item relations in our terms] ... become a most useful perspective to study the various dimensions of socio-temporal order' (Zerubavel, 1976, 87).

(ii) Theoretical approaches

(a) The Spectrum of Social Time (1964).

Georges Gurvitch, a Russian-born (1894) French sociologist has been identified as the leading representative of the phenomenological branch of sociological formalism by Martindale (1960), but Gurvitch has denied ever having been a 'partisan of the phenomenology of Husserl or of ever having been a formalist' (Korenbaum, 1964; citing letters from Gurvitch). He 'prefers to be known as a representative of dialectical realism' and in the *Spectrum of Social Time* he applies 'depth level analysis' to the nature of social time. Depth level analysis (which is outlined below) is 'designed to deal with the open moving drama of social reality, forever in process and changing'. It utilizes the concept of levels as 'a systematic way to penetrate from the surface of readily observed, objectified social data to the very deepest, most obscure and veiled layers of social reality' (Korenbaum, 1964).

Depth level analysis

Gurvitch classifies his method as 'hyper-empirical dialectic' (Gurvitch, 1953) deriving his depth levels from Proudhon, Marx, and Bergson. He distinguishes between micro-sociology with its concern for manifestations of sociability of kinds of relations and macro-sociology, the study of groups and inclusive or global societies, which are clusters of the micro-social elements.

Eight criteria for classifying social structures are utilized:

Criteria and social structure

(a) The hierarchy of groupings
(b) The combinations of forms of sociability
(c) The depth levels which are accented
(d) Division of labour and wealth accumulation
(e) The hierarchy of social control
(f) The cultural products which cement the inclusive social structure
(g) The scale of temporalities

Fourteen types of inclusive society are distinguished by Gurvitch and there are also six historical types of social structures, e.g. 'feudal' (in Europe from the tenth to fourteenth centuries) and 'competitive capitalism' (which was predominant in the nineteenth and early twentieth centuries in Europe) and four 'contemporary' societies. They may be distinguished in terms of the eight criteria listed above and subjected to depth level analysis.

Gurvitch defines social time as the 'convergency and divergency of movements of the total social phenomena, giving birth to time and elapsing in time' (1964, 30). But varieties of social time are necessarily identified because social structures are unstable, 'they are involved in perpetual movement' (1964, 11). The life of social frameworks, their products and particularly their structures unfold in their own time (Gurvitch, 1964, 12, 13).

Gurvitch social time definition

76

constructs eight different kinds of social time as frames of reference for sociological analysis and maps a category of social time to each of them. The eight social times proposed by Gurvitch are:

i. Enduring time (ecological).

> The time of long duration. It is the most continuous of social times. Here the past is projected in the present and in the future. Peasant 'classes' or small scale cultures above all 'actualize' this time.

ii. Deceptive time.

> Under an apparent calm, sharp crises produce a time of surprises. This time belongs to large cities. The organized level of social life unfolds in this time. It is discontinuous due to 'catastrophic' interruptions.

iii. Erratic time.

> The time of irregular pulsations between the appearance and disappearance of rhythms, the time of uncertainty. The time of social roles, of unstructured groupings, of classes in the process of formation.

iv. Cyclical time.

> A time in which past, present, and future turn in a circle. This time visibly prevails in the mystic-ecstatic groupings, for instance of churches and sects.

v. Retarded time.

> This is time too long awaited. The time of social symbols 'outmoded at the very moment they are crystallized. Community sociability emphasizes retarded time through the value it places on symbols. 'Closed' groups move in retarded time; mobility, licensed professions, etc.

vi. Alternating time.

> This time alternates between delay and advance. Realization of past and future compete in the present. Discontinuity is stronger than continuity. This is the time of economic groupings.

vii. Time pushing forward.

> Future become the present. This is the time of aspirations towards the ideal, the time of active masses and 'communions' in revolt. The time of proletarian classes. Perhaps the time which trade unions operate in?

viii. Explosive time.

> This is the time of acts of collective creation. In centralized and pluralistic collectivism there is an attempt to make this time dominant.

The depth levels which are applicable to the study of social data are as follows:

(1) The level of the ecological surface, including natural and technological environments which limit the social sphere; the corresponding time is 'ecological time', emphasizing continuity.

(2) The organizational level in which organization exercises external constraints over individuals and groups. The ecological surface level and the organizational level may be in conflict but there is always relation. The time of the organizational level is 'deceptive time'. Here continuity and discontinuity collide and never achieve an effective compromise. Advanced urban societies, where complex organization systems operate, occupy this level.

(3) The level for conduct with relative regularity, includes behaviour outside organizations but which is nevertheless more or less regular involving rituals, routines, mores, fashions, and fads. Models, rules, signals, and signs guide, conduct and demand obedience through their validity. Patterns and rules are sometimes oriented to routine immobility, sometimes towards flexibility and adjustment to circumstances and sometimes toward innovating initiatives. Continuity and discontinuity exist in conflict. Past and future compete for representation. Deceptive and 'erratic time' prevails.

(4) The level of social roles, and

(5) The level of collective attitudes operate under the erratic time of irregular pulsations (see Figures 1.15, 1.16), and uncertainty. Discontinuity is dominant over continuity. As depth levels these are more spontaneous than the depth level of relatively regular behaviour. The erratic time of these levels is 'wedded' to other kinds of time from which it seems to get 'support' and an element of continuity.

(6) The level of symbols, ideas, collective action, and cultural products, and

(7) The level of the collective mentality, dependent on other levels.

Symbols 'push individuals and collectivities to participate in action even while they obstruct participation' (Korenbaum, 1964). Collective values may become encrusted and incarnated in the 'status quo' or they may claim to be eternal, predestined, and perfect; in which case they become the strongest obstacle to human intervention in social reality. 'Symbols, ideas, and collective values unfold in a time in which delay and advance struggle, but never achieve alteration nor compromise ... Intellectual ideas ... tend to endure to an excessive degree and to move in a time held back on itself. That means that the symbols are often outmoded at the very moment when they crystallize. ... Most of the new ideas are awaited too long a time and by the time they are clarified they are not always of their time. Sometimes however the future may be rendered effectively present for instance in the discovery of major new ideas'[7] (Gurvitch, 1964, 51). The level of the collective mentality is the deepest layer and has three aspects:

i. Mental states (e.g. memory or suffering)
ii. Opinions
iii. Mental acts (intuitions and judgements, cf. Doob and Section 2.3 (vi)).

**Group classifi-
cation by time**

The micro-social framework and its time manifestation of sociability, for instance in mixed interpersonal and intergroup relations, also receives detailed discussion from Gurvitch. He proposes criteria for the classification of groups which are directly related to time, such as duration (temporary, durable, and permanent groupings) and rhythm (groupings of slow, medium, or precipitated cadence).[8]

The time of groups may also be classified according to their dispersion, mode of admission, function, and orientation. Groups which meet[9] periodically may do so at fairly long intervals or fairly frequently; there is therefore a dispersion in locational universe time, notionally similar to dispersion over territory. Such locations may or may not be appropriately matched by psychological and social times and this is precisely why they are so important to behaviour. The fact of a 'projected' or scheduled meeting is evidence of group specific coherence achieved through the satisfactory mapping of universe, life, and social time. It is 'hard' information (Chapter 1).

**Alternating time
Cyclical time**

Ecological time

Erratic time

The particular type of social time which Gurvitch associates with the (time) dispersion of groups is alternating time (vi) and it is essentially discontinuous. Groups which meet rarely but regularly combine cyclical and alternating time. With *mode of admission* as a basis for classification there may be open, closed, or conditional access. The open grouping tends to lack internal coherence and 'leaves the doors wide open to all and sundry social times'. Closed groups incline to time of long duration, or ecological time (i). Kinship groupings are a familiar functional grouping, they move in time of long duration limited by erratic time (iii), or uncertainty and changing social roles. Economic groupings are also functional groupings but tend to produce (*sic*) a time where future and present are in conflict. Orientation may be divisive (e.g. social classes) or unifying (e.g. political states). Time consciousness is greater in the divisive than in the unifying groupings, however. Each social class possesses its own dynamic or direction, rate, and path to change. It is now clear that it may also produce its own time scales.[10]

At the 'global-society' scale Gurvitch investigates time scales of various types, for instance the archaic or primitive (the Australian aborigines would have a primitive social time), and the historical (for instance as in the Greek city states). He also treats the time scales of modern societies. Each type of society is treated by depth level analysis against the categories of social time outlined above.

Most of the important notions which go to make up Gurvitch's 'dialectic on social time' have been summarized. But because of the complexity of his scheme, an illustration of the interplay of depth levels involved in a real world situation is given below. This extract is taken from Korenbaum's Introduction to Gurvitch (1964).

'Modern industrial societies apply numerical designations to both demarcations of point in time and to measures of duration, whereas primitive societies use concrete activities or observable phenomena. It is rather obvious that for primitive societies time designations are communicable within very circumscribed local areas. Hallowell (1937), for instance, could not use the typical western meal time referent to make appointments with his Salteaux Indian informants. Meals for them are an individual matter rather than social and are eaten whenever convenient for the individual. He was forced to say "Come as soon as you have lifted your nets", an activity that regularly took place the first thing in the morning but recognizable as a time designation to a very limited local group.

What is not so obvious is the extent to which local areas in modern industrial societies have attempted to control their own time designations. Numerical labels are often credited with an almost magic-like power to impose co-ordination. However, local groups have often and long resisted attempts to co-ordinate time designations as required by industrial efficiency. As late as 1860, three hundred local times were observed in the United States. Philadelphia was five minutes slower than New York time and five minutes faster than Baltimore time (Cowan 1958, 45). On the "day of two noons", Sunday, November 18, 1882, standard time was adopted by the railroads, but it was not until 1918 that Congress made standard time the rule for the entire nation.

That we have not yet achieved coordination in this year of nineteen hundred sixty three is apparent in this quotation from the May 10th issue of *Time Magazine*. "It's time to do something about time. This was the consensus of a parade of witnesses representing transportation, communication, finance and farm who testified before a Senate committee called to consider three bills for reforming the US's unhappy clock chaos...

Of the 28 states that observe D.S.T. (daylight saving time), only half impose it on a statewide basis, and they all turn it on and off whenever they feel like it. Compounding the confusion are the country's four time zones. In Indiana, for instance, the boundary between Eastern Standard and Central Standard Time splits the state from north to south. In parts of northern Idaho, Daylight Saving Time is observed on a door-to-door basis. And passengers on the 35-mile bus route between Steubenville, Ohio and Moundsville, W. Va., would, if they wanted to keep local time for all the stops on the way, have to adjust their watches no less than seven times.

Most of the witnesses of last week's Senate Commerce Committee hearing cited the wastefulness and expense of the U.S. time snarl...".

Resistance to time co-ordination still continues. The latent conflict between rural and urban elements is expressed when Daylight Saving Time is shunned as "city time" and in the outcry against any change as tampering with "God's time."

We can illustrate the interplay of depth levels involved in this problem. (1) At the *ecological surface* the geographical breadth of the United States, the location of cities and towns that are economically dependent on each other, the daily migration of workers from outlying suburbs to city centers, and the network of communication systems thrust very specific problems on time coordination. (2) At the *organization level* the requirements of national coordination conflict with democratic organization which implies choice based on local preference. The economic organization and the political organization must come to terms on this issue. (3) *Signs and signals* are involved in that the hour designations direct the scheduling of daily conduct. Signs and signals are not an individual matter but have to be arrived at by social convention. In this case, confusion results when individuals must behave according to signs and signals which originate from a number of different locality centres. (4) *The network of social roles* come into play when, for instance, the conduct of the father role is scheduled according to suburb time while the corresponding occupational role must be scheduled according to city time. (5) *Collective attitudes* are involved in our attitudes toward "clock time". The "clock time" has to be changed in order to take better advantage of daylight. It is much more difficult to schedule everything one hour earlier than it is to move the hands of the clock, because we are bound by our collective attitudes toward clock time. (6) Daylight Saving Time symbolizes to the farmer the dominance of city centres and industrialization in our economy and acts as a "call to arms" for immediate, non-reflective resistance. (7) *Collective mentality* is reflected in the degree to which notion of the power of human intervention can overcome the mental state of powerlessness to solve complex issues. In the microsociological sphere, we can examine the formation of we-nesses among individuals and groups to support the issue of unification of time, as well as the scissions that will arise. At the macro-sociological level one could examine the formation of the Committee for Time Uniformity, the leadership of its chairman Robert Ramspeck, the effectiveness of its operation as a pressure group, whether it is a dispersed group, how it attracts its membership, how one gains admission, etc.' (Korenbaum, 1964. Reproduced by permission of D. Reidel Publishing Company).

(b) Social system and time and space (1969).

Kolaja's theory is based on a cyclical view of human society presented through the idea of recurrent behaviour. The notions are rather complex and as the title suggests, the work is about space and time. However we shall concentrate on the temporal elements here. The spaces which Kolaja deals with are social and culture space. Physical or geographic space is mentioned from time to time but it is usually only by way of distinction from the former spaces.

Physical space

Social space

Culture space

For Kolaja physical space is that space commonly defined by distances betwen earth-bound objects. Social space is represented by a number of persons or organizations in relation. Culture space is represented by the number of items that are available in regard to a particular problem [project or task]. It is 'also significantly a function of time because it takes time generally speaking to examine certain alternatives that are available'[11] (1969, 72). Thus an innovation, contributing first to culture space, may filter into social space and ultimately be established in physical (or geographical) space. The explanation for the location of any phenomena in geographical space would therefore seem to have its roots in culture space-time because of the operation of a *principle of finitude* that it takes some absolute quantity of time to consider alternatives: *where* in a real sense depends on *when* and *how much* time is available. A spatial location having been chosen or imposed indicates above all else that others were not selected or delegated and in addition that there was time for their consideration as well (see Chapter 6 for an expansion of this theme).

For Kolaja, social structures and social systems are 'summary terms for recurrent phenomena' but he acknowledges structure to be a more general concept than system because, whereas structures can contain systems the reverse is not true. He isolates four dimensions (or categories) which are crucial to a rounded view of social time and social space, and which are existential dimensions. These 'dimensions' are derived from categories of usage in anthropology and sociology.

1. *The existential locus* — the ontological problem of 'where society is'. Anthropologists have been the forerunners in this approach, but there have been differences between them regarding how structure exists as 'something out there' (Radcliffe-Brown, Evans-Pritchard, and Nadel) or as something to be found only in the heads of observed persons or of researchers (Leach, or Lévi-Strauss). Kolaja suggests that for 'sociologists today, traditional questions regarding ''where society is'' appear to be only historically relevant' (p.41), but recent discussions in sociology would disagree.

2. *Parts and their extensiveness* — this dimension involves questions concerning the nature of the smallest unit in a society, its 'elementary particle'. Are these simply persons, (Radcliffe-Brown) or groups and explicitly not persons (Evans-Pritchard) or are they interpersonal relations (cf. Nadel, 1958) or the ideas that persons have about the distribution of social power (Leach)? There is a difference of viewpoint about 'smallest units'. The most extreme view perhaps being Sorokin's notion that a 'search for the smallest unit (is) a mechanistic imitation of natural science' (Kolaja, 1969, 43), whereas the concept of relations (von Wiese's 'Benziehungslehre') has gained increasing adoption. These relations (among small groups) is a focus that Kolaja finds useful because it provides a 'certain link and inspiration (for) a theory of recurrent behaviour'. **Parts and extensiveness**

3. *The modus of the transaction relationship* is a dimension which Kolaja adopts as a phenomenal one 'experienced by persons who produce recurrent behaviours' (1969, 44). The nexus among the parts in a transaction relationship is in the present. Spatial concepts are more important to this dimension than are the temporal. **Transaction relationship**

4. *Degree of recurrence* or its absence is the dimension which draws attention to the conceptual difficulties involved in the relation between change, recurrence, and structure. Because Kolaja's major focus is on recurrent behaviours, he gives no further discussion to non-recurrences, i.e. the social change concept (1969, 47). **Recurrence and absence**

82

Social change

The concept of recurrence as used by Kolaja is based on two other notions:

(i) The number of items (discrete units that can be differentiated) which are involved in a recurrent situation, and

(ii) The density of intervals, or our perception of the series of events which constitute a particular phenomenon.

Interruption and interval

Although primarily a function of time, recurrence also involves space to the extent that certain of the items in any phenomenon are spatial. If the notion of an identical number of items being necessary for recurrence is replaced by the notion of a similar number of items, then a phenomenon may recur up to some limit, before it changes character. The time during which a phenomenon is interrupted, namely the interval, gives us the notion of interval density based upon memory. An interval density of (0) means that the phenomenon is never interrupted, it is continuous rather than recurrent. An interval density of (1), on the other hand, implies that the phenomenon does not occur at all, there is no matching of items. Thus recurrent behaviours tend toward zero interval density but may lie anywhere between a value of 0 and 1. But due to the very fact of being recurrent and structured they are usually only on the threshold of attention and are therefore not intensive.[12] With

Innovating behaviour

unstructured behaviour, Kolaja postulates that one response is to try and make the component acts recurrent. This leads to innovating behaviour (IB). Recurrent behaviour (RB) is familiar behaviour, but it is not habitual because the person is aware of it

Recurrent behaviour

and distinguishes it from a wider series of recurrent behaviours for however short a time period. A person is not aware of habit.

In terms of the existential locus, recurrent, and innovatory behaviour (RB and IB respectively) can be classified (a) by the actor himself, and are called *personal* RB and IB. (b) when classified by other persons they are called *group* RB and IB.

RB chains

The basic part of RB and IB is the item, the differentiable quality of a perceptual unit. Although RB are by definition a part of the time dimension, a characteristic is that they become 'dateless'; all that remains is the structure. RB is defined in relation to past or future behaviours and not in relation to present behaviour. Any single individual has a whole suite or CHAIN of RB.

The difference between personal RB or IB and group RB or IB is that group RB/IB allows the group to anticipate the RB of an individual. A personal RB will project along the time dimension more than the group RB. The latter needs the (social and cultural) 'space' dimension to a greater extent because reference to another person's behaviour 'involves a spatial relationship, i.e. the relation between the observer and that person' (1969, 80). It demands appropriate relative location in social space.

There are many other kinds of RB. There are physiologically

derived RB. There are also socially derived RB. Kolaja argues that the latter are likely to be more time-precise. RB are also hierarchically ordered; thus we have socially formalized RB (e.g. work times). Where a change is likely to occur on all criteria simultaneously, formalization is a normative restriction that occurs among spatial parts as well as temporal units.

Different patterns of configurations of RB are possible as the result of our ability to change their rank and order in a cycle. Pattern may also be changed if the number of chains is altered. A chain may then appear only in every nth cycle; but it must re-occur within memorable time or else it becomes IB. There are certain implications for the theory of complex social organization, for instance, 'Persons occupying higher positions within an organization are persons who ought to be able to see RB within higher cycle levels' (1969, 68) and can see IB as part of a higher level RB (for instance business decisions as part of a business cycle).

Five levels of cycles[13] are suggested within which each group RB operates:

1. The family cycle of 24 hours. This is the most basic, frequent and regular set based on physiologically derived RB. This will be recognized as the biological time component of life-time discussed in 2.3 above. It is also similar to Gurvitch's notion of ecological time which was continuous. For Gurvitch, you will recall, it was the time of small scale or traditional societies. The 'family' is also seen as occupying traditional social and cultural space, as for instance in the social area 'paradigm' (Shevky and Bell, 1955).

2. The work cycle, although it also operates in a 24-hour cycle, is less regular due to interruptions. Generally speaking only five of the seven days in a week are included. This cycle is impelled by social items (career ambitions, trade unions, etc.) rather than the biological items of the family cycle. Here we would find deceptive time operating: 'urban' as opposed to sub-urban or traditional life styles.

3. The voluntary group cycle is generally a cycle of weekly or monthly activities. Examples are church or club attendances. Cyclical time operates, interval density is relatively high.

4. The anniversary cycle covers all events happening once within the period of approximately a year. The year need not be the formal calendar year of course. Interval density is relatively low.

5. Life cycle 'covers events of individual entries and departures from life' (1969, 78). The notion of life cycle was discussed briefly in 1.3 when we considered certain terminological issues related to the study of time (Young and Ziman, 1971).

Image RB/IB

When referring to past or future recurrent behaviours the notion of Image RB (or RB_i) is introduced. Its general meaning relates to whether or not the behaviour is overt. If it is not overt it is an image RB. Image RB (RB_i) may be personal or group related. For RB and IB whether personal or group, overt or covert, there are probabilities of recurrence and therefore of cycle patterns. With a high level or interdependence between activities, as in most work situations, RB predictability is high. However, without occasional shifts in the cycles of RB (or RB_i) there is a deterioration in social relations and life becomes humanly meaningless.[14]

Federation of RB

The loose interlocking of RB in a social system is defined as a federation by Kolaja. Beginning his discussion with the process of interaction between two individuals he seeks to understand the influence of the pattern of IB and RB (including IB_i and RB_i) of one person upon another. To achieve this he estimates the probability of recurrence and shows that there is a range of interaction probabilities dependent upon the susceptibility (our term)[15] of one individual to the conduct of another. The time order of the items which make up the chain of behaviours in an interval is a factor in the development of interlocked behaviours. Once interlocked or federated the image RB becomes important in allowing the individual to pursue his personal IB and RB. He or she has an understanding of what others are doing, have done and might do.

Susceptibility

To develop the notion of federated RB in groups and aggregates Kolaja uses MacIver's three categories of action. These categories are:

i. Distributive
ii. Conjunctural
iii. Collective

Distributive actions

Distributive actions do not involve others. For example, the private activities of individuals within a family; here the chains of distributive RB run parallel to each other and this type of action makes up the largest block of waking time consumed by persons (Kolaja, 1969, 92). Taking the family to be a part of the larger 'containing society', all those activities within the family household, usually termed private activities, could be classed as distributive. Age itself has a distributive character. Old people, for instance, tend to produce less future image RB, than do the young. Perhaps the main reason why this is not reflected in social institutions is because of the very large middle-age group which is usually in a dominant majority position of power. With aging societies this may become an important issue, amenable to satisfactory solution only if social time is given studied attention.

Conjunctural actions

The conjunctural actions of individuals are not organized, but participants are aware of the consequences of certain actions, whether their own or those of others. This is not the situation in the distributive category; actions are in parallel. However, action is still

individual, with reflection 'on the cumulative effect of...
individual actions' (1969, 92). Mass media (e.g. TV) generates
conjunctural actions but participation is passive and distributive.
In totalitarian political systems, for instance, compulsory
attendance at meetings of 'The Party' operates in this category.

With organization and a certain division of roles we have
collective actions: a federation of loosely interlocked RB with
awareness of the consequences of action. Each of these action
types, with associated IB and RB are the 'harbingers' of social
time.

Distributive actions (and their times) are most closely associated
with physical space, conjunctural actions with social space and
collective actions with culture space.

Collective actions

(c) Time as conflict (1975, 1977, 1978)

Fraser aims toward the development of a theory of time per se. It is
not specifically designated as a theory about social time. His
conceptualisation of time as conflict covers natural and social
science philosophy. It is selected for discussion here because we
consider it to be a seminal contribution to the study of social time.

It was the German biologist von Uexkull who first proposed the
concept of Umweltlehre: specific worlds or perspective universes.[16]
In Fraser's theory of time as conflict an Umwelt is a synthesis of
Merkwelt and Wirkwelt. The Merkwelt is that world of specific
receptors which determines which signals can be received and the
Wirkwelt is the world of specific effectors which determines the
sum of responses available once these signals are received.
Translated from the original concern of von Uexkull for animal
ecology, the Umwelt becomes the 'perceptional' world in which an
individual must live.

**Umwelt
Merkwelt and
Wirkwelt**

**Receptors and
effectors**

The theory of time as conflict considers the tension between
series of archetypal Umwelts of increasing temporality and degree
of causation. Six hierarchically ordered Umwelt are isolated. They
are listed below in order of increasing temporal complexity. The
significance of time increases, in all systems, (think about 'places'
if you like) as the number of items which may adopt a relation with
other items increases.

**Temporality and
degree of
causation**

1. In the Atemporal Umwelt no meaning can be given to causality
   because the direction of time cannot be gauged. An example
   would be the world of pure electromagnetic radiation.
2. The Prototemporal Umwelt comprises indistinguishable
   entities, countable but not orderable — some 'states of mind'
   are in this form of temporality. Connectedness is weak, laws are
   entirely probabilistic. However, spatial and temporal features
   can often be distinguished.
3. The Eotemporal Umwelt is one of pure succession, or its
   equivalent pure asymmetry. There is a dyadic before-after

Umwelt 'levels'

relationship. Causation is deterministic, as in action-reaction.

4. The Biotemporal Umwelt arises from biological morphogenesis and is distinguishable by the separation of two aspects of causation: connectedness and intentionality. The triadic form of past, present and future time appears with the emergence of the present defined in terms of needs and is therefore distinguished from deterministic causation. Future and past become increasingly polarised, causation is multiple as well as final.

5. The Nootemporal Umwelt is the temporality commanded by individual social man and is associated with certain unresolvable conflicts of his faculties or capacities and in particular with memory. This is the Umwelt of personal identity and of signs and signals that are symbolic transformations of experience. Future and past are sharply distinguished. Causality is enlarged to include free will.

6. Sociotemporal Umwelt is the time of global, collective man.

Fraser (1977) has written that, 'The large variety of times with which you deal in the context of human geography constitutes a 'fine tuning' or modulation of the two highest levels of time: nootemporality (that of the individual mind) and sociotemporality. Since the mind and society are immensely rich in creative capacities, this modulation is a whole world [umwelt] in itself: the world of man.' By way of example, think of small scale and large scale societies as Umwelt operating in minimally overlapping social times. Between the poles of small and large scale, however, a hierarchy of social times exists. Advanced scale urban societies are especially associated with the nootemporal (highly specialized individuals) and the sociotemporal (telecommunication-linked networks required for commercial and military survival). At both these upper hierarchical levels survival is dependent on intended connectedness, the ability to plan ahead. Such planning involves sophisticated connectivities or lawful necessities. The small scale society, say a pygmy tribe, is able to survive in the eotemporal and biotemporal range, without any loss of orientation. But large scale societies could not because they rely on the connectedness between nootemporal and sociotemporal.

Intended connectedness

Sophisticated connectivities

Transcendencies

Transcendencies refer to the flow of symbols through communication channels. Tracing movement through the hierarchical levels, it appears that 'individuals in communities are connected not by physiological processes (as are neurons) but rather by the flow of symbols... thus they tend to form images of things and conditions external to the group images, of the group itself,[17] as well as images with no prior correspondence in reality. The aggregate of these symbols comprises the substance of, and determines the next higher integrative level above the noetic: that of man in societies.' Transcendency thus 'involves the revolutionary new use of already existing structures and functions...' (Fraser, 1975, 441).

At the nootemporal and sociotemporal levels there is increased separation of past and future. The space between time past and time future increases as hierarchical level rises. Multiple causation sharpens the separation. Multiple causation and the growing interdependence of individuals and elements in complex societies, revealed through intended connectedness, sets greater demands on the anticipation of dramatic regularities, catastrophes, destructive events (from earthquakes to nuclear explosion to floods to epidemics) because any single such irregularity, however spatially remote, is likely to have an extensive spatial and temporal impact in the contemporary world of interlocking communication; of syndicated news and nested levels of political co-operation.

It is the biotemporal, nootemporal, and sociotemporal Umwelt that 'contain' social man except in certain states of isolation or in small scale systems where the eotemporal may also be manifest, albeit in veiled form. At a coarser level of conceptualisation we have, in universe time, life times and social time a similar notion of the many-element structure of time which faces human geographic studies.

(iii) Social time and social scale

There are many classifications of society based on social time characteristics. Hall (1966) has suggested a simple dichotomy within European cultures based on the 'time dispersion'[18] pattern of activities, their time use (cf. Chapter 4 following). Thus he suggests that south Europeans, who use their time in a somewhat casual, unstructured way (in Kolaja's terms where the interval density is low), are more heavily reliant on item relations in the present than in the future. Item or event order is imprecise and the duration of events is relatively flexible or elastic. Such a culture space is polychronic. Hall speculates that there are built space correlates of this characteristic: the plaza is a space of congested, unstructured use. The association of social or cultural time morphology with the built morphology of settlement is rather similar to Durkheim's view that the concept of time is reflected in the organization of space. On the other hand, the north Europeans see time as serially structured, here interval density is higher, event order is more inelastic: this is a sign of a monochronic culture. The North American derivative is also monochronic. Main Street is the built space reflection.

By considering time as a resource, as a means to the satisfaction of a want but not explicitly as social time, Linder (1970) proposes a classification of global culture-types into three broad categories. They are:

1. The time surplus cultures where productivity is so low that a certain proportion of time yields nothing whatsoever. Such cultures have little need of precision in time reckoning. Image

Time dispersion and time use

Elastic

Polychronic

Inelastic

Monochronic

Time surplus

RB may, however, imply scarcity, for these cultures of course!

2. The time affluent cultures have an adequate time supply and occupy a middle position; a process of economic growth has started and extended stretches of unfilled time have largely disappeared. But certain slacks in the use of time still remain. Linder, following Mead, cites aspects of modern Greek culture as an example in this category.

3. The time famine cultures are typically those of urban industrial societies. As these societies perceive it (their image RB in Kolaja's terms) they have no remaining slacks in the supply of the time resource. All that remains is for the further colonization of time (Melbin, 1978, and Chapter 8 following). There is likely to be much stress and social disruption between the levels of the social system as the times appropriate to those levels (Gurvitch, cf. above) undergo change.

You will recognize that there is a similarity between Linder's time surplus-time famine cultures and the small scale-large scale polarity incorporated in the theory of increasing societal scale which played such an important role in the development of social area analysis adopted by Shevky and Williams (1949) and later by Shevky and Bell (1955).

Lynch (1972) has also suggested a culture scale classification. It is based on seven factors, each with an inherently temporal structure.

1. Grain is the size or precision of the intervals into which time is divided by a culture in order to satisfy its needs and wants.
2. Period is the length of time within which events or items of social consequence occur.
3. Amplitude is the degree of change within a cycle.
4. Rate is the speed at which changes occur. In Moore's terms social time is an accelerating time, and this would be so especially in time famine cultures (Moore, 1963).
5. Synchronization is the degree to which the cycles and changes are in phase, or begin and end together.
6. Regularity is the degree to which the previous characteristics themselves remain stable and unchanging.
7. Orientation is the degree to which attention is focussed on past, present, and future. This you will recognize as the time perspective component in life-times discussed in 2.3 above.

'Advanced' scale urban industrial society is fine grained, has short period, high amplitude and rapid rate. It is highly synchronized, is regular and is both near and distant future oriented (evidenced, for instance, in near and distant bond redemption times!) On the other hand 'retarded', traditional, or small scale societies are coarse grained, have long period and low amplitude. They have slow rate with very loose synchronization.

They have regularity but are past and present oriented.

Fortes (1970) distinguishes time as duration, continuity and discontinuity, genetic or growth processes, and historical sequence and lays different stress on them according to the culture studied (Barnes, 1971). Moore's (1963) scheme is based on the notion of degree of urbanisation in which time scheduling is seen to have an important role. Hawley (1950) considered societies to be differentiable in terms of their modes of meeting periodic adversities; how did they adjust? Three types of adjustment were identified. Movement, as in the nomadic way of life (Chapter 7 following); storage through the generation of a surplus, and circulation of commodities and capital. All three modes of adjustment to adversity are space-time strategies, but the space and time dimensions should include paraspace and paratime and cannot be couched simply in terms of physical space and time. This is so because the items in relation are not necessarily either earth-bound (space) or orbital (time)!

Movement
Storage
Circulation

Paraspace and paratime

(iv) The small scale societies: the evolution of time

We have seen that small-scale cultures have what Linder has called a time surplus. Consider the time budget of a nomadic Australian aborigine group (Table 2.3). The variety of activity is noticeably limited, at least from an advanced scale society perspective.

Table 2.3. Sample time budget in a time-surplus culture (Mountford, 1960)

Activity	Time Allocated (hr)	
	Women	Men
Sleep-lying	9·0	12·0
Sitting-talking	8·3	5·0
Prepare } instruments Repair }		1·0
Preparing and cooking kangaroo		0·3
Collecting food	4·3	
Preparing and cooking food	2·0	
Singing and dancing		1·3
Hunting		4·0

Lee (1969) has shown that the Dobe Bushman of South Africa have similar time budgets to that of the Australian aborigine, as do the Hadza of Lake Nyasa (Sahlins, 1972). Some time-affluent cultures are nomadic. Their movement ensures greater reliability of food supply. Nomadism involves the principle of regulation of space and time: the timing of space use and the spacing of time use.

From our standpoint in urban, advanced society, the time budget of the nomadic Australian aborigine or Trobriand Islander, seems

The dreaming

to include a lot of slack or unused time, measured in universe time. Their experiential and social time, however, is likely to be very different to ours: the Australian aboriginal views the world through a cosmology called The Dreaming where the creative past is united with the present and future. The culture heroes who dwell in the 'Dreaming' not only made this world and its creatures, but are still continuously concerned with its activities. On initiation the Aboriginal becomes a part of this experiential 'time zone' and shares the life that comes from the 'Dreaming'. Trobriand

Trobriand islanders

Islanders apparently conceive of all past events (real or mythical) as included in a 'universal' present. How can such cosmologies be explained?

A certain ecological determinism is probably involved — thus Tuan (1974, 80) in a review of Turnbull's work (1961) likens the attenuated time sense of the Pygmies to the lush continually

Pygmies

growing forest in which they exist, where seasonal change never occurs. Much of the time sense of such peoples is also a matter of

Time

the symbolic attributes attached to the environment. This is clearly revealed in Whorf's study of the Hopi Indian (1968). The Hopi

Hopi Indians
Manifested
Manifesting

have two different concepts of space and time. The first is all that is manifested. This time is objective and comprises all that is or had been accessible to the senses. The second is all that is manifesting. This is a subjective concept and comprises all that will be acceptable to the senses. There is no existence in the way that a painting-by-numbers waits only for a colour. Thus the only tense which might be considered in the Hopi language could be called 'expective' — something about to be manifest or at the moment of inception. It includes part of 'our' present and part of 'our' future (Whorf, 1956).

Space

In terms of space a similar situation prevails; only subjective space is produced. What happens at another village only actually happens when a person hears about it. And the surrounding countryside is so besotted with Hopi myth that its mystic significance is linked with the space dimension — there is no differentiation. Differences exist only in terms of validity or

Hopi Indians

intensity of an event and interestingly there is no concept of velocity. This is indeed an extraordinary condition for western societies to comprehend.

We must also look to language for some of the explanations. In English and European languages events are stable objects. The collocation, the fluidity of passing experience is lost in the language. Contrast this with the Hopi culture in which the days are not totalled (e.g. the 10th day) but are referred to in their order (i.e. by the 11th day). Different days are re-appearances of the same entity rather than distinct slices of time.

Information is another factor. Thus Ortiz (1972) has pointed out that for most members of a small scale culture there is only a piecemeal and very limited amount of access to the totality of information held. Thus amongst the Pueblo indians only the priests would have access to the complete store of cultural knowledge (White, 1960).

Pueblo Indians

We still have the fact that for the Pueblo and the Hopi, time is not an important factor, except in a seasonal context and as a progenitor of ritual drama. Time is a rhythmic order, an 'all at once' phenomenon; festivities mark a new cycle and the re-enactment of myth provides for periodic re-entry into the past. The Polynesians, for instance, when setting out on an adventurous journey, denied its novelty, because it was seen as a re-enactment of a myth.

Myth

Polynesians

Successful agriculture, as opposed to nomadism, demands more accurate time and space measurement, though their joint status may not in fact be as complex. In space, certain areas may become 'property' and need to be marked out as such. The schedules demanded by agriculture derive from the temporal behaviour of crops and animals and the need to produce surplus food to store over unproductive seasons and as an insurance against unpredictable circumstances. Many events in an agricultural society are fixed in time and over space, but have the property of recurrence. The tasks associated with these events form distinct cycles of activities. In agricultural societies these cycles of activities and the recurrence of certain prominent physical events become the basis of time measurement. Two cycles are particularly important, the daily and the yearly.

Figure 2.19 shows the daily or circadian rhythm of activities on typical summer and winter days amongst the male Kabyle, an agricultural tribe in Algeria, who also keep animal stock (Bourdieu, 1977). Whereas the Kabyle have no formal hourly structure to the day, based upon time pieces, there is, nevertheless, a remarkably formal structure which has emerged around the five Moslem prayers.

Kabyle

The pattern for the dry season day is shown in Figure 2.20(a) as a cycle of physical events, by which its passage is marked. The events which mark time are a mixture of religious activities, the position of the sun and the occurrence of secular activities such as mealtimes. For the Kabyle the time structure of each day is a quite natural association of religion, obligatory tasks, and the position of the sun: in effect a combination of social, life, and universe times.

Time marked

As with the daily rhythm, so the yearly or circannual rhythm of activities has a formal time structure. Figure 2.20(b) shows the cycle of women's activity associated with the 'farming year'. It is strongly based around weaving and pottery. Cooking has its own circannual and circadian periodicity and is paced by the foods which are available at different times of the year. It is also related to markers; the occurrence of feast days, Figure 2.20(c), (d). The feast days are the first day of spring, midsummer, and midwinter. Special meals are also prepared for the activities of ploughing and harvesting, which in turn have a religious significance.

The Kabyle calendar is a function of religious activities, the

92

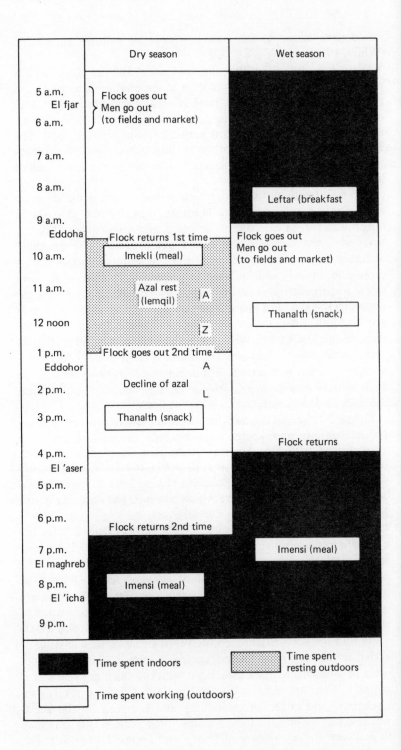

Figure 2.19. Daily rhythms in summer and winter. (After Bourdieu, 1977, 149. Reproduced by permission of Cambridge University Press)

agricultural years and many other things. It is shown in Figure 2.20(d). It has 'eight' months, six of equal length, and two of greater length. The problem here is that 'The calendar cannot be understood unless it is set down on paper (but) it is impossible to understand how it works unless one fully realizes that it exists only on paper.' (Bourdieu, 1977, 98). In other words, the calendar is really a remembered and culturally transmitted oral narrative. The year starts about the 1st September *or* on the 15th August ('the door of the year') *or* on the first day of ploughing and continues from there as an often *ad hoc* mixture of religious observance, agricultural event and the Julian calendar which has also now been mixed into it, as Algeria has modernized. It is what Bourdieu has called 'a practical taxonomy' which provides a set of keys to the archive of Kabyle social and environmental knowledge.

These features of the Kabyle calendar are a mixture of two time types, similar to those which Evans-Pritchard, in his work *The Nuer* (1940), has called oecological and structural time. Oecological time is related to the environment, its vagaries and seasonalities. By contrast structural time relates to the social system and involves the longer time cycles of the ordained passage of man through the social system (rites de passage) and myth.

The Kabyle calendar is representative of most time systems in agricultural societies in its principles if not its detail. It is not so far removed from the English feudal calendar with its saints' days, its unlucky days and its field rituals, as Thomas (1971, 738) has documented: 'In the mid-seventeenth century the Yorkshire farmer, Henry Best, drew heavily upon the church calendar in his review of the year's work. He knew that lambs conceived at Michaelmas would be born before Candlemas; that the ploughing should be over by Andrewmass; that ewes should go to tup at St Luke; that Servants were fined at Martinmas; and that long fields should not be grazed for more than a fortnight after Lady Day. Certain days thus became traditionally appropriate for certain activities. One let blood on St Stephen's Day, weaned lambs on St Philip and St James, and paid the rent on Lady Day. The sixteenth century agricultural writer, Thomas Tusser, offered scores of such maxims to his readers,

> 'Set garlic and beans at St Edmund the King,
> Pore saffron between the two St Mary's days,
> Have done sowing wheat before Hallomas Eve.'

Such almanacs were the chief source of printed knowledge about time and activity (Neuberg, 1977).

Cope (1919) has split calendars into three types:
(i) Purely descriptive — this type consists simply of descriptive designations for the lunar periods, the count commencing with some important natural event.

94

(ii) Astronomical — this calendar type recognizes the descriptive system but mixes it with astronomical observations of the solstice and also, perhaps, the rising of the constellations (the Pleiades is a favourite). Months of about equal duration are more common.

(iii) Numerical — this calendar type comprises those calendars in which numerical designations have replaced descriptive terms and obviously requires a number system.

Cope mapped these three types of calendar system for North American Indian tribes. Figure 2.21, A and B, provides a fascinating glimpse of calendrical variation over space at the turn

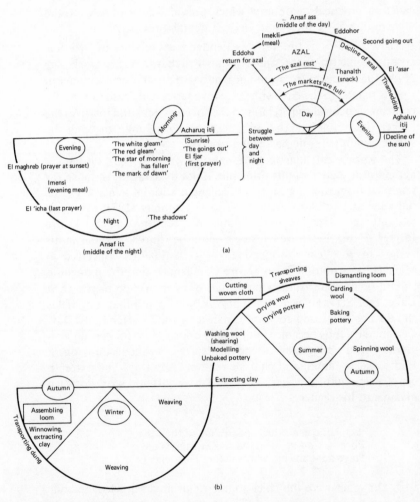

Figure 2.20. Cycles among the Kabyle.

(a) Structure of the dry season day. (b) Cycle of women's activities. (c) Cooking cycle. (d) Abstract calendar.

(Adapted from Bourdieu, 1977. Reprinted by permission of Cambridge University, Press)

of the century, when Western calendar systems had not completely infiltrated Indian cultures.

Figure 2.22 shows a comparison of various calendar systems which will be in operation for the first three months of 1979,[19] and underlines Cope's point that most calendar systems do operate on some common principles because, as we have now seen, they have more or less common roots in the shared experience of agricultural scheduling.

The type of calendar used by a society influences that society's perspective on time. An agricultural calendar like that of the Kabyle is not built for looking ahead but is both a practical guide to

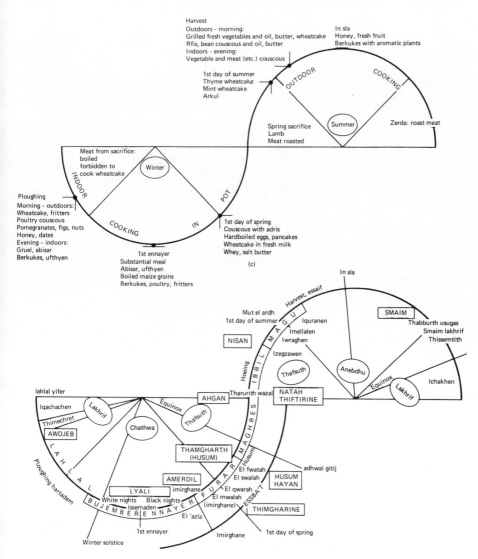

96

Figure 2.21. Calendar differences of North American Indian tribes. (Cope, 1929)

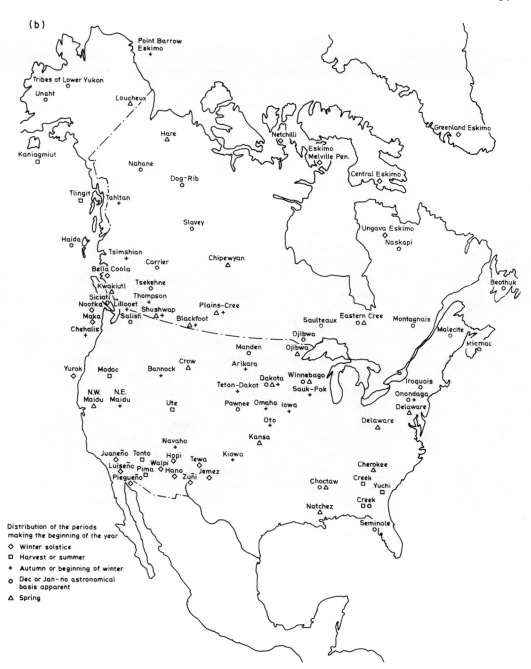

(b)

Point Barrow
Eskimo
+

Tribes of Lower Yukon

Unaht ◇

Loucheux △

Hare △

Netchilli ◇

Eskimo
Melville Pen. ◇

Greenland Eskimo ◇

Kaniagmiut ◻

Nahane ○

Dog-Rib ○

Central Eskimo ◇

Tlingit + Tahltan +

Slavey ○

Ungava Eskimo ◇
Naskapi ○

Haida ◇

Tsimshian +

Carrier ○

Chipewyan △

Beothuk

Bella Coola ◇

Tsekehne ○

Thompson

Plains-Cree △ +

Saulteaux ○ △

Eastern Cree ○ △

Montagnais

Kwakiutl ◇
Siciatl △
Nootka ◇ Lillooet + ○
Maka ◇ Salish △ + Shushwap △ +
Chehalis ◇

Blackfoot △ +

Ojibwa ○
Ojibwa ◇

Malecite ◇

Micmac

Yurok ◇

Modoc ◻

Bannock +

Crow △

Arikara △

Manden ○

Dakota ○ △ +
Teton-Dakot ○ △

Winnebago ○ △
Sauk-Fok +

Iroquois △

Onondaga ○ +

Delaware ◇

N.W.
Maidu △
N.E.
Maidu +

Ute ◻

Pawnee ○

Omaha ◇
Oto +

Iowa ○

Delaware △

Kansa ○

Navaho ○

Juaneño Tonto ◻
Luiseño ◇ Hopi ◇
Piegueño ◇ Pima ◇ Walpi ◇
Hano Zuñi ◇
Tewa ◇
Jemez ◇

Kiowa +

Cherokee △

Creek ◻
Yuchi ○

Choctaw ○ △

Creek ◻ ○

Natchez △

Seminole ○

Distribution of the periods
making the beginning of the year

◇ Winter solstice

◻ Harvest or summer

+ Autumn or beginning of winter

○ Dec or Jan - no astronomical
   basis apparent

△ Spring

farming and an almanac of religion and myth. Therefore the future is not emphasised in practical terms. This way of seeing things may be reinforced by language; thus in the Kamba and Kukuyu languages there are only two future tenses — action in two to six months and action in the foreseeable future, where the foreseeable future is regarded as two years! Long term negotiation, say between an Englishman and a Kikuyu untouched by western civilization against the background of the Gregorian calendar will not make a lot of progress! There is 'no such time as 1984, for instance, viewed from 1978'. If you try to explain an event beyond two years you are neither understood, nor taken seriously! (Mbiti, 1968). Pocock (1964) has stressed the necessity of appreciating this cross-cultural difference in time reckoning, especially in complex international political agreements.

Nowadays, of course, in most agricultural areas in the world, the traditional agricultural calendar runs side by side with other calendars. Chapman (1978) has documented this in his study of the use of agricultural almanacs. In Bihar, a State of Northern India, the farmers use these almanacs as their main source of agricultural knowledge. The almanacs mainly use an agricultural calendar based on 27 divisions of the year into periods of 13 to 16 days. These periods are ideal time slices for the organization of agricultural activity. But use of this calendar coexists with the monthly Bihari calendar and a monthly Bengali calendar. The traditional calendar is used for precise instructions in the almanacs but the monthly calendar may be used when more general sayings are mentioned or in general comments on the weather. Chapman found that the local calendar allowed a much better understanding of the local weather variation. This is important to farmers in a monsoon area. In particular the structuring of the calendar was sympathetic to weather variation. The local calendar picked up most of the variation in rainfall, whereas the non-local Bengali and Bihari calendars did not. Thus the final point must be that the agricultural calendar is a source of knowledge over and above the simple function it sems to have of dividing up the year. It is part and parcel of the social and environmental milieu in which a people live. The new calendars do not fulfil this role and, in terms of 'development', more thought needs to be given to introducing innovations which are sympathetic with local calendars.

Thus we can see that coordination within a social system is the vital factor in the articulation of the concept of time as a social entity. The need to plan is important in agriculture but it becomes more so when man starts to move towards an urban civilization, and towards faster and more complex commodity and capital circulation. The combination of specialization of roles and of new social organizations in the city leads to the spread of precise measurement of time and to a heightened awareness of the

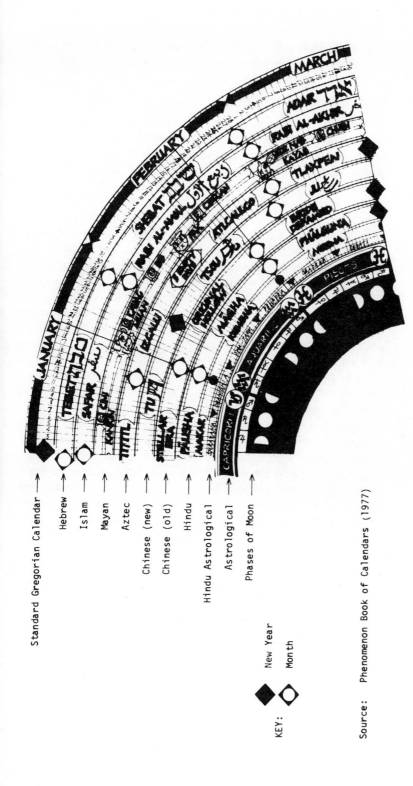

Figure 2.22. A comparison of calendar systems in operation January 1st to March 20th (spring equinox, northern hemisphere), 1979. (Reproduced by permission of Phenomenon Publishers Ltd.)

significance of finitude, of boundaries and limits in what can be carried out in one day. Many of these boundaries and limits are then used as ways of marking the passage of time in different ways.

(v) The large scale societies

We will give rather less space to the large scale societies in this section than we gave to the small scale systems because most of the rest of this book refers to them in one way or another. This section should therefore be considered as a brief overview, emphasizing generality of perspective.

With Weber's emphasis on the rationalization of time in the West (Weber, 1958; 1920), we find one of the firmest statements about time as a scarce resource; together with attitudes to time, as linear and convertible into monetary terms. Industrial society seems dominated by time consciousness and it has been suggested that this 'represents a transformation of the impelling metaphor FROM a view of time as basically cyclical and subservient to spatial concepts TO a view of progressive time and timing in which spatial concepts are subordinated. A state dominated by spatial concepts is oriented toward law and order and qualitative judgements. Space is its own product. A state dominated by linear time is oriented toward productivity and quantitative judgements. This is a measurement of another object' (Ilchman, 1970, 138). Hence Taylorism and the tyranny of the clock.

Time rationalization

Urban societies of advanced scale display all the signs of time famine. To accomplish the wide range of tasks wanted and needed by their members, various adjustments are made such as shift work programmes, part-time work, role specialization, staggered work hours and flextime systems. Advance bookings and appointments bring a semblance of order to an otherwise 'turbulent' space-time environment. As societal scale increases an inverse spare-time law seems to operate! The more complex the society the smaller the units of time which become critical to the attainment of a goal. As Linder (1970) has argued, the consumer society demands time to consume and with fairly finite lumps of it to allocate, time for consumption must be taken away from time for production or from time required for allocation to physiologically determined needs like eating and sleeping. Yet these two universally necessary activities themselves consume about ten hours in every twenty four: 3650 hours a year.

Unlike the situation in small scale, traditional societies, the largest settlements in advanced scale systems display considerable variation within. These variations are a key to the nature of regional variations in social times within urban places. Universe time in the city is a standard time only to the extent that it marks the boundaries of public events. Time perspectives and the appropriate time locations of many actions, as well as their order, become increasingly more important at smaller more human scales, especially following secondary judgement.

A person's age is still one of the most complex metrics of all. Measured in universe time units, it relates to life-time item relations and apart from its manifestation in the physical appearance of a person, it is also a marker to the widest possible range of social groupings. When to go to school, when to vote, etc. But the 'oldest' people in cities are the lonely people because they are by definition cut off from the rich variety of times which might otherwise be available to pace life.

Heterogeneity is a factor in the making of an urban environment (Wirth, 1938). But the large city is not, because of this, unstructured. Vertical and horizontal structuring has been identified by many researchers and is intuitively known to exist by every adult urbanite. Power to control the use of space is vested in the hands of individuals, governments and classes, in private and in public enterprises. But power to control the use of time is less well understood. Just as the space which is most directly controlled by authorities of one complexion or another is physical space, so too the time which is directly amenable to immediate 'control' is physical or universe time. We can think of the manipulation of interest rates and minimum credit periods or waiting times for public housing or the licencing hours of certain urban functions. Access to time has become a major cause of urban social stratification, especially in terms of social class. The higher the social status niche one occupies the greater the access one seems to have to the future. As the existence of a future to plan for is perhaps the single most important factor in present activity, then the ability to colonise this time region at various scales (length of waiting time for public housing, or a job or a dental appointment) becomes a major source of power, yet it is one almost wholly ignored by sociologists and human geographers.

However, Schwarz has some rather interesting observations to make on the ecological distribution of service units (especially medical facilities) as they affect the time costs of clientele. He argues for the 'linkage between class, status, time, and space' (1978, 1203) in relation to access and delay within a 'resource-availability theory'. Caplowitz's (1963) thesis, 'The Poor Pay More', which must be applicable to time as well as money 'is found to hold not because of the consumption habits of the poor or the selling practices of the marketplaces, but because of the spatial distribution of marketplaces in large cities. Delay is one expression of an overarching macroscopic linkage between class, status, and space' (Schwarz, 1978, 1219). Delay is a temporal symptom of social pathology (or ecopathology). How times are utilized and by whom, indicates the nature of social time. The extent and stability of a personal future (a time perspective) is a clue to social class and therefore to power, but it is also a component in a more catholic issue: the very quality of (urban) life.

The existence of groups and individuals with short time horizons or with unstable, uncertainty-dominated futures, gives further opportunity to high 'time' status groups because the need for the security of a promising future perspective is a basic human need, and people without it are susceptible.

Social times appear to have many paradoxes associated with them. Unlike money, we cannot always equate access to 'large quantities' of it as a good thing. In terms of measurement, it is still necessary to associate locations and durations in social time with universe time values. 'After lunch' as location, and 'between lunch and tea' as duration is not always sufficiently precise for the arrangement of activities with others. But there are other situations in which the universe time measure is quite inadequate as a means for comparing the duration of socially distinct events. Four weeks of unemployment is not comparable with four weeks of holiday; the universe time scale is not a suitable base for purposes of comparison because it is not constructed on item relations which are common to the meanings of work and leisure.

So universe time, on its own, is an inadequate measure of time in large scale societies. Take for instance man-hours, working-week, financial-year; we seem to require some sort of time which gets closer to the dynamics and item relations of the social system in the way that Rummel (1970) has attempted in the political context. Abu-Lughod (1969) and many other urban social scale analysts have pointed out that increased scale brings with it an increase in the number of dimensions of social space, an increase in the heterogeneity of social space. Why should there not be an increase in the dimensionality of time and in this particular, in social time?

If you live in a large town or city and do not wear or own a watch, but aim to take part in the social life of the community, then there are probably enough regularly recurrent signals (see Gurvitch) to enable you to fulfil your needs. But innovative behaviour (IB), which is also a characteristic of urban lifestyles, becomes much more difficult. Urban cultures need accurate time measurement because they are 'experimental' and innovative cultures and many of the innovations adopted, whether as material artefacts or in social organization are directly concerned with gaining advantages in time allocation and use.

For instance, recently, flexible working hours have been introduced in many countries, and Table 2.4 shows an example of a flextime sheet. The objective of flextime is to allow the individual employee to allocate time in such a way that work obligations and non-work obligations can be fulfilled with a minimum of disruption. In terms of social time there is an opportunity for time 'saving' due to a widening of the range of choice of alternative time

NAME J. CITIZEN   DEPARTMENT ADMINISTRATION PERIOD FROM 30/5/77 TO 24/6/77

STANDARD WORKING HOURS PER DAY __

I declare that the details completed hereunder by me are true in every respect. (Signed)... J. Citizen ......

	DAY	LATE	START TIME	LUNCH OUT	LUNCH IN	TIME TAKEN	FINISH TIME	ACTUAL TIME WORKED HRS	MIN	REMARKS	LEAVE TYPE	HRS	MIN	PUBLIC HOLIDAY SUPER E'T	CREDIT ER	MIN	DEBIT HR	MIN	CUM HR	MIN
1	MON	30	5	8 45	12 30	1 30	1 00	5 00	7 15							15				16
2	TUES	31	5	8 45	12 30	1 30	1 00	5 00	7 15							15				30
3	WED	1	6	8 40	1 00	1 30		30	4 40	7 30						30			1 00	
4	THURS	2	6	8 45	12 30	1 30	1 00	5 05	7 20							20			1 20	
5	FRI	3	6	8 40				12 20	3 40			1			OUS			3 20	2 00	
6	MON	6	6	8 40	12 30	1 00		30	4 45	7 35						35			1 25	
7	TUES	7	6	8 45	12 30	1 00		30	4 45	7 30						30			55	
8	WED	8	6	8 30	1 00	1 30		30	3 00	6 00	Study time 3-4pm	6	1 00							
9	THURS	9	6	8 45	12 30	1 00		30	4 45	5 30	Study time 1-3pm	6	2 00			30			25	
10	FRI	10	6									3	7 00							
11	MON	13	6									3	7 00					30	55	
12	TUES	14	6	9 15	12 30	1 30	1 00	4 45	6 30											
13	WED	15	6	8 30	1 00	1 30		30	3 00	6 00	Study time 3-4pm	6	1 00						25	
14	THURS	16	6	8 45	12 30	1 00		30	4 45	5 30	Study time 1-3pm	6	2 00			30			10	
15	FRI	17	6	9 00	12 30	1 45		45	5 00	7 15						15			10	
16	MON	20	6	8 40	12 30	1 10		40	5 10	7 50						50			40	
17	TUES	21	6	8 45				12 00	3 15		Went home sick	2	4 00			15			55	
18	WED	22	6	8 30	1 00	1 30		30	3 00	6 00	Study time 3-4pm	6	1 00					1 25		
19	THURS	23	6	8 45	12 30	1 00		30	4 45	5 30	Study time 1-3	6	2 00			30			25	
20	FRI	24	6	8 45				12 10	3 25			1			OUS			3 35	2 10	
								110	50			27	00		5	15	7	25		

* TYPE 1. Flexleave 6. Study time
2. Sick leave 7. Examination leave
3. Annual leave 8. Public holiday
4. Long service leave 9. Concession day
5. Short leave 10. Other

Net Credit/Debit this period 2 10
Hours brought forward - previous period
Hours carried forward to next period ... 2 10

All calculations checked by A Checker , Supervisor ACP Supervisor Date 27/6

Table 2.4 An example of a 'typical' flextime worksheet

locations and durations as well as the sequence or activities. But, following the adoption of flextime, the pattern of time use and activity order for any individual seems to be that after an initial period of widely fluctuating behaviour (space timings and time spacings) there is a convergence towards the pre-flextime routine. This is an indicator of the invested power of social times and the need for collaboration (Sorokin and Merton, 1937) which is a hallmark of urban industrial societies.

Urban society is also being forced to colonize the night hours. Note the graph in Figure 2.23 recording the times at which the House of Commons rose, in 1945 and in 1976, and the graph of the number of sittings that ended at or after midnight between 1945-1976 shows an example of this colonization.

In Spain a recent move by the Spanish Government to adjust the traditional social times of Spaniards, because of their unfavourable impact on industrial efficiency and commercial linkages, in a rapidly urbanizing and integrating Europe, has received an unsympathetic response as the Spain correspondent of *The Economist* reveals below. (We shall return to the colonization of the night in Chapter 8.)

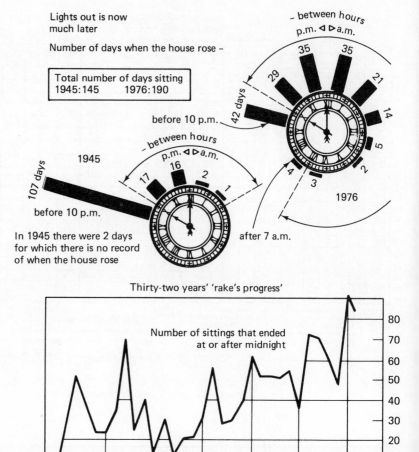

Figure 2.23. Time colonization and the time structure of the British Parliament 1945-1976. (*The Economist,* October 22, 1977. Reproduced by permission of *The Economist*)

The next chapter completes Part I and with it the introduction of the main concepts and principles. Chapter 2 has illustrated an expanded perspective on the everyday notion of time and the times associated with this view have been called paratimes. The discussion in Chapter 1 gave time a necessary role in the study of place and in Chapter 2, with the introduction of the notion of paratimes, we found that the contact between space and times required that we be aware of the operation of other times than

Early to bed? It's a flop
FROM OUR SPAIN CORRESPONDENT

No wonder somebody has been taking potshots at the Spanish prime minister's residence. Having wound up a dictatorship, legalised political parties and trade unions and won an election, Mr Suarez is now tackling something really big: he is trying to persuade Spaniards to go to bed early.

Like Dr Johnson, Spaniards suspect the worst of people who go to bed before midnight. Midnight is when Spain's air becomes breathable, when its wines are most drinkable, and when public entertainments are just getting into their swing. Climate, social custom and long working hours (often in two or more jobs) have combined to make a night-bird of the average urban Spaniard. He considers it abnormal to go to bed one day and get up the next.

Economists and efficiency experts have been urging the authorities for years to impose a "European" timetable on Spain. Foreign clients sometimes find businessmen who have lunch in the afternoon and work in the evening hard to get in touch with. Drowsiness is one cause of Spain's high industrial accident rate. All that nightlife guzzles electricity. But even General Franco, backed though he was by a pullulation of police services and a termitary of technocrats, failed to change his subjects' noctambulant habits.

Mr Suarez was impelled to act by the size of Spain's oil bill. On December 3rd he introduced a long list of restrictions. Cinemas and circuses must close at midnight, and cafes and restaurants at 1am. Basques must stop playing pelota, and greyhounds stop racing, half an hour after midnight. Street lights are being dimmed and fountains and illuminations must be switched off before 11pm. Television goes to bed soon after the chickens.

A few of these measures may be self-defeating. Television's early close-down sends some restless viewers out for a stroll. Frustrated greyhound fans gamble in cafes. Deprived pelota players probably join Eta. So as not to appear irredeemably puritanical, the government has decided to allow cinemagoers, pelota players, greyhounds and the rest an extra half-hour on Saturdays and fiesta-eves (of which there are a lot), and a full extra hour every night during the summer. Even so, many Spaniards are cross. They know that once governments get a grip on anything, they rarely let go.

Some critics are uneasy about the coincidence of the government's go-to-bed campaign with its more tolerant attitude towards adultery and abortion. Indeed they suspect that it is no coincidence but a single package. They could be right. It would be sloppy economics to send people to bed to save fuel if the result were a population explosion.

(*The Economist*).

common clock time, even though we still conduct our lives predominantly in universe time. Chapter 3 presents further ideas on the way space and time may be related and suggests that times may be adjusted in order to effect changes in the nature and use of geographic space. The chapter concludes with a short discussion on the role of cyclic events in giving persistence to urban places. Perhaps the understanding of persistence should be a precedent aim to the understanding of change.

Notes

1. This movement is, of course, related to functional activity space and not to territorial space.
2. Professor Doob is a social psychologist at Yale University. His major work, *Patterning of Time* (1971) occupies nearly 500 pages. Clearly it is a hard task to do justice to it here. The synopsis which we offer must therefore only be treated as a guide to the principles he works with. Reference should be made to Doob (1978).
3. Dispositions, or predispositions as Chapin calls them, are an important factor in the development of the transductive model for human activity analysis (Chapin, 1974, 1978) to be outlined in Chapter 5.
4. The summary is from Zerubavel (1976). In Chapter 8.6 the images of time shared by social status groups are discussed with an empirical example and should be related back to this notion of intersubjectivity.
5. Our attention to Neisser's notion about mediation in the knowledge of reality comes from Pocock and Hudson (1978), though the association of ideas is rather different in our use of the term. (Neisser, 1967, 3, and Pocock and Hudson, 1978, 20).
6. Social space is paraspace to human geography. Urban and social geographers might well note that this paper by Sorokin and Merton was dated 1937; nearly two decades before the social area framework appeared and three decades before *N*-dimensional social space was 'in' urban geography its relation to time was already being studied.
7. Gioscia's vectorial scheme, discussed in 2.3(v) is rather similar. For instance new ideas awaited too long (Gurvitch) would be aligned on the metachronic (ahead)-anachronic (behind) vector.
8. Cadence is a notion we take up again in later discussions, especially in Chapter 8.
9. The notion of a meeting appears in the Time Geographic paradigm to be discussed in Chapter 6. Every meeting occurs in a space-time locational and experiential context. A phenomenological perspective on meeting is provided by Seamon (1979).
10. In urban social geography the notion of *N*-dimensional social space is now well established. Social status, family status and ethnic status for instance, are accepted as pervasive dimensions or urban social structure in industrial societies. Apart from the social times associated with each dimension we can differentiate with some success the time scales appropriate to life styles of individuals and groups located at different positions within a dimension, e.g. high status and low status groups, cf. Chapter 8.
11. This is an important concept in the time-geographic approach, Chapter 6 following.
12. There is possibly some congruence between this view and the deliberated choice model of human activity (Chapter 5 following), which has been developed by Cullen.
13. Cycles are a recurrent theme throughout this text, especially in Chapters 8 and 9 following.
14. This poignant piece of graffiti appeared on a bridge in London: 'Is there a life before death?' (Lynch, 1972). The writer was looking for a shift in RB!
15. We use the term susceptibility because it is one which we have already introduced and which we will be using in the next chapter in a rather more generalized way as an important factor in the mechanism or process of activity persistence or routine which gives urban places

coherence from day to day. Change being induced by innovation, requires a susceptible context, but for day to day living, all cannot be flux.

16. The usual interpretation of Umwelt would be 'environment' but this is 'too open' according to Fraser, so he uses the German term without further translation.

17. You may note some similarities of concept here between Fraser's approach and the concepts put forward by Kolaja on group and personal IB, RB, and IB_i and RB_i, Gurvitch's level 3 for conduct with relative regularity and Doob's 'potentials'.

18. Refer back to Gurvitch's classification of groups according to time dispersion.

19. We are writing in November 1978!

3
Spacing, timing, persistence, and place

3.1 Preliminaries

So far we have considered the notion that there are a number of qualitatively different times and that each is effective in the patterning of human behaviour. By operating together they provide the basis for a more or less coherent environment for daily living. The places which we live in, whether private or public, and their co-ordination into systems of places (regions) result from coping strategies which are developed to handle the scarcity of geographic space and universe time, and also to cope with the decline in coherence which all of us experience from time to time. These coping strategies enable completion of the tasks which face us in day-to-day living and in that way they make us less susceptible to unforeseen and undesirable events.

Coherent

Susceptible

There are scarcities associated with all resources and with each of the factors of production, and there is scarcity of time. Resources, including time, are appraised differently by producers and consumers; individuals, groups and associations; households and families, and so scarcities of time and space are category specific. How do times work to give coherence to space and what sort of room is there for manoeuvre? How adjustable is space-time, or place structure? With all the time in the world at our disposal we occupy either hell or utopia and who knows where they are?

Coherence

Place

Coherence is the harmonious connexion of the several parts of a system. The impelling factors which operate to bring coherence (in a sense liveability) to places are known as pacemakers (Parkes and Thrift, 1975, 1977, and 1979).

Liveability
Pacemakers

Relocation of the times of recurrent events makes it possible to cope with spatial problems, such as congestion, which might otherwise be treated using only spatial solutions, (e.g. an alteration in the road network geometry). As well, but not conversely, a particular existing patterning of time may be adjusted by inducing change in the way in which periods of time are perceived to be available. Flextime is an inducement to spacing universe time in alternative ways, in order to satisfy personal wants and needs which are more precisely located in lifetime or social time.

The terms timing space and spacing time may seem rather confusing. There is, however, both a conceptual and an operational difference between them. Timing space means the maintenance or adjustment of the space context by time manipulation, and is usually bound by universe time referents. The excerpt from a daily newspaper (1978) indicates the times at which various activities in Bradford in the UK (which are fixed in space) are located in clock time. The times at which cinemas, theatres and restaurants are open, so often displayed in newspapers, indicates when spaces in the city are available for use. With spacing time the emphasis shifts to the time domain entirely. What is the pattern in time for various activities?

Timing space

Spacing time

Each week similar charts to our example appear in newspapers in most of the major cities in Britain and elsewhere, and of course other media are also used to transmit temporal signals of this sort. From one week to another there is little or no change in these timings. There is a weekly recurrence of events, cycles of one week in periodicity which gives a sense of stability, of persistence, to urban space, inducing a sense of place. Personal and image RB develop about the rhythm of the system. Innovations in social services, recreation opportunities and transport facilities take time to be adopted and their adoption is always some function of their ability to find a suitable phase relation with existing recurrences. Disharmonies prevail because of ill-conceived universe time locations which conflict with the social rhythms of different social spaces.

Persistence

Any space in 'the most general and abstract sense of the word, is simply a class or set of items in relation' (Shackle, 1972). When we think of geographic space we usually have a bounded territory of some sort in mind, delimited either formally (e.g. a political state) or functionally in terms of relations among selected items (economic, political, etc.). Formal territories or geographic areas usually have a given proper name to identify them in an unambiguous manner which is appropriate to a particular time period, New York or Leningrad.

Formal regions

Functional regions are less readily or frequently labelled although some of the economic planning regions which have appeared in many countries in recent years, particularly those based on daily urban systems, are an exception. Formal or functional regions can be adjusted in various ways or may be sustained by operating on the items which are either contained within their boundaries (i.e. formal) or on the items which determine their boundaries (i.e. functional).[1]

Functional regions

Clockwise

Market days
Bradford: Monday-Saturday; *Bingley:* Wednesday and Friday; *Ilkley:* Tuesday and Saturday; *Keighley:* Wednesday, Friday and Saturday; *Shipley:* Friday and Saturday.

Half-day closing
Bradford: Wednesday; *Basildon:* Tuesday; *Bingley:* Tuesday; *Ilkley:* Wednesday; *Keighley:* Tuesday; *Shipley:* Wednesday.

24-hour petrol
Petrol stations giving 24 hour service: Texaco Service Station, Leeds Road, Bradford; West Bowling Service Station, Manchester Road, Bradford; Hartshead Moor (M62) Service Station; Snaygill Service Station, Keighley Road, Shipton (£1 and 50p pumps); Central Garage, Bradford Road, Cleckheaton (£1 and 50p pumps).

Hospital Visiting
Bradford Royal Infirmary — Daily 7.15-8.15 p.m. also Saturday and Sunday, 2.30-3.30 p.m. Patients children under supervision Sunday 2.30-3.30 p.m. Maternity unit: husbands only daily 7.30-8.30 p.m.; also ante-natal wards daily 4-5 p.m.; post-natal wards daily 4-5 p.m. (husbands only on day of delivery). Patients own children 4-5 p.m. must be with an adult.

Advice
Citizens Advice Bureau — Thorpe Chambers, 12a Ivegate, Bradford, (telephone Bradford 25325); 5 Sandywood Street, Keighley, (Keighley 605454); 2 Mortimer Street, Greenside, Cleckheaton (Cleckheaton 877607); Voluntary Services Centre, Market Square, Pudsey (Pudsey 578000).

Bradford Association of Widows — for widows, single or divorced women. Meetings: Room 1, Central Library, Prince's Way, Bradford. First Friday in month at 7.30 p.m. Telephone Bradford 673026 for details.

Bradford Association for the Elderly — Advice, information or aid. 1a Upper Piccadilly, Bradford. Monday, Tuesday, Wednesdays and Thursdays 10 a.m. to 4 p.m. Telephone Bradford 20225.

Legal Advice Centre — Citizens Advice Bureau, Thorpe Chamers, 12a Ivegate, Bradford 4. Telephone Bradford 25325. Thursdays and Fridays 10 a.m. to 12 noon. By appointment only.

Lighting-up Times
4.26 p.m. to 7.21 a.m.

Social Services
Metropolitan District Social Services, Provincial House, Tyrrel Street, Bradford. BD1 1NP. Telephone Bradford 29577 (*during office hours*).

Samaritans
Telephone Bradford 494949 (any time) or Bradford 494940 (between 10 a.m. and 10 p.m.

Tourist News
Regional Tourist Information Centres, Central Library, Prince's Way, Bradford (Telephone Bradford 33081, extension 451) Monday to Friday 9 a.m. to 9 p.m.: Saturday 9 a.m. to 5 p.m.: Mill Hey.

London Trains
From Bradford Exchange, Manchester 21.34 (Sunday 21.28): Halifax 22.52 (Sunday 21.28): Leeds 22.43 (Sunday 22.23): New Pudsey 22.04 (no Sunday transpt). London, last through train 21.20. (Saturday 21.50. (Sunday 22.00).

(Bradford Evening Telegraph 1978)

Before and after

One way or another it is the order of item relations as before and after in association with the geometry of the same items which provides a basis for geographic realizations. If, for example, item relations are adjusted by changing their location and duration in universe time, (e.g. banks open until midnight) then we have a change in the timing of space and a different or new geographic region or place is generated even though the locational space co-ordinates remain unadjusted.

However, adjustment of the universe time domain does not necessarily bring about a change or even permit us to gain an understanding of how any change which does result (i.e. a particular system of conduct) came about. To achieve this it is necessary to be aware of the interrelations among time and paratimes in the lifeworld of individuals and of such cohesive social groups as the family or licensed professions. Any change in terms of locational time is treated as new information and is related to other space and time elements, at every level of the social system (cf. Figure 1.23). The effective environment then becomes a partial function of the coincidental and often unappreciated experiential times of the society.

Geographic realizations

Paratimes

3.2 Timing spaces

Who is able to make the sort of adjustments to time locations which have an impact on space use? The greatest power to time space is clearly with government agencies and commercial and industrial organizations. These are superstructural bodies. The result is that it is rather hard for the individual to time more than his backyard and for some people who live in multi-storey dwellings, rented from state or local housing authorities, even this freedom to time their 'Entrances and Exits', is denied due to fear of hooliganism or the effects of vandalism which make lifts inoperable or stairwells unsafe and offensive.

Superstructural

The impact which the superstructure's timing of space has on the spatial organization of society will depend very much on the time intervals which are involved. In general the longer the period, the greater the impact since this permits (or even forces) a potentially larger number of individual timing strategies to be developed. The cutting below from a British newspaper (in February 1978) illustrates the power of the educational superstructure to time space over a long period, and if adopted, perhaps to time space into a future which will extend over a number of generations.

'FLEXI' SCHOOLING

Flexible hours may soon be introduced in Northumberland schools to cut costs by reducing the demand for buses at peak times.

(Daily Telegraph, 16.2.78. Reproduced by permission of the Daily Telegraph.)

Land use planning

Time use planning

Cadence

Oscillatory pacemaker

Zeitgeber

Activity density

At present timing of space tends to be achieved indirectly and unintentionally because statutory planning emphasis in many countries tends to have been set in laws and by-laws which are couched in spatial terms: we have land use planning but seldom time use planning, although some transportation planning may be an exception to this.

Residential areas of the city often experience an increase or a decrease in the cadence of their activities due to alterations in the timing of events. School holidays, for instance, act as a low frequency oscillatory pacemaker with the cycles of their own system (i.e. three terms) acting as zeitgeber to the rhythms of the domestic space. The opening of a new shopping centre tends to be associated with the notion that a new physical structure must have been built. Following the ceremonial cutting of ribbons 'there it lies' a new place to shop! To some degree this is usually so, but it is equally the case that a change in the hours of an existing retail centre, as in late night shopping, may be as effective as the opening of a new centre. There is now a retail facility **when** there was none before.

Within the city there may be differences in the timing of space as the result of changes in the function within buildings, or as the result of rehabilitation and renewal of the city's physical fabric. With rehabilitation of old and architecturally valuable buildings it is experiential time which operates on space,[2] (see Chapter 9).

Through powers of spatial zonation planners have control over the speed at which areas run. This is the activity density of an area, the number of different events which occur in a unit area in some unit time. Figure 3.1 illustrates some results from a space timing study undertaken in Australia (Jones, 1977). Data was derived from newspaper announcements, sampled for alternate weeks over a two-year period. The figure shows in a straightforward manner the differential timing of space throughout an urban area with a population of around 275,000 people. A kilometre square grid was used as the net in which to catch activities and the surface shows variations in the total number of hours for which an area was 'switched on' during the year. The CBD quite clearly dominates the activity environment of the city, reflecting its urban status. The scale at which the computer drawn maps have been set understates the amount of activity in suburban areas but part of the principle of space timing is brought out quite well.

By zoning an area as a conservation space the planner not only freezes the physical character of a space but ideally also its temporal character. He aims to maintain the original essences of the area, although as the plate and news cutting below suggest, this is not always well accepted (Figure 3.2 and Newscutting).

Museumization (Relph, 1976), on the other hand, attempts to recreate the places of the past in as much detail as money can provide but the authenticity is achieved only if the appropriate

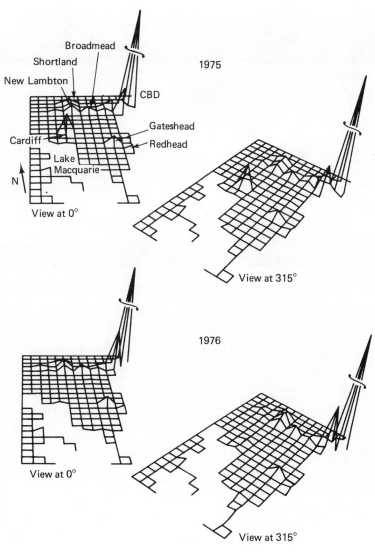

Broadmead
Shortland
New Lambton
CBD
Gateshead
Cardiff
Redhead
Lake
Macquarie
N
View at 0°

1975

View at 315°

1976

View at 0°

View at 315°

Figure 3.1. An aspect of space timing: total activity hours in Newcastle NSW, 1975 and 1976. (Jones, 1977)

times are fused into the spaces and structures of the museum — the time must feel right — and this can best be achieved by the sense of activity density which is induced.

Other examples of space timing are not hard to find. For instance, architectural styles in buildings may impose a powerful association between a time — Georgian, and a space — Bath! An 'effective bond' (Tuan, 1974) is established when space and time are right and the product is place. The process of 'place-making' depends heavily on the collocation of spatial and temporal elements and in particular the 'right' experiential time elements will be important. Some geographers and planners have recently

114

The village fish and chip shop at Saltaire where residents claim they are being ordered to turn back the clock and recreate dark Victorian homes.

SLUM FEAR AT UNIQUE VILLAGE

Daily Telegraph Reporter

RESIDENTS of a 100-year-old model village claim they are being ordered to turn their houses into Victorian slums.

Bradford Metropolitan Council yesterday issued a leaflet containing a policy statement of development 'do's' and 'dont's' for Saltaire, near Bradford, following demands from the residents.

Saltaire, built between 1853 and 1873 by Sir Titus Salt, is classed as an outstanding conservation area.

The council has announced it is to clamp down on modernisation schemes. They want a return to original style sash windows, panelled doors and Welsh slate roofs.

The council insists that planning permission will not be given to remove chimney stacks and cast iron drainpipes. Residents will also be given a standard colour chart to be used for paintwork.

Expensive work

A leading protester, Mr Reginald Hodgson, who runs a local bakery and corner shop, said: "They have got a fight on their hands. We are the people who live here and we are proud of the village. We will look after it and not be told by a big council what to do."

Another protester, Mr Roland Teale, of Albert Road, said: "This is all very well but how are we, as ordinary people, expected to pay for expensive work like this."

Council Development Control Panel Chairman, Councillor Smith Midgeley said: "There is nowhere else like this village in the world. It must be preserved. Over the past few years changes have been made for the worst."

He added: "Unfortunately the only way we can stop drastic alterations is by using these powers. It means restricting people's freedoms to alter the homes."

Saltaire has more than 800 homes, a textile mill, alms houses, a hospital and schools —but no public house or off-licence. Sir Titus, a devout congregationalist, insisted that alcohol should never be sold within its boundaries.

Figure 3.2. Turning back the clock of a neighbourhood. (*Daily Telegraph*, October 1978. Reproduced by permission of the *Daily Telegraph*)

published studies which give some support to these ideas (Lynch, 1976; Rapaport, 1977; Tranter and Parkes, 1977).

Space is also timed when a plan is proposed for the development of land, for instance as a freeway. The plan is an 'advance warning system' which may cause individuals to alter their personal action spaces and activity spaces and their associated times.

Such planned changes in space (i.e. of the class or set of items in relation) will impinge on image RB and image IB (future behaviours), but to different degrees according to the Big Time[3]

Figure 3.3. Museumization: timed space; late 19th and early 20th century from 1978. Beamish, Co. Durham, England

horizon of the project: is it one year or five years? Different degrees of future-discounting are involved according to the length of the planning time interval and the time perspectives of the individuals and groups affected. As we saw in the previous chapter, these time perspectives may vary quite substantially according to age, social class and sex. The constraints on Small Time changes, which people may wish to make as a personal adjustment strategy to the new scheme, will also eventually bring about changes in the timing of space. The durations of activities also time space, a factor in the (geographical) significance of time budget (Chapter 4) studies which has not been properly investigated to date.

In addition, the location of an activity (x) within a sequence of activities $(a,b,c,d, \ldots, (x),y,z)$ times space. This seems to occur not only because an activity may have a necessary location at a particular point in the sequence but also because social convention or habit has located it there. Certain obligatory activities (Chapin,

Obligatory activity

116

1974; and Chapter 5 following) time space and the most obvious of these are sleep and eating. The mean velocity of communications can also time space: if the velocity is low (bullock carts in India) then space is likely to be slow. If the velocity is high (telephone) then the space is likely to be fast, but cross-culture comparisons do require that a suitably comparable time metric is used. (Some of these principles are taken up in Chapter 7).

3.3 Spacing Times

Spacing time

With the notion of spacing time the focus shifts to the interval between events[4] and the recurrence of events. Spacing becomes the adjustment of these event (or item) relations and therefore of time. In biological or psychological time, drug inductions are a good example of time spacing agents (Fischer, 1968). The items may be activity episodes or activity sequence strings. At an everyday operational level spacing time becomes the ordering of universe time intervals so that they are available for allocation to events, in advance; timetabling of various sorts is the obvious example. Some activities require a fixed amount of time for their completion (a football match), they are inelastic in duration and impose constraints on the amount of time spacing which is possible, others are more 'flexible' but may impose constraints on other activities because they make rigid demands on their sequential location.

An alteration in the spacing of time may result from the relocation of activities in geographical space or as the result of changing values and fashions in social space, such as Sunday church service attendance. The particular structure that time spacing reveals at any one time within a society is therefore not necessarily an actively premeditated condition. When events change their geographical location they will always cause respacing of time which extends to events beyond those in question. There is therefore an opportunity time to most events, and an opportunity space for many events.

Opportunity time

Any evaluation of the impact of geographical relocations should incorporate an assessment of the time spacing changes which are likely to be induced and the times considered should not be restricted to universe time. There should be some attempt made to consider the impact of the relaxation time in relation to life-times and social times: how are obligatory physiological needs likely to be affected, how will time perspectives be affected, what changes in social RB will be induced?

Time spacing involves the recognition of limits. There are limits in every time domain and there are limits set by the length of cycles, which may relate to the duration of certain events or conditions. For instance, universe time limits exist as hours of daylight, biological time limits exist as aging or the recovery time from illness, retirement is a social time limit. Thus it becomes necessary to be aware of limits which forbid an unbridled freedom in the spacing of time. In general time is spaced towards and away from the appropriate limits which are associated with a project and the process always involves the utilisation of markers (key events and key times). How tasks are actually strung together to form projects (i.e. their duration, sequence structure and frequency) will in turn result from the behaviour of the relevant pacemakers. Time spacing is therefore a rather complicated process and involves much more than simply moving the time of an event. The facility to space-time in turn varies between individuals, groups and superstructural agencies.

Limits

Project

Markers

The individual spaces time but the manner in which he/she does so will depend on the profile of predisposing and preconditioning factors which he or she possess. Perhaps the most important of these factors are once again age and sex in conjunction with the joint impress of the communities (or social) spaces which the individual occupies at the moment, and which he or she has occupied in the past or anticipates occupying at some time in the future.

Predisposing and preconditioning factors

These space factors will be recognized as co-terminal with the temporal perspectives within the socialization process outlined in Doob's scheme of potentials.

Medical researchers at Johns Hopkins University (USA) found error in time estimates among their patients from poorer backgrounds which were not found in those from higher socio-economic backgrounds (Luce, 1972). These findings link up with cross-cultural study which has shown that people who have not been brought up to a life strictly regulated by the clock will tend to free run to a greater extent than those who have. (Free running behaviour is a form of time spacing). Recent research suggests that exogenous socio-economic factors may play an important role as pacemakers to the inner time-reckoning systems which influence human behaviour. Relations like these suggest that if sub-areas of cities are occupied by individuals with similar socio-cultural attributes, which are related to life-time and social time, then in low status space, times may well be spaced rather differently to the way in which they are spaced in high status space (Parkes, 1973, 1974 and Chapter 8.3 following). The resultant behaviours which are observed are therefore in some way dependent on the scope for time spacing in each individual group's space-time and we have seen already that this characteristic may itself be a suitable basis for group classification and discrimination (Gurvitch 1964). The family is the most 'familiar' group and various aspects of group time spacing have been well illustrated by Young and Willmott (1973). Differential timings among the members of a group are adjusted by rules and other displays of authority. But, conflicts do

Free run

occur, and their basis is the set of dispositions and preconditioning factors (such as age and sex) which themselves have endogenous timing properties. However, there are also effects which are the result of long socio-cultural roots such as the 'appropriate role' of women in the family. Why else is it that work hours for women are the ones which are most frequently cited as being inappropriate?

Long intervals of each day require, often by custom, the presence together of every member of a family group, or every member of a club committee. The same space and the same time must be occupied by all members. This sort of group spacing of time, multiplied for instance through all similar family groups in the city is a considerable influence on the timing of activities of many different sorts.

Loosely we may think of aggregates as large collections of individuals or groups with a common interest, crowds in a shopping mall or at a football match for instance. In this way the time spacing of the family group acts as an important pacemaker for urban rhythms in general, and our understanding of space-time rhythms in the city will therefore depend on our understanding of the spacing of time in family groups, each of which occupy locations in different parts of social space.

A regular, pulse-like timing system is adopted when events depend on a large participation for their economic and social success. Football matches every Saturday through the season, festivals such as Mardi Gras in New Orleans, or major international exhibitions such as the Boat Show in Earls Court, London, are examples. Each have time spaced with high fixity in order that they may be clearly marked and therefore in turn act as marker pacemakers for the scheduling of other activities (the design and production schedule of manufacturers for instance), and appropriate time spacing strategies within groups such as 'the family' who attend the event.

Groups can also assert their identity by appropriate spacing of time. Thus Boal (1976) distinguishes between behavioural and structural assimilation of ethnic groups. Both forms, however, vary in rate of assimilation and ascribed assimilation according to origin of the group and numbers and this distinction in the pace of assimilation becomes a manifest feature of group difference.[5] Characteristically ethnic groups derive much of their community by celebrating festivals and days of national significance which are out of phase with the days of festival and national meaning in the host community. They therefore space time differently, and this is what distinguishes their social space distance. Because such behaviours are manifest, for instance 'in the street', the material environment periodically reflects them and excites physical space in a distinguishable but often temporary manner.

The difficulty with such notions lies in gaining a fuller understanding of the difference between temporary and permanent: changeable and persistent: in effect it confronts us with one of the most tantalizingly difficult concepts associated with time, that of interval density; continuity and discontinuity. For ethnic groups, when certain critical markers disappear from the calendar and clock of the group they move closer to the host culture. The time perspectives among the old, however, can cause loneliness and distress, and every large city in western society which has experienced unprecedented change in its social and physical fabric over the past century will contain pockets of dreams that cannot be realised. The ethnic group which has been discontinuing its regular ceremonies, moves towards the host community not only because of the positive adoption of new traits but also because of the discontinuation of old ones. Gradually, through the processes of assimilation and acculturation, the traditional social times of the ethnic group become increasingly discontinuous and interval density decreases. In the silent language of space and time (Hall, 1966) there are many idiosyncratic traits which members of a culture group share and take for granted but which distinguish them as clearly as their passports or their spoken words. The way that groups handle time is one such trait and has a lot to do with the nature of space. It is a pointer to the resistances they wittingly or unwittingly transfer between themselves and the host community. As Hall points out, time of day may indicate the importance of the occasion as well as on what level an interaction between persons is to take place. In America advance notice of an event is often referred to as lead time, an expression which is significant in a culture where schedules are important (Hall, 1966, 26).

Temporary and permanent

Changeable and persistent

Finally, superstructures space time according to their own needs to control the individuals and groups within their domain. They may cover almost the whole scale of time relevant to the organization of social activities even including the spacing of time of future generations. Typically they are concerned with bundles of Big Time such as are involved in the generation of development plans for hospital location and health care services, education programmes, transport networks, etc. Successful realization of such plans, however, depends on their impact on Small Time respacing and this is where lifetimes and social times become a necessary component of study because it is within rather short intervals of time that the greatest impact on liveability is felt: the eight hours of a 'good night's sleep', the ten minutes waiting for a bus, or the closing times of public libraries. After all, in the long run we are all dead!

Superstructures

Time is spaced in relation to the rhythm or cadence of activities. Such rhythm is determined by various pacemakers having different levels of fixity. Some pacemakers are periodic features of lifetime such as sleep, others are irregular but generally recurrent (or fluctuating) such as childbirth or illness and these can influence time spacing very significantly. Adjustments may occur before or after the marker is reached.

Time is spaced by the introduction of new information. As the scale of society increases and as the specialization of roles becomes more marked in some social areas and less marked in others, it is necessary that the urban system (or place) should have some way of coping or adjusting (Hawley, 1950), so that its ecological relations can be maintained or changed under control. There is the somewhat paradoxical situation of increased independence among many dimensions of social life (Abu-Lughod, 1969; Timms, 1971, 1976; Parkes, 1973) and an increasing interdependence among individuals playing specialized roles from day to day. The inter-dependence relies on the capacity to achieve synchronization of action and the maintenance of necessary phase relations.

Information

Ecological relations

Synchronization

3.4 The accordion effect

The co-ordinated complex of changing interdependencies among activities, spaces, and times is partially achieved through the operation of a mechanism called the accordion effect (Parkes and Thrift, 1975). This effect involves the 'use' of time and paratimes and the impact of markers and pacemakers and helps to explain changes in the behaviour of one system (which may be an individual) in relation to another, for instance the workplace and the residence place. As no empirical work[6] has been undertaken, based on this notion, it is only possible to illustrate the mechanism with a hypothetical example.

The accordion effect is the process of timing and retiming spaces and of spacing and re-spacing times forward and backward from a new marker following reception of new information. It involves the replacement of some initial space-time event sequence with a new one centred around the co-ordinates of the old marker. The accordion effect is a product of the principle of stress minimisation in which the impact of unexpected events (themselves a recurrent characteristic of advanced or urban scale social systems) is accommodated by time spacing, i.e. by organizing time appropriately.

Accordion effect

The principle operates at all space levels of the social system and at all time scales. Many of the social pathologies which plague urban societies are a consequence of the inefficient operation of the accordion effect, i.e. an inability to respond adequately to new information. Some social deprivations are associated with just those individuals and groups who have been unable to cope with the demands of time spacing and the concomitant controls and influences exerted by established space timings. The need for retraining due to technological change, inflexible working hours, the time location of medical services, time fixities imposed by domestic routines, all these are factors susceptible to an accordion effect. The locations of clinics, hospitals, schools, day care centres, parks, and playgrounds are all space-timing and time-spacing factors in the everyday life of someone in the city: they have to be coped with and brought into synchronization according to needs and wants. The fixity of their spacings and timings act as constraints within which, however, there are also limits. The accordion effect is indicative of the means of resolution of the disruptions of daily life imposed by constraints and limitations.

Figure 3.4 illustrates the accordion effect schematically and will be referred to during the discussion which now follows. Consider an office worker j. He leaves his home location L_1 for his work location L_2 at 0800 hours. This familiar clock time event is also located in a decision time dimension at d_1. (The meaning will become clear as the discussion develops). At d_1 the decision to leave for work is made but so is the decision to leave work at 1700, located at d_2. (Journey to work is an RB with a very high image RB within a family, in this way the decision to leave home and the decision to leave work at a particular time are made together.) Between d_1 and d_2 the residence system or place and the work system or workplace are synchronous. Events (items) which occur at L_1 are essentially independent of the events which occur at L_2. The residence and workplace systems are synchronous rather than synchronized because their common characteristic is simply that they occur in the same clock time interval. In effect L_1 and L_2 are dislocated in activity space-time. Although there may be a joint location in clock time there is no common ground among the event sequences which occur in each domain unless some new event (IB) occurs which causes a need for synchronization between the spaces L_1 and L_2. The important factor to appreciate is not that something unexpected has occurred at the work (space) or at the residence (space) but rather that the IB affects the significance of the d_2 time location because it (d_2) acts as a marker pacemaker which now allows the residential system to pace its events according to the choices and constraints within its domain.

Decision time

Synchronous activity

Let us suppose, for the purposes of illustration, that at some time in the day beyond d_1 and in his office location (L_2) an instruction (new information in the authority domain) is received by j from the superstructure level (management) which leads to the expectation that d_2 is no longer a realistic marker. This instruction in effect becomes d_3. In decision time it is after d_2 but now becomes antecedent to d_2 in clock time, occurring at 1300, whereas d_2 referred to 1700. For j the workplace, now located on L_2d_3 is immediately put into disharmony with L_1 (home), where his family only have the interval $d_1 \rightarrow d_2$ to associate with. Telephone communication with L_1 may allow some rapid readjustment of the

Antecedent

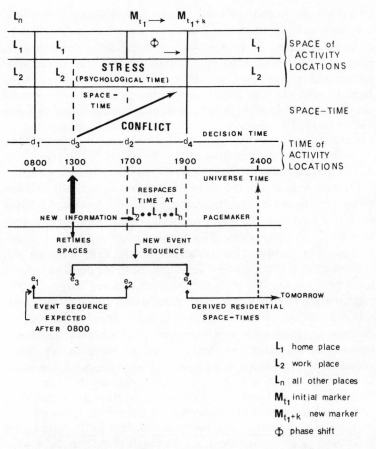

L_1 home place
L_2 work place
L_n all other places
M_{t_1} initial marker
M_{t_1+k} new marker
Φ phase shift

Figure 3.4. The accordion effect. (After Parkes and Thrift, 1975. Reproduced by permission of Pion Ltd.)

residence system to reduce disruption if a new marker M_{t1+k} can be set up at d_4, at say 1900 clock time.

To comply with the demands set by the new information, and the particular authority constraint of management, j's strategy will be to respace time between d_3 and d_4 putting tasks (items) into a new relation: spending more clock time on some and less on others within the limits set by the work environment. From the instant of d_3 (1300) not only is j's time spaced differently but because of the extension of work onto d_4 (1900), space is also retimed, the office at L_2 is expanded in space-time. From the point of view of the city as a whole there is an increase in size (or room), by at least the product of the geometric area or volume of L_2 and the extension of working time from d_1 to d_4[7].

j's behaviour in space and in time has been constrained by the authority domain and this leads to coupling problems (cf. Chapter 6 following) with the residence space-time because as j has to move from d_2 to d_4 there is always the possibility that L_1 cannot be synchronized with L_2 at and following d_4. Some sort of phase synchronization of L_1 and L_2 clearly becomes desirable at or shortly after d_3 when the instruction was received. However, this is a time when under normal circumstances whilst the two systems may have been synchronous in universe time they are wholly dysynchronized in social time, the residence time system and the work time system being assessed in terms of quite independent item relations once the influence of the original decision ($d_1 \rightarrow d_2$) time locations had been set.

Coupling constraint

Social time

For some period before d_2 and now up to d_4 there is a likelihood of phase shifting as activity spaces are rescheduled. This is illustrated as the area Φ in Figure 3.4. New markers are set up and new spaces can be activated as the office (L_2) comes under the influence of the new spaced time (or region) $d_3 \rightarrow d_4$ as opposed to the $d_1 \rightarrow d_2$ region on which behaviour had initially been constructed. In this example the generation of social stress and space-time tensions (Nystuen, 1963) (in a family group) has its cause between d_3 and d_4 and will tend to increase in intensity up to d_4 when it will be suppressed if d_4 was precisely estimated at d_3 and communicated to the homeplace (L_1) in time for adjustments to be made.[8]

Note the area above the lowest of the horizontal lines in Figure 3.4. This represents the new event sequence at the office L_2 given the event instants e_1, e_3, e_2, e_4 mapped one to one with d_1, d_3, d_2, d_4 at clock times 0800, 1300, 1700, 1900 which, although independent of what is 'going on at home' sooner or later impel a retiming of the L_1 space to produce a derived space-time residential environment between 1900 and 2400 as shown in 3.4 which is a function of the initially internalized time of L_2. There is no simple forward movement of time and of activity organization, some actions are brought back in time and others brought forward: what is yet to happen determines what is about to happen and what has happened determines only to some degree what will happen. If the IB which has been set up by the instruction at d_3 becomes or is expected to become the basis for a new RB chain (recurrent behaviour profile in 2.4) then new coping strategies are initiated which may result for instance in a change of job, change of residence, adoption of flextime work schedules etc.

The accordion effect demonstrates how space timing and time spacing ideas allow us to investigate quite complex space-time situations and help us to come to grips with the processes which operate in the construction of urban places. On most days we return home to the same neighbours, and leave home for the same colleagues or enemies at work. Such persistence is the characteristic of urban activity space-times or places. We will now consider this concept with the aid of two principles: the principle of susceptibility and the principle of entrainment.

Persistence

Susceptibility

Entrainment

3.5   On cycles, activities, and persistence

It was probably in about 1838 that rhythmic ups and downs in business activity were first noticed (Dewey and Mandino, 1973). In 1849 periodic activity was noticed in the frequency of sunspots. Shortly after this discovery rhythms were detected in the size of populations of animals. And by 1875 an American economist recognized the existence of rhythmicity in prices. These disparate phenomena were all connected by the fact that they exhibited periodicity and formed cycles. By 1930 the study of cycles was well accepted and, in economics in particular, where the study of business cycles was *en vogue,* the explanation and prediction of cycles had become a major theme.

In 1931 an international conference was held in Matamek, Quebec, on the subject of biological cycles. This event was to be a turning point in the geographic study of cycles. Two of the papers read at the Conference were by the already eminent geographer, Ellsworth Huntington, and he was also to become the rapporteur for the Conference. The Conference led to the establishment of the Foundation for the Study of Cycles. Huntington was on the board of directors, along with such noted scholars as the biologist Huxley, the astronomer Shapley, and the economist Mitchell (Martin, 1973). The 1930's and 1940's were to be characterized by a feverish study of cycles of all sorts. The 'Great Depression' graphically illustrated the effects of simultaneous troughs in many economic cycles and, for this reason, the study of cycles was particularly pre-eminent in Economics, for instance in the work of Schumpeter and Kondratieff, and we will take up some of this work in Chapter 9. But in other subjects as well cycles were being studied. Geography was amongst them, especially through the work of Huntington. He warned, however, that 'It must be recognized that there are innumerable pitfalls in this whole subject (of cycles). It is easy to see cycles where none exist. It is also hard to pick out those which undoubtedly exist' (Huntington, 1945, 470). Huntington's major concern was with 'civilization' in the large. As part of this project he had already tried to measure the contribution of different types of people to the progress of 'civilization' in his book *The Pulse of Asia* (1907). This work eventually resulted in one of the classics of the cycles literature, *Season of Birth — Its Relation to Human Abilities,* published in 1938. His conclusions are best summarized in his own words:

> 'The season at which people are born has far greater importance than is generally supposed. At certain seasons the number of babies is unusually large, and the proportion of girls is high. The children born at these times have a low death rate in infancy, and the survivors live to more than the average age. In addition, the births of persons who achieve distinction rise to high proportions. Such conditions indicate not only that

reproduction is stimulated at certain seasons, but that children then born are more vigorous than those born at other times. There is also a little hint that mentally defective children are born in unusual numbers at this same time. Such children, however, as well as those that are malformed show a more decided tendency towards birth at the time of year when evidences of vigour are least numerous. At the most favourable season, however, criminals and persons who suffer from insanity and tuberculosis as well as those who achieve eminence, show an unusual number of births. This, too, is evidence that reproduction is especially stimulated at certain seasons.'

(Huntington, 1938, 1)

These conclusions were the result of the painstaking collection of data from an enormous number of sources, including the F.B.I. and Bureau of Patents as well as the Census! Figure 3.5 shows one of the series of graphs Huntington was able to draw up showing the variation in month of birth, and it also shows one of the cycles to which this was thought to be related — seasonal incidence of insanity, suicide, and crime (c). Huntington's identification of these relations has since been verified by many other studies, although they have yet to be adequately explained (Huntington's explanation was climatic — the human variations in season of birth are an inheritance from earlier times when few children survived unless born at the best season).

Huntington's work on cycles was to find its final form in *Mainsprings of Civilization* published in 1945, 'an attempt to analyse the role of biological inheritance and physical environment in influencing the course of history' (Huntington, 1945, v). This massive survey again relied to some extent on cycles as an explanation of the progress of civilization and includes Huntington's now 'notorious' hypothesis that atmospheric ozone causes an increase in intellectual activity at certain seasons of the year. However, that part of the book that does not simply smack of eugenics remains as an interesting document.

His known emphasis on the cycles of weather, electromagnetic fields of the solar system, and variations in ozone and other 'respects of the atmosphere' should not be taken simply as an example of rather naive determinism. He was well aware that 'only a minute fraction of the entire field (could) be covered' in his own work (1945, 460), and the three kinds of physical conditions referred to above were only examples of some of the influential cyclic conditions which affect civilization. 'The matter becomes still more complicated when we examine the delayed as well as the immediate effect of all these (physical cyclic) conditions, together with those of many other social, political, economic and psychological conditions which vary in cycles and influence (entrain) the general flow of business and the conditions of civilisation' (Huntington, 1945, 464, our parenthesis). While much of

126

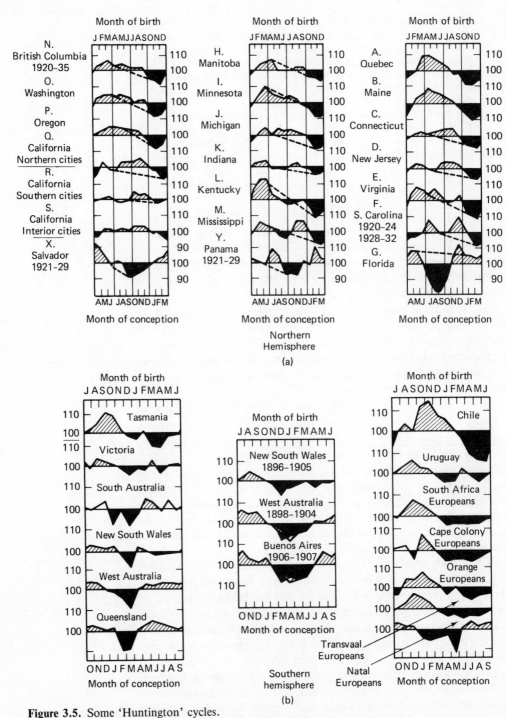

Figure 3.5. Some 'Huntington' cycles.
(a) Months of birth in United States and Canada, 1926-1932, and Central America, 1921-1929.
(b) Seasonal distribution of births in the southern hemisphere, 1921-1929.
(c) Seasonal incidence of insanity, sexual crimes, and suicide.
(Huntington 1945, from Mentor Books, 1959)

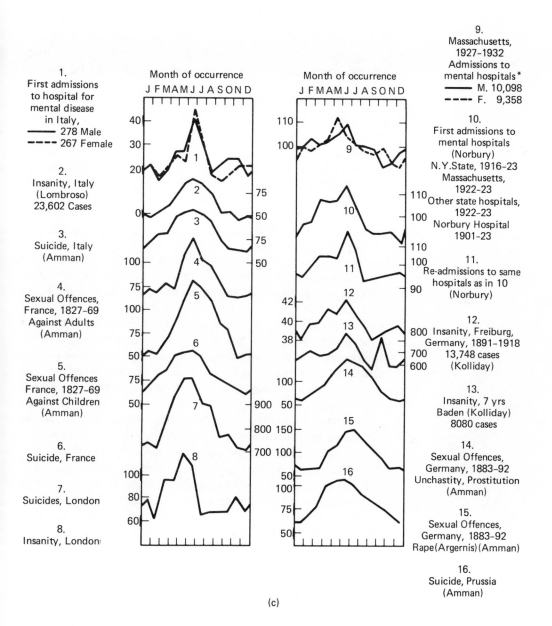

(c)

Huntington's work on cycles must now be treated very cautiously indeed, there is much else that is still demanding careful attention. One wonders what Huntington would have been able to infer if modern time series analysis techniques had been available to him! (See for instance the special 1977 issue of Part A of the *Journal of the Royal Statistical Society* on reanalysing of old cycle data using new techniques).

As one of the more controversial figures of geography, Huntington's contribution has not fitted well into modern orthodoxy. Clark (1968, 85) has called him 'the figure whom most contemporary geographers might most avidly choose to forget'. Yet, paradoxically, he has been the geographer who has had most influence on subjects other than geography. It may be time to rethink his role.

After the crescendo of work in the 1940s the study of cycles became rather less fashionable but it is now enjoying a considerable renaissance, particularly as the result of a renewed interest in biology and economics (and, it appears, particularly in Marxist economics (e.g. Mandel, 1975). There are many reasons why human geographers should participate in this renaissance in cycle research, and there is some evidence that this is now happening, especially in so far as the idea of rhythm can be incorporated into a theory of experience of the lifeworld (Schutz and Luckmann, 1973).

Urban space

Urban space is usually defined by a set of item relations selected on the basis of whim, theory, and empirical evidence. It is conceivable that the whole world might be defined as an urban space because of its persistent display of linked causes and effects, as in Doxiadis's notion of Ecumenopolis. But consider more familiar 'place' scales. Any atlas of Australia which contains maps of the settlements in New South Wales will locate Newcastle about 150 kilometres north of Sydney. Though Newcastle may occupy more space now than it did a year ago it remains essentially the same place from one day to another or from one week to another

Coherent system

and from one year to another because it is a more or less coherent system of synchronized and synchronous facilities and actions as outlined in our discussion of the accordion effect. It becomes a different place not simply because events or functions have relocated in space, as this is a trivial commentary, but because they have also shifted in times. Any move in space is also a move in time, except in terms of geometrics, with which we are not concerned. A retiming of the spaces themselves may also have occurred and this also constitutes a geographical change. The manifest difference brought about by the construction of new buildings or new roads is superficial evidence of change because in the end for us as human geographers it is the liveability of space which is of consequence and this is dependent on space timing and time spacing. Consider Buttimer's view that 'to record behaviour in an isometric grid representing space and time is only an opening onto the horizons of lived space and time. Neither geodesic space nor clock/calendar time is appropriate for the measurement of experience.[9] The notion of rhythm may offer a beginning step towards such a measure. Lifeworld experience coud be described as the orchestration of various time-space rhythms: those of physiological and cultural dimensions of life, those of different work styles and those of our physical functional environments.' (Buttimer, 1976, 289).

Through sets or families of cyclic, rhythmic events, whether experiential or locational, the territory of recurrent human actions is largely determined. Persistence of a pattern of event (item) relations, which are often strictly cyclic, gives structure to spaces, but this structure is itself temporary; it is only a relatively permanent set of item relations within the larger structure of society. This poses some rather difficult problems for the study of place when a Big Time perspective is adopted. As geographers we are left trying to grapple with a situation rather like that faced by a psychologist attempting to interpret the results of a Rorschach test while the ink is still flowing! The time frame chosen becomes the most important determinant of explanation at every scale and consideration should always be given to the mapping of appropriate times and spaces. The mapping of time frame scales to space scales may be illustrated in a rather simple way by the diagrams in Figures 3.6 and 3.7.[10]

Space-time scale mapping

SPATIAL SCALE

TEMPORAL SCALE	Room	Building	Neighbourhood	City	Metropolis	Region	Nation	World
Minute	3	3	2	1				
Hour	3	3	3	2	2			
Day	2	3	3	3	3	1		
Week	1	2	2	2	3	1	1	
Month		1	2	2	2	1	1	1
Season			1	1	1	2	2	2
Year			1	2	3	3	3	3
Life			1	2	3	3	3	3

Figure 3.6. Space and time scales. The numbers in the cells simply indicate the strength of association: for instance the minute or hour and the room have a sensible scale relation, the 'score' of 3 indicates this. However, the minute and the world is clearly a mismatch for all practical human geographical purposes. Minute by minute records of socio-cultural processes at the world scale are quite incongruent at the present time

130

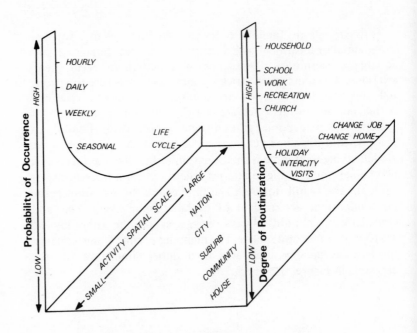

Figure 3.7. Time scales and activities. (After Whitelaw, 1972, 04. Reproduced by permission of the editor and author)

Corporeality

Human corporeality (van Paasen, 1976) and human mentality limit action in both space and time. Movement (or more generally mobility, which is literally the ability to move) is the only evidence required to establish their joint status (refer back to Figures 1.1, 1.2, 1.3), and it is movement which establishes geographic relations among items. The same notion can, of course, be applied to the conjucture of social space with time as in the concept of social mobility and the pauses in mobility at one's place in society.

The lifeworld (Buttimer, 1976) which we occupy is predicated on activity which is based on the tacit assumption of a tomorrow and this usually means a tomorrow of routine, planned or expected events, of almost automatic decision-taking (Seamon, 1979). There are really very few surprises for us in a single lifetime let alone in a day. This preferred orderliness depends on the stability of relations among families of cycles, and upon recurrent events that might, however, have a less systematic structure. Such orderliness (rhythm in a loose sense) requires that there be certain dominant items which act as zeitgeber and pacemakers which impel or influence personal, small group, and even societal-level 'sense of time' and which therefore determine the specific (universe) or appropriate (life and social) time of all events. The biological and psychological times of human life are sometimes referred to as the times inside but they are influenced, as we have seen, by the times outside, the universe and social times of the community environment: the socio-temporal Umwelt (Chapter 2.4(c)).

Zeitgeber

In the strictest sense times and spaces are related through their common item sets. For time the items are observed in sequential and recurrent relation and for space they are observed in juxtaposition. In everyday life they are brought together by movements and are given a joint and specific complexion according to the situation and context in which they are set.

Situation and content

Context is an amalgam of factors: physical and biological, mental and societal which together fully encompass a situation in physical and experiential terms. Situation is the capacity (in space and time) to realize any one of a number of the many possible interactions which exist in the context of a specific lifeworld. This capacity is not fully determined in the universe time of the calendar and the clock, though such time may be used as a convenient referrant. Nor is capacity adequately defined simply by objective items and elements which are surveyable onto maps and atlases, clocks, and calendars, because there also exists a complex of spaces and times which are subjectively determined, and which appear to be differentially structured according to levels in the social system (cf. Althusser, 1970).

Capacity

In much the same way as individual conduct is a product of the level of coherence among the times inside and the times outside and of empathy with the contextual and situational factors which influence the content of experience, so too place is a product of its internal and external timings with regard to events as they recur within the context and situational constraints of calendar and clock, life and social times. A reference back to the meaning of 'context' will show that calendar and clock times belong with physical factors, life times with biological and mental factors and social times with societal factors.

Place

To illustrate the notions of persistence and coherence, situation and context, pacemaking and place, consider how a city appears when viewed from the top of a tall office block, as darkness approaches (Parkes and Thrift, 1979). Lights switch on and shine from buildings and other items of urban furniture. The major roads appear as rivers of light; fixed lights on the banks, traffic, as flow. Intermittently the flow is halted as traffic lights intervene, some synchronized with the rhythms of the traffic others responding to the less regular, less predictable signals of pedestrians. The view also reveals that some spaces, excluding the channels of the transport network have a higher proportion of their buildings with lights switched on than do others. Then as clock and social time passes lightened areas darken and darkened areas brighten. Some spaces remain much the same, however, unpenetrated by the demands of society, uninfluenced by its changing role from a work-time space to a rest-time space. Such a night-time view repeated from the same vantage point over a number of 'days' reveals a rhythmic situation in a time context which is a high order socio-temporal lifeworld, its most important periods set to a circadian or circaseptan cycle (cf. Table 2.2).

However, within the circadian period there may also exist various ultradian frequencies or periodicities like shiftwork in factories and hospitals, so that the city is characterized by the dependent relations between these fairly high frequency cycles and a range of lower frequency cycles. However, from one day to another there is not much change except when the weekly rhythms are considered, in which case the social times of the 'weekend' exert their influence.

Tools

The growing similarity among urban places around the world is often remarked upon. But such reference to the 'sameness' of places usually refers directly to the appearance of the material, built environment. And yet standardization of technology is making similar tools available to an ever increasing proportion of the world's population. These tools extend capability (among varying abilities) up to more or less common limits, so inducing a convergence of material socio-cultural traits. The timing of operations which are associated with production of the built environment is also being standardized by the capacity of similar technologies. The integration of similar machines into industrial and commercial complexes is accompanied by a further convergence of secondary, tertiary, and quaternary organization of the working day. Not only does the 'heart' of one city look much like the 'heart' of another, but it beats with a similar, commonly entrained rhythm.

Differences among urban places do exist of course, but apart from superficial differences in building design, now usually limited to residential buildings, these differences are based on the timing of events of public significance which act as markers or low frequency oscillatory pacemakers, and which can have a periodicity of as much as a year perhaps. For example, Edinburgh in Scotland is persistently different to Glasgow from year to year because in August it holds a major international festival of arts and drama. This festival acts as a pacemaker and time giver (zeitgeber) contributing to the specific urban personality which Edinburgh has for its own, for some period once every year, though to which it is 'geared' for a number of months before and after the festival. Here we find place distinction through the operation of a persistent low frequency cycle.

Inner-time domain

The notion of time structure and its contribution to urban personality can be illustrated further by reference to high frequency pacemakers. Consider again an industrial city such as Newcastle, Australia, with its population of around 270,000. Ecologically the city system is dominated by a large integrated steel plant. The plant's impact is visible on the morphology of the surrounding streets, buildings and wastelands. Its impact is also identifiable in the price of housing and in the social structure of the environs, but the greatest social geographic impact comes through its pacemaking properties. The inherent rhythm of steel manufacture entrains social and territorial subspaces of the urban area to its own dominating cycles. Associated metal fabricators, service facilities, retail trade rhythms (fortnightly in harmony with pay periods) and domestic routines each in turn imprint the times of this ecological dominant onto their local spaces: so the steel plant, brighly lit at night in defiance of the universal cycles, captures 'susceptible' times and spaces of the city and entrains them to its recurrent high frequency (circadian and ultradian) cyclic needs.

Susceptible
Entrains

Big Time effects like those associated with trade cycles and the more or less irregular perturbations induced by labour relations conflicts are also evident but in each instance the steel plant itself is a pacemaker, occupying territorial, and functional space. Specifically, it is the times inside the steel plant system which are the critical determinants of day to day and week to week liveability for a large proportion of the people in Australia's sixth largest city and it is from this time base that the city ultimately derives many of its idiographic place qualities. Many of the others emanate from similar if less dramatic properties associated with other places of employment and of the interaction among times within them and within the social mosaic of individuals and groups in the population at large (refer back to Figure 3.4).

Time inside

Situational information is held by each and every individual about the times at which events occur indicating his or her capacity to participate, and arises from experience of having participated in certain activities in the past or from recurrent reinforcement of information about the existence of certain events through the media and other communication channels. Day by day a programme of routine activities is followed and the configuration of this programme is largely determined by the strength of marking which is associated with each activity (its clock time and its social time). The fixity (Cullen and Godson, 1975) and elasticity[11] of activities in relation to the clock and to other activities is a constraint on the exercise of choice and is a contextual factor related to situation. The 'daily' routine of any place is, in the final analysis, a function of the operation of a series of entrainments to pacemakers. The improvement of our understanding of the morphogenesis of place will ultimately depend on our ability to explain how places stay the same from day to day rather than on how they change. The key to such an understanding lies in the study of susceptibility to entrainment and the properties of pacemakers. For instance, over the past one hundred and fifty years of industrial urbanism it has been the case that the workplace has been the dominant zeitgeber for (high frequency and low frequency) residential or domestic routines. However, in the future it seems likely that the residential space of cities will impel rhythm adjustment towards the workplace as the status relations and roles of women and men and their children change and as the labour fraction in any workplace decreases against the machine. Flextime

Elasticity

schemes provide one of the clearest examples of the beginnings of this reversal of the direction of space timing and time spacing. A recent report on flextime in Reading, England, supports some of these concepts (Shapcott and Steadman, 1977).

We end this chapter with two brief empirical examples of entrainment to new time systems; one successful and one unsuccessful.

London time and local times

Elsewhere (Parkes and Thrift, 1979) we have shown how the introduction of Greenwich Mean Time acted as a successful entraining factor in the space-timing of British towns, in conjunction with the need for coherent timetabling spurred on by the development of the railway system in Britain. Until about the middle of the nineteenth century town times in Britain had been set to suit the internal needs of the particular settlement and were not systematically related to longitude. Figure 3.8 illustrates the position of towns in relation to London Time in 1852.[12] The subsequent adoption of Greenwich Mean Time by these towns is a good example of the diffusion (see Chapter 7) of a new entraining agent. The history of the diffusion can be found in Thrift (1977) and Howse (1979). Over a space of about thirty years from 1852 until the passing of an Act of Parliament which made Greenwich Time universal in Britain in 1880 the towns of Britain were gradually entrained to the aptly named 'London' Time. This entrainment was not just a matter of travellers arriving at stations, sure that the time of the timetable was the same as local time, but also had far-reaching social consequences. 'Our time' was replaced by 'their time', although popular resentment was not as great as at the time of the changeover to the new Julian calendar when the 'New Style' calendar caused rioting in Britain. (Many thought that ten days had been taken off their lives, others viewed the new calendar as a Papist plot, some saw it as both). The onset of Greenwich time heralded a new era in which social habits were timed to the minute, instead of the hour. For instance, in 'society' the idea of 'calling cards' started to decline (Davidoff, 1973) being replaced by exact appointments.

An example of an unsuccessful entrainment is provided by the French Republican Calendar. This was introduced during the French Revolution and lasted only one year, from 1793 to 1794 (Zerubavel, 1977). The calendar was segmented into twelve months of thirty days, and the day was divided into ten hours of one hundred minutes. Such a grand change was too much to take, even for convinced Republicans, and watches at the time had a small reference dial showing the 'right' time (Figure 3.9).

These two examples, one of a successful adoption of an innovation and one of an unsuccessful adoption, show that ultimately the phenomena of entrainment, persistence and susceptibility are deeply rooted in prevailing social structures. The surface levels of everyday life and everyday places at one and the

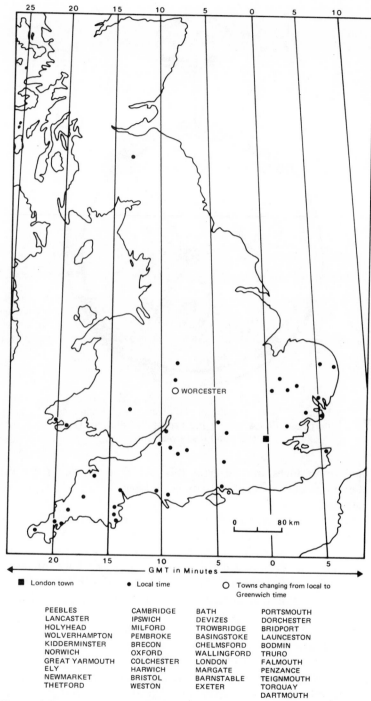

Figure 3.8. Towns operating on local time in England, Scotland, and Wales, 1852. GMT as pacemaker with the railways. Operation of the entrainment principle and evidence of persistence. (Parkes and Thrift, 1979)

Figure 3.9. A sketch of a silver watch made by
Berthond Frères during the French Revolution. (Note
the small traditional dial)

same time conceal and advertise this fact. By studying the process
of entrainment we can hope to move one step closer to the syntax of
these structures.

We have now completed the first part of this introduction to the
role of time in issues of interest to social geographers. The concern
so far has been with the introduction of certain concepts and
principles. In Part II which begins with Chapter 4 we introduce
approaches and methods for the application of a time perspective
to the study of human geography and to urban places in particular.
The use of time has been studied by evaluation of time budgets. We
shall first discuss some of the characteristics of time budgets and
then consider the structure of human activities. This is followed, in
Chapter 6, by an introduction to the time-geographic paradigm
developed by Hägerstrand, Carlstein, and associated researchers in
the Department of Geography at the University of Lund in Sweden.
Time, space and movement are then illustrated by periodic
markets, diffusion principles, and the notions of convergence and
divergence of space and time locations. From this interplace
perspective we turn to a space-time ecological view of the city and

follow with some illustrations based on empirical work including space-time urban imagery, the rhythms of urban social pathologies or ecopathologies and a discussion of the process of time colonization. Chapter 9 then addresses some issues concerning time depth in the environment, the rhythms of building construction which are a condition for the existence of a sense of time. The Appendix is a guide to some analytical methods which will have been referred to in the text. A rigorous approach is not adopted; many admirable books exist for those wishing to follow up our account.

Notes

1. If a functional region has been determined in terms of the 'daily urban system' notion of journey-to-work-patterns, then an improvement in travel times due to freeway construction or improved rail services will adjust these spatial limits as a function of timing, see Chapter 7.
2. The plates in Lynch's study of Boston time (Lynch, 1972) are well worth referring to for some good examples of this concept.
3. The terms Big Time and Small Time have been introduced elsewhere (Parkes, 1973, 1974; and Parkes and Thrift, 1975). Big Time refers to time units of one year or more, Small Time to any shorter intervals, like the hour, day, or week.
4. In the strictest sense there can be no 'interval between sequential events'. The concept of interval density and time density is beyond our province but reference to Lucas (1973) is recommended. Gioscia (1972) also has some interesting comments on the notion of instant and process which you will recall from the section on social time, Chapter 2.
5. Refer back to Gurvitch's notion of time categorization of groups.
6. The term accordion effect has been used by Calkins (1968) in a social-psychological study of the behaviour of in-patients in a mental institution. The particular use which we propose for the notion is, however, somewhat different.
7. Although the product of (area × time) has no metric values in urban geography at present we shall see in Chapter 7 when we discuss convergence and divergence in space-time that Forer (1978) does have some suggestions to make for their combination.
8. Studies of human activity structure and stress have been carried out by Cullen *et al.* (1972, 1975, 1978) and will be discussed in Chapter 5 following. The accordion effect which we are discussing is notionally similar to Elliot Jaques' concept of Time Span (Jaques, 1964) derived from longitudinal studies of the Glacier Metal Co. and known as the Glacier Project. Goodman (1967) provides an interesting critique of this concept and a discussion of Jaques' decision time and time span capacity as characteristics in the exercise of discretion. The latter is similar to the idea of limits which we have mentioned as a dictate of behaviour, a better spatial image is perhaps created by the term bounds: as in 'out of bounds'.
9. These are elements in the psychological time component of what we have called lifetimes.
10. Isard (1971) has attempted to formalize certain concepts of scale transformation in relation to time. He remarks that 'the choice of a relevant transformation of time resembles the choice of a relevant map' (p.19).

11. Elasticity will be discussed in the next chapter. For the moment it may be thought of simply as the amount of variability in time location that an activity can accommodate. For instance sleep would be inelastic, reading a book, elastic.

12. From an historical urban geographic viewpoint it may well be worth investigating the impact of local times on the structure of the central place network. In 1852 it appears that the majority of towns which were still operating in local time were situated in the areas where the classical central place system has also been found (cf. Bracey, 1962; Dickinson, 1964).

PART TWO

Practice

4
What is the use of time?

'In the world of common-sense experience the only close rival of money as a pervasive and awkward scarcity is time.'

(Moore, 1964, 4).

Wasting time

THE LATEST piece of dizziness from California tops most: spending part of your spare time seeking advice on what to do with the rest of it.

"Leisure counselling" is a serious, booming business. People worried about what to do in their spare time can now go to a counsellor and, for about £25, get an interview and psychological tests which tell them what hobbies, sports or evening classes are best suited to their personality.

Sunday Times, 24.9.78

4.1 Preliminaries

'To understand the city, one must understand patterns of activities in which urban residents singly and collectively engage' (Ottensmann, 1972, 1). Since all human activities occur coincidentally in space and time we need to find ways of describing and explaining their space and time patterning. In urban social geography the emphasis to date seems to have been weighted towards the description of the space patterning of activity. This might have been acceptable were it not for the fact that when attempts have been made to explain this patterning the role of time use in the urban system has usually been ignored. This is especially surprising given the recent interest among urban social geographers in the study of behaviour.

So far in our discussion about time we have been trying to handle some rather abstract concepts (Chapters 1-3) and we have also attempted to isolate some principles around which we can develop operational methods. However, we have only been able to measure and classify time in a limited number of ways as yet. It is especially **Social time** unfortunate, as others have observed, that these difficulties are most pronounced in trying to handle the intriguing concept of social time (Ottensmann, 1972, 4), because in dealing with man's use of time and daily activities we are concerned with just those 'items' (activities)-in-relation which result from specialization of role and function and the widening area of social interaction which we saw to be the basis of the notion of social time (cf. Chapters 2-4, **Items** also Sorokin and Merton, 1937).[1] However, 'if the variable natures of time are taken into account, (time) loses some of its convenience for activity study purposes: considering its paramount importance to industrial societies, clock time may provide an adequately "objective" standard' (Anderson, 1971, 355).

Heirich (1964) has considered the use of time in the study of social change and although a discussion of social change is not our objective here it is worth noting one or two of Heirich's points because they lead to issues to be dealt with later. He proposed that **Time as social** the concept of time could be used in 'at least four distinct ways' and **change** the first three of these seem to be readily applicable to the study of time use in a more general manner:

(i) Time as a social factor
(ii) Time as a causal link between other elements
(iii) Time as a quantitative measure
(iv) Time as a qualitative measure of the interplay between elements.

As a *resource* and as *social meaning,* time is a social factor, 'in and of itself'. As a resource it 'becomes a variable in its own right' because its quantity is finite. Marx conceived of time as a resource in his equations defining surplus value. 'Exploitation consists of extracting more time from a labourer than is required to produce goods having the value of his wages' (Heirich, 1964, 387). Marx **Time as resource** was 'saddened (and angered) because the worker (former peasant) now [had] no choice but to sell his time, i.e. his labour per hour. Tyranny [had] been removed only to be supplanted by a new form **Expectations** of subjugation' (Gioscia, 1972, 76). Time is scarce because we perceive alternative uses for it. Within a culture, the value of time as a resource varies according to the stratification level being considered, [because] roles differ in expectations and possibilities for the use of time ... but whatever the value placed on time by a person or group, it will be allocated for a variety of purposes. Thus relative time uses can be one indicator of the changing importance of various activities' (Heirich, 1964).

Possibilities

Time as social meaning relates to the attitudes toward time which can in themselves be dynamic factors because 'specific moments of time acquire a social meaning of their own'. The Wall Street crash of October 1929 evokes memories that still affect the actions of many people (Heirich, 1964, 387, Sorokin and Merton 1937) and at higher frequency, within a day, certain times are given social meaning which also prescribes behaviour: meal times are thought of as private times when one should not visit without being invited.

Time as social meaning

As a causal factor Heirich distinguishes between time as setting and time as sequence. As setting we have space-time relationships, as sequence we have time-time relations. With time as setting the questions are not simply about coincident relations (correlations) such as: 'Is A present whenever B is present? Do variations in A occur with variations in B — or occur following an appropriate lag? Rather the questions are about the configuration of many patterns within a particular time period' (p.388). With a knowledge of sequence the possible causal links between events are heightened when correlation is also established but causality is only implied. We establish 'negative causal links, not positive ones'. From the identification of a sequence we can say that if activity (B) follows activity (A) it probably did not cause activity (A) but this may not always be so in human environments — two hours spent in a cinema (B) will follow the trip to the cinema (A) but will also be considered to have caused the trip (A).

Time as causal factor

Setting and space-time

Sequence and time-time

As a quantitative measure of relationships the type of time scale to be used becomes important. If an interval scale is available, as with clock time, then durations and rates may be computed. Both these measures are used in the analysis of human time budgets, and they also allow the development of other measures for evaluating social activities as we shall see.

Durations

Rates

Time budgets

Discussion in this chapter will now proceed as follows. Some theories of time allocation will be considered in a short review of the approaches taken by economists and other social scientists. The time budget, as the principal instrument in time use studies, will then be outlined. Problems of and approaches to data collection and classification are followed by a brief outline of some models and methods of measurement of data collected in time budget surveys. Your attention will be drawn to the so-called space-time budget but as this really amounts to no more than a conventional time budget with the spatial location of activities also recorded, the discussion will assume the term time budget to be a generic term which includes the space-time budget. The chapter will conclude with some examples.

4.2 Time allocations[2]

> 'Technology has mastered the art of saving time, but not the art of spending it'
>
> (Soule, 1955, 100).

Allocation

Use of any scarce commodity involves allocation decisions. When a commodity is scarce constraints are put on its use, through such devices as rationing and prices. As a scarce resource time acts as a constraint on the amount of adjustment which is possible, by individuals and 'collectivities' of one type or another to achieve a satisfying 'quality' or standard of life. The use of time and its allocation are not one and the same thing. Thus 'time allocation studies differ from simple time use studies in that they enquire into the mechanisms of allocating time as well as the consequences and opportunity costs of different alternatives' (Thrift, 1977, 418).

We would expect to find time allocation studies well established in economics but they are in fact rather scarce![3] There are at least two reasons for this. The first is that the factors of production and issues of exchange related to Man 'in the everyday business of his life' (as Alfred Marshall saw the objective of economic analysis) have been studied only when considered to be in short supply. Marshall was troubled by the concept of time and its incorporation into the economic analysis of human behaviour, 'for the element of time, which is the centre of the chief difficulty of almost every economic problem, is itself continuous. Nature knows no absolute partition of time' (Marshall, 1898). The second reason is that the traditional economic theory of the allocation of time was constructed to explain the supply of labour (Ware, 1977) and it is really only quite recently that theories of the allocation of time have been considered in terms of other aspects of social organization such as the household (Becker, 1965).

As far as we are aware, adding time to the traditional factors of production was first suggested by Soule (1955). 'Still, however, economic theorists have not absorbed the concept of time formally into their basic thinking, as physical theorists have done. Specifically, economists have not regarded time as a scarce resource, co-ordinate with land, labour and capital, and have not developed their theory of resource allocation on this four dimensional framework ... (but) ... as often, popular language anticipates scientific theory by its intuitive, common-sense phrases.

Time, land, labour and capital

The wage earner does not refer to the employer as buying his labour; what the employer buys is his time' (Soule, 1955, 89).

The explicit interest of economists in developing theories about the allocation of time and their interest in time as a resource per se re-appears a few years later in the work of Harrod (1958), who drew attention to the limitations imposed on the productivity of service industries due to the non-availability of time. But the best known works on time allocation are possibly Becker's 1965 study and his subsequent work with Ghez (Ghez and Becker, 1975) which have been developed in the framework of neoclassical equilibrium analysis and utility theory. Becker, unlike the earlier neo-classical economists, explicitly recognized that consumption takes time and also considered production activity in the household, where he saw time being combined with market goods to produce more basic commodities which entered directly into the utility function (or satisfaction level). His more general analysis of the allocation of time stressed the relative importance of time as varying among commodities (Ware, 1977, 4). Some years later (1970), Linder also published a theory of time allocation in which the time taken to consume was of central importance. These theories were a considerable advance on the earlier labour-production-consumption theories which treated consumption as instantaneous, although production 'took time'. With de Serpa (1971) and Evans (1972) the notion of bounded and unbounded activities is also introduced, as a result of technological or institutional constraints which put a lower or upper bound on the time required to undertake an activity. Sometimes an individual may be able to substitute his own money for someone else's time and in that way avoid undertaking an activity. These activities might be called individual unbounded activities, but others are physiologically obligatory or socially obligatory and cannot be avoided. Time must be allocated to them. Evans' view is that the 'individual chooses his most preferred set of activities subject to his time and budget constraints, and the technical relationship between intermediate activities and final activities. Intermediate activities are those activities which are not undertaken for their own sake but are technically or institutionally necessary in order to undertake final consumption activities which yield positive satisfaction. For example travelling to the cinema would be an intermediate activity associated with a final consumption activity of cinema attendance' (Ware, 1977, 8).

The implications of the Becker, Linder, de Serpa, and Evans theories, can be treated as a whole for our purposes and may be summarized quite simply. Because production and consumption cannot be separated and the household is itself a small production unit (producing commodities or utilities) and consumption and production both use scarce time, increases in productivity raise the

Basic commodities

Utility function

Bounded and unbounded activities

value placed on time and this induces a more goods-intensive consumption: i.e. a consumption of the goods which have been produced (Hirsch, 1977). There is a growth in conspicuous consumption. Activities which do not consume hardware are rejected in favour of those that do leading to a 'self-service economy' (Gershuny, 1978). Thus, Seeley, Sim, and Loosley (1963, 65-66) pointed out that church services had been shortened in their 'upper class' study area in Canada, the pseudonymous suburb Crestwood Heights, in order to give more time to a conspicuous consumption.

Benjamin Franklin's advice to a young tradesman was:

'Remember, time is money.'

But contrary to the soothing harmonies of equilibrium economics it patently is not. To start with, all individuals have an equal time income of 24 hours in a day. The only differences relate to length of life itself, but even here the expectations in a single society are more or less equal. The time allocation problem arises because the number of alternative possibilities and the time required for their completion consumes more time than is usually available. There is a scarcity which forces the individual to choose among alternative activities which have associated constraints and which are also related to a varying set of preconditioning and predisposing factors. One of the ways in which he copes with this scarcity is by ordering activities into some sort of recurrent, even cyclic, pattern. Routine profiles of time expenditure are adopted. But while there is usually room for the substitution of one time location with another, time itself cannot be used as a medium of exchange in the way that money can. However, although time cannot be stored or banked it can be used more or less efficiently and it is here that spacing time (Chapter 3) becomes an important operational principle.

Preconditioning and predisposing factors

The allocation of time to activities occurs in two ways. First of all in terms of when an activity is located and secondly for how long it is located. The equivalent spatial conditions of the activity are where it is located and over what area. The occupation of a particular time location means that some other activity cannot occupy that location (but there are some usually 'trivial' exceptions to the 'rule' that only one thing can happen at a time). This means that the time cost of an activity is measurable as the time taken away from some other activity, in other words as an opportunity time cost (Matzner, 1974). This is a useful notion for time budget analysis, so long as the survey and accounting framework has been designed to accommodate the notion of alternatives.

Opportunity time costs

An allied notion is that of activity and time elasticity. In the study of time use by social scientists the idea of elasticity is probably due to Szalai (1964). He introduced it as the compressibility of an activity. 'How much time was spent on an

Activity elasticity and time elasticity

activity and how much had to be spent on it?' The notion is directly related to that of elasticity in economics. We shall discuss ways of evaluating it in a moment.

Meier's seminal study of time budgets as a basis for the development of social accounting frameworks (1959) has stimulated an interest in time allocation surveys as a method 'to develop indices of the richness of life in urban communities' (Meier, 1959, 27). Here the time budget becomes 'a simple extension of the census ... (with) .. a continuous accounting' being made of a sample population (*ibid.*, 29). The episode has been used as a unit of analysis in the study of people's patterns of activity in order to understand how the 'productive machine and the urban environment interact', and also to relate 'spatial aspects of activities to time patterns'. Such an approach is based on the 'notion that every urban resident's life consists of a flow or a series of more or less interconnected events which have an essential order and follow predictable sequences' (Chapin and Logan, 1968, 305). Events (or activities) are here seen as the output of a set of choices based on motivations which have a particular status relative to the norms of the social system. The resultant 'flow or sequence of activities is called an activity routine, and when the activity make-up of a routine is typical to a class of population, the 'choices' reflected in the activities are said to consist of an activity pattern (Chapin and Logan, 1968, 310). The scarcity of the time resource induces an allocation process which mediates between the choices which can be made and so in turn determines the activity pattern.

Episode

Activities as outputs

Activity routine

Activity pattern

An attempt to build a social indicator based on time allocation has been made by Hobson and Mann (1975). Their Lambda index is based on 'the manner in which individuals allocate their time among various life activities'. As a weighted sum indicator, it attempts to take into account both objective and subjective aspects of time allocation which are related to the discrepancies between the amount of time actually allocated to an activity and the amount of time preferred for allocation. This example relates time allocation to social indicator research and in so doing implicitly emphasizes the absence of time elements in the geographical literature on social indicators and in welfare approaches to human geography in general.

4.3 The time budget

'A day has 24, a week 168, a year about 8,800 hours. From day to day, week to week, year to year, this amount of time falls to the share of the poor and the rich, the young and the old, men and women — in short to everybody on this globe of ours, irrespective of nationality, mother-tongue, colour, political, or religious convinction and of his position in society .... The question how he spends these 24 hours, how he must and how he may spend it is not

> so simple and not so equitably solved . . . . It is primarily his own and his fellow men's social activity that man quantifies and co-ordinates by means of time'
>
> (Szalai, 1964, 105).

The time budget is a means to the study of the time location, frequency, sequence and duration of social activities. The space-time budget is therefore only an extension of time budget analysis which includes information on the spatial location and separation characteristics of activities (cf. Szalai *et al.*, 1972).

The time budget is usually a log or diary of the sequence and duration of activities engaged in by an individual over a specified period, most typically the 24-hour 'day', and its function is to show how an individual's time is consumed or utilized. Time budget surveys involve the collection of individual logs and diaries from members of a population and can be used in the analysis of the main trends (a longitudinal diary) and subgroup differences in the allocation of time over a number of months or years. (cf. Converse, 1968; Michelson and Reed, 1975; and Szalai *et al.*, 1972).

(i) Antecedents

Family budgets

Time and motion

Just when time budgets were first introduced into the literature of social science is not clear but since the advent of industrial wage labour, productive human activity has been measured and largely paid for against 'the clock'. Szalai suggests that time budgets were essentially an extension of the family budgets undertaken by Frederic Le Play during the last quarter of the 19th century. In 1892 Engels also drew attention to 'certain regularities in the housekeeping statistics' based on family budgets, and in the USA, Taylor's time and motion studies (1911) are generally considered to have been among the first time budget studies though more restricted than those to be undertaken by Strumlin in 1924, on Moscow workers. A comparative study by Prudensky (1959) using the Strumlin data, with fresh data from Moscow in 1959, allows a rare opportunity for comparison of time budgets over a long priod. See Table 4.1, based on Szalai (1966).

By 1934 Lundberg, Komarovsky, and McInery had published work on the time budgets of housewives in the USA and at the same period Sorokin and Merton considered the notion of social time as well as the time budget.[4]

In 1939 Sorokin and Berger published the results of extensive time budget studies at both the city and the national scale in the United States, but as Meier (1959) has remarked it still seems to be necessary to point out the utility of these approaches to governments, planners and all organizations that are concerned with the activities that people undertake so that they may be used to the benefit of society and the more efficient use of all scarce resources, including time.

Table 4.1. Comparative time budgets for Moscow workers 1924-1959 (from Ware 1977, mimeo)

Time not at own disposal	Strumlin 1924		Prudensky 1959	
	m.	f.	m.	f.
(i) Productive work for the community (with overtime)	7·83	7·64	7·17	7·20
Private	0·45	0·57	0·78	0·62
Total	8·28	8·21	7·95	7·82
(ii) Housework (in the family)				
Preparation of meals	0·48	2·56	0·23	1·41
Care of children	0·16	0·53	0·43	0·65
Other	1·08	1·71	1·04	1·85
Total	1·72	4·80	1·70	3·91
(iii) Lost time				
To and from the place of work	0·95	0·86	1·93	1·30
Shopping and waiting in queues	0·22	0·20	0·37	0·65
Total	1·17	1·06	2·30	1·95
Total of time spent for indispensible activities ((i)-(iii) total)	11·17	14·07	11·95	13·68
(iv) Indispensible necessities				
Meals on the job and at home	1·55	1·27	1·18	0·93
Sleep at night and during the day	7·74	6·83	7·48	6·97
Total	9·29	8·10	8·66	7·90
((i)-(iv) total)	20·46	22·17	20·61	21·58
Time at own disposal				
Study and individual cultural activity	1·86	0·68	1·68	1·15
Recreational and entertainment	1·68	1·15	1·71	1·27
Total	3·54	1·83	3·39	2·42
Percentage of time fund	14·7%	7·6%	14·1%	10·1%
Total hours (sum of the budget)	24	24	24	24

Values are shown in **hours** and decimal parts.

The work of Lundberg, Komarovsky, and McInery (1934) in the USA contributed significantly to the development of time budget methodologies as did the work of the Mass Observation Unit at the Case Institute. But with the intervention of war there was no resurgence of interest until the late 1950's and then it came from the publication of Meier's (1959) work. Broadcasting corporations such as the BBC in Britain and the NBC in the USA conducted massive time use surveys, as did the Japanese. (BBC, 1965; Nakanashi, 1966). A review of this work will be found in Szalai (1972) and Thrift (1977a). Figure 4.1 is an adaptation of one of the summary charts produced by the BBC following their 1965 time budget study of British households. This particular time budget study has been carried out many times since 1939 and provides a good opportunity for longitudinal study (White, 1978; BBC, 1978). The latest (1975) survey takes up this challenge (BBC, 1978).

In the USSR and other 'eastern block' countries there has been an intense utilization of the time budget as a means to social and economic control (planning?). Research has been so widespread that there now exist standardized activity codes and specialized indices such as the index of activity elasticity and structural difference outlined later (Chapter 4) developed by the Central Statistical Office in Budapest, Hungary (1965).

A logical outcome for all this rather sporadic and unrelated research was a cross-cultural, multi-national time budget study. The idea was first suggested by McCormick in 1939 and reached fruition with the establishment of the Multi-national Comparative Time-budget Research Project in 1965. The project consisted of a comparison of the time budgets of people from twelve countries. It involved nearly 30,000 interviews and was launched by the European Co-ordination Centre for Research and Documentation in Social Sciences (Vienna). Computations were carried out at the Survey Research Center of the University of Michigan, the Institute for Comparative Social Research of the University of Cologne, the French National Institute of Statistics and Economic Studies and the Sociological Institute of the Free University of Brussels, as well as at each of the Universities and Institutes directly involved in the surveys. A huge programme, it was finally brought to fuition and published in considerable detail in *The Use of Time* (Szalai *et al.*, 1972). We shall draw on some of the results in a moment.

When a time budget is taken the description of activities in terms of time parameters is the first objective but in many studies there is also a desire to know something about their spatial locations and urban geographers and planners are likely to be particularly concerned with such data. This approach which 'includes the spatial co-ordinates of activity locations' (Anderson, 1971, 353) is

Space-time budget

known as a space-time budget. There are 'patterns of time and space use' and the 'definition of spatial loci of activities permits the analyst not only to identify broad service areas associated with particular community facilities, but also to determine to what

Activity communities and ecological time

extent different life styles support different activity communities' (Chapin and Logan, 1968, 312). The ecological time of Gurvitch's scheme is the time of activity communities. For the urban geographer or planner time budgets which are used as a means to the analysis of activities can also be used as a means to the analysis of the 'human use of urban space' by the simple strategy of recording an *x-y* co-ordinate for each activity and then adopting one or a number of the range of methods of spatial analysis which exist in human geography.

What is involved in a time budget?

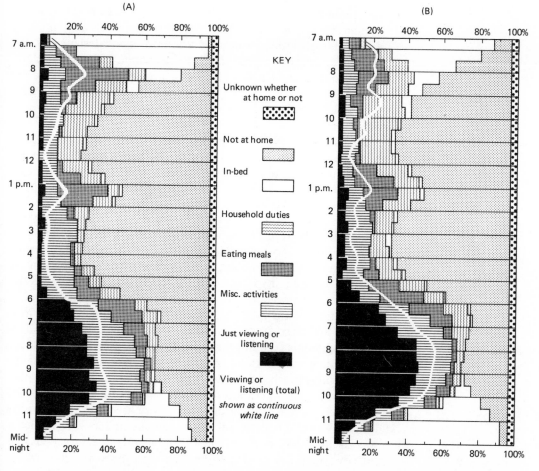

(A)

(B)

KEY

Unknown whether
at home or not

Not at home

In-bed

Household duties

Eating meals

Misc. activities

Just viewing or
listening

Viewing or
listening (total)
*shown as continuous
white line*

Figure 4.1. Time use charts from a BBC survey in 1965. (a) Upper middle class (or high social status) groups. (b) Working class (or low social status) group. (After BBC, 1965. Reproduced by permission of the BBC)

(ii) Data collection

At the outset it is necessary to clarify objectives and also to determine the sort of analyses which are to be undertaken. Amongst the sort of questions that need to be asked are: What is the scale, in time and space, that activity occurs in? Is description all that is required or is there to be an inferential objective? What is the size and composition of the behavioural unit — is it the individual or a household? Is there a specific interest in particular population sub-groups (e.g. the elderly)? (Abrams, 1978).

A number of basic conditions affect all social systems. These include firstly, numbers; for instance the demographic dimensions of societies vary in number. Secondly space; for example the ecological concentration and dispersion of the human population, and finally time; as revealed in the temporal boundaries of life and

**Social systems
numbers**

Space and time

the sequential ordering of action. Time budgets provide an accounting system for each of these dimensions but they must be designed to elicit their relevant characteristics and designed so that their effects may be noted during the evaluation and analysis phase which follows the field work. The particular form which a time budget survey takes will be determined by its objectives. But there are a number of basic data objectives which all time budgets should aim to satisfy.

First some predetermined period of clock time, usually the length of some cyclic period, is selected say 24 hours, or of a calendar period such as a week. The period chosen should relate to the scale of the time subdivisions which are of interest and also to the way in which activities are to be classified (cf. 4.3 (iii) following). For instance, if a 24 hour budget is required it is usually necessary for activities to be recorded at a time scale which includes activities which consume less than 60 minutes. In fact recording of time expenditure should probably involve measurement to the nearest minute. Secondly if spatial co-ordinates are also required then the bounds of the space scale must also be established. The data analyst must anticipate the need for definition of functional spaces, how they are to be recorded and how they are to be identified after data collection. Some kind of grid system is usually used for spatial data but the time scale must be congruent with the space scale (Figure 3.6). A survey record such as, '1132 shopping in London' would, for most purposes, be a mismatch of minutes with a very large and ill-defined space scale. Thirdly, the activity itself must be recorded in some way that is standardized throughout the survey or in such a way as to be amenable to classification after the survey. Is the activity which is recorded to be limited to those activities which occur at predetermined time intervals, such as on the hour and every fifteen minutes following or is a free formatted time to be used relying on the subject to decide the density of the activity record? Typical time budget pro-forma and extracts of instructions are shown in Table 4.2.

Some time budget studies are designed to test hypotheses of one sort of another, while others are concerned simply to record information about a population rather as a Census does. Either way the design of the study will have established whether the interest is in a single cross-sectional approach or whether a longitudinal approach is required. The latter are extremely expensive in time, labour, and money, and the attendant problem of maintaining subject co-operation is also considerable (Michelson, 1976).

Whatever the objective a time budget should aim to generate

data which allows summary evaluation of at least four features of activity structure:

   (i) Starting time of an activity
  (ii) Duration of an activity
 (iii) Frequency of occurrence
 (iv) Sequential order of activities in the period concerned

If a spatial record of activity is also required then there is a fifth requirement:

  (v) Activity location.

Although, in the strictest sense, a time budget only tries to establish a picture of time use, it is possible to inject motivational or choice-constraint information into the record with little extra effort. This has been successfully achieved for instance by Chapin (1974), Cullen and Godson (1975), Parkes and Wallis (1976, 1978), Michelson (1977), and Robb (1977). One simple and effective approach requires that the subject should record whether each activity was fixed or unfixed, planned or unplanned, in terms of its time location and in terms of its space location. Such a record also provides an indication as to whether the space-time 'environment' of a recorded activity is set in a hard or soft information field (cf Chapter 1). Some examples of this approach are given later in this chapter.

In time budget surveys, as in any other survey, there are decisions to be made about the size and structure of a sample (see Table 4.14). Should the data be collected in the presence of the subject? How much explanation of the objectives of the study should be given to the subject? What should be done about non-response, and how is an 'inadequate' interview or diary to be determined?

These are general problems associated with social survey design, problems compounded in time budget surveys because they inevitably demand high levels of co-operation from subjects, take a relatively long time to complete and are expensive. The importance of careful sample design and the advantages of statistically acceptable *small* samples cannot be overstressed. Such small sample strategies also allow 'pre' and 'post' survey discussion with subjects. Just as a sample census with enumerators at hand tends to produce more 'accurate' results for each subject than a 'normal census', so too a small time budget sample can pay off in the long run.

INSTITUTE OF COMMUNITY STUDIES

STUDY OF EVERYDAY ACTIVITIES

DIARY
for

Weekday:

Week-end

Weekday Evenings

We are grateful to you for agreeing to complete the diary. These notes may help you. The diary is in two parts. For the first, which covers the next weekday and next weekend, we would like you to keep a full record of all your activities. For the second, which covers four weekday evenings, we need to know only about away-from-home activities.

FULL WEEKDAY AND WEEKEND
We would like you to keep the full diary for the first weekday after your interview and for the first weekend after it.

* What Did You Do? First record in this column each of your activities through the day.
Please try to record each activity separately and on a different line.
Then record the time you started and completed each activity.
A few examples of things you might write in are: 'got dressed', 'worked', 'cooked dinner', 'watched TV', 'had friends in for a chat', 'ate breakfast'.

* With Whom? If you were doing any activity with someone else you should write in who it was.
If you were watching television with your son you would write 'son' in this column. Some other examples of what you might write in are: 'wife and child', '3 friends', '2 relatives', '8 work colleagues'.

* Same Time as Usual? Then, for the weekday, we would like you to record for each activity whether you did it at the same time as usual on weekdays. If you did, please write 'S'. If you did not, please write 'D'. Do the same thing for the weekend. If you did the activity at the same time as usual on a Saturday (or for Sunday activities the same time as usual on a Sunday) record an 'S', and if not a 'D'.

* 'Work' or 'Leisure'. You remember that one of the questions we asked you in the interview was about what you thought made the difference between work and leisure. We would be grateful if you would go through all the activities you record in the diary, and put a 'W' in the column against those activities you think of as work as an 'L' against those you think of as leisure.
If you think that some of the activities are a mixture of both, put an 'M' in the column. And if you think some activities are neither work nor leisure nor a mixture of both, put an 'N' in the column.
Everyone will have different ideas about whether an activity is work or leisure, a mixture or neither. There are no right answers. Please just label the activities *how you feel*. It is what *you* think that counts.

Please try to fill in the diary at several times during the day, as it is so easy to forget things if you leave it all till the end of the day.

WEEKDAY EVENINGS
For the other weekday evenings we need information only on those activities you do *away from your home*. We would like you to start keeping the record from 6p.m. and continue until you go to bed.

* It may happen that on some or all evenings you will stay at home all the time. If so, just write 'at home all evening'.

* If you are at work in the time between 6p.m. and midnight, just write in 'at work'. There is no need to give more details.

Name:

Tel. No.
(if any):

Address:

RETURNING THE DIARY
When you have completed the final section of the diary, please post it to us in the stamped addressed envelope. We will then post the £1 fee to you.

THANK YOU FOR YOUR HELP

(a)

Table 4.2 Extracts of the instructions and typical 'diary' pro forma

155

TIME	WHAT DID YOU DO? (no more than one activity on each line)	TIME BEGUN	TIME ENDED	WITH WHOM?	WAS ACTIVITY DONE AT SAME TIME AS USUAL ON WEEKDAYS ? S = SAME D = DIFFERENT	W = WORK L = LEISURE M = MIXED N = NEITHER
Midnight		Midnight				
1a.m.						
2a.m.						
3a.m.						
4a.m.						
5a.m.						
6a.m.						
7a.m.						
8a.m.						
9a.m.						
10a.m.						
11a.m.						
Noon						
1p.m.						

(a)

KINDS OF ACTIVITIES YOU MAY DO DURING THE DAY

(but please use your own words to describe what you are doing)

TRAVEL: All the trips you make, both at home and at work.

WORK: Actual work; work breaks; delays or sitting around at work; work meetings or instruction periods; meals at work; overtime; work brought home.

HOUSEWORK: Preparing meals and snacks; doing dishes; arranging and straightening things; laundry and mending; cleaning house (inside and outside); care of yard and animals; repairs.

CHILD CARE: Baby care; dressing; helping with homework; reading to; playing with; supervising; medical care.

SHOPPING: Groceries, clothes, appliances, or home furnishings; repair shops; other services (for example: barber, hairdresser, doctor, post office).

PERSONAL LIFE: Eating meals and snacks; dressing; care of health or appearance; helping neighbour or friends; sleep or naps.

EDUCATION: Attending classes or lectures; training and correspondence courses; homework; reading for the job.

ORGANIZATIONS: Club meetings or activity; volunteer work; going to church services; other church work.

GOING OUT: Visiting (or dinner with) friends, neighbour or relatives; parties, dances, nightclubs or bars; sports events and fairs; concerts, movies, plays, or museums.

ACTIVE LEISURE: Sports or exercise; playing cards or other games; pleasure trips and walking; hobbies, knitting, painting, or playing music.

PASSIVE LEISURE: Conversations; radio, TV, records; reading books, magazines or newspapers; writing letters; planning; thinking or relaxing.

(b)

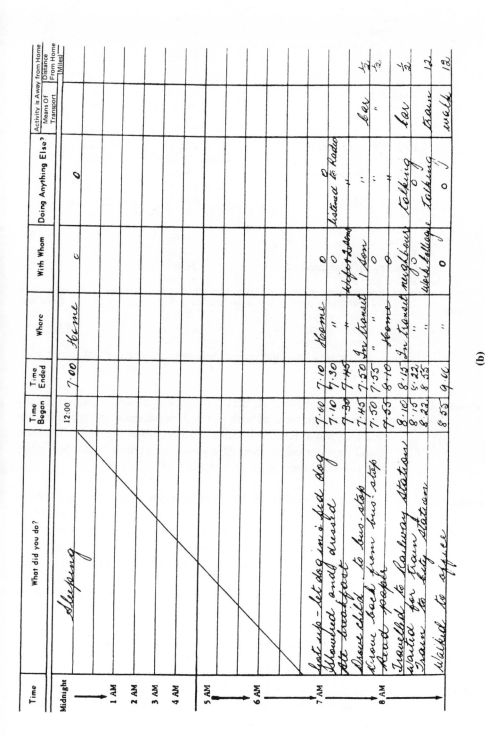

Time	What did you do?	Time Began	Time Ended	Where	With Whom	Doing Anything Else?	Activity is Away from Home	
							Means Of Transport	Distance From Home (Miles)
Midnight	sleeping	12:00	7:00	Home	c	0		
1 AM								
2 AM								
3 AM								
4 AM								
5 AM								
6 AM								
7 AM	Got up – let dog in/fed dog	7:00	7:10	Home	0	0		
	Showered and dressed	7:10	7:30	"	0	Listened to Radio		
	Ate breakfast	7:30	7:45	"	wife + children	"		
	Drove child to bus-stop	7:45	7:50	In transit	1 son	"		
	Drove back from bus-stop	7:50	7:55	"	0	"		
	Read paper	7:55	8:10	Home	0	"		
8 AM	Travelled to Railway Station	8:10	8:15	In transit	neighbour	Talking	bus	1/2
	Waited for train	8:15	8:22	"	0	0	bus	1/2
	Train to City Station	8:22	8:55	"	work colleague	Talking	train	12
	Walked to office	8:55	9:00	"	0	0	walk	12

(b)

Table 4.2 contd.

Time	N.T	UN.T	Activity	N.A	UN.A	Location of Activity	Distance	N.L	UN.L
7.00 am		√	Awoke	√		Home	—	√	
7.15		√	Showered etc.	√		Home	—	√	
7.45		√	Breakfast with family	√		Home	—	√	
8.20		√	Doing dishes	√		Home	—	√	
8.35		√	Housework	√		Home	—	√	
11.30		√	Travel		√	Black St. White St.	½ mile		√
11.33		√	Shopping		√	Brown St.	½ mile		√
1.00 pm		√	Travel		√	White St. Black St.	½ mile		√
1.05		√	Lunch	√		Home	—	√	
2.10	√		Travel		√	Black St. Green St.	¼ mile		√
2.15	√		Doctor's appointment	√		Blue St.	¼ mile	√	
3.00		√	Travel		√	Green St. Black St.	¼ mile		√
3.03		√	Watering garden	√		Home	—	√	
3.50		√	Travel		√	Black St. Yellow St.	⅛ mile		√
4.00		√	Visiting friends		√	Red St.	⅛ mile		√
5.00		√	Travel		√	Yellow St. Black St.	⅛ mile		√
5.15		√	Preparing evening meal	√		Home	—	√	
6.15		√	Evening meal	√		Home	—	√	
7.00		√	Cleaning up after meal	√		Home	—	√	
9.30		√	Retiring to bed	√		Home	—	√	

Table 4.2 contd.   N.T Necessary Time   UN.T Unnecessary Time   A Activity   L Location

In the very broadest terms there are four methods of time budget recording.

Recall method A   (a) *Recall method A:* here the activities of some specified period in the past are recalled with as much precision with regard to time and location as is apposite for the objectives of the study. The most usual past period to select is a yesterday period. Szalai has called this the fresh diary approach (Szalai, 1972, 37).

Recall method B   (b) *Recall method B:* here the activities of some normal period are recalled. In this method the analyst has already determined the activities of interest and requires the subject to supply the normal time location, duration and possibly also the space location for each activity. In some cases the subject may be asked to construct a normal 'day' (beginning at some prescribed time) by supplying the information on predetermined moments. He or she will also be asked to supply the duration of the activity which is recorded and any other relevant information, such as the time and space fixity of the activity. One of the advantages of method B is that the budget can be focussed around project-specific details.

Diary method   (c) *Diary method:* this is in fact something of a misnomer because there are a number of diary methods in use. The essence of the method is to get subjects to keep a diary for a specified period into the future, recording activities either according to predetermined time locations or in a free formatted manner. This method may or may not require interviewer intervention; depending almost entirely on the availability of money! Obviously if it is possible, ex post evaluation of a diary with subjects present, is desirable as a check on reliability. The diary method is fraught with problems, however, especially in the free formatted approach, where records often include varying levels of detail and where activities considered to be of a private or personal nature are often omitted. The diary method also requires very detailed explanation of the range of information required. For instance, how to record two activities occurring at the same time, how to record the company in which the activity occurred; was the activity planned in advance, and so on.

Unless these and other details are pointed out in advance a diary is likely to produce rather limited results, and especially in free formatted style becomes very hard to code.

(d) *Game based method:* there are various reasons why such an approach is used and according to the reason, various sorts of games can be devised. It is unusual to use the gaming method on its own: it is more usual to use it in a post-diary or post-interview situation to gain further information, especially about 'changes in the contingencies of the choice-making environment' (Chapin, 1974, 75). Chapin's trading stamp game is perhaps the best known example of a game-based method and a part of his game format sheet is reproduced below (Chapin, 1974, 75-85). 'A subject is asked to allocate a limited amount of time to selected 'free time' activities according to a few simple rules, considering his present family situation and other constraints ... after the subject has finished making his time allocation on the basis of initial instructions, the interviewer introduces a change in the contingencies: 'Now let us assume that you would have more free time during the week — a half day free on Tuesday and a half day on Thursday — but the same income you have now — please show how you would allocate these eight additional hours.' A game like this 'functions somewhat in the manner of a primitive form of linear programming where the respondent seeks some optimal combination of activity choices in the allocation of free time, subject to a set of constraints.' Another gaming approach has been developed by Jones (1978) at the Transport Studies Unit at Oxford University. Once again the time budget is a component in the whole scheme but in order to assess the impact of transportation changes on household activity patterns, subjects are asked to participate in a game played on an ingeniously simple board (Figure 4.2) on which they are asked to 'create a physical representation of their daily activity-travel patterns, which can then be modified under guidance from the interviewer to explore response to different hypothetical external changes (e.g. revised bus schedules, staggered

Game method

160

working hours)'. Each household member is given a board to play the game with and it seems that young (children) and old (usually rather difficult population categories from which to get survey data) participate with high levels of enjoyment and this, in itself, is a most important factor. The game is called HATS (Household Activity Travel Simulator).

(iii) Classification procedures

Activity classification

Prior to data collection and bearing in mind the considerations outlined above, every time budget study is faced with the extremely difficult problem of classification of activities. There is an almost unlimited number of classification schemes available to the analyst. But the present stage in development of time budget studies demands that only one or two of these schemes are adopted. The reason for this is quite simple: without some standardization comparative studies become impossible; theory generation is harder and discipline is not introduced into research strategies.

Classification is a difficult problem in every empirical science. In the study of time and human activities this is no less true. Both time and activity need to be classed and as we have seen they are not always independent of each other, for instance in the notion of social time. In the final analysis the object of the study will determine the appropriate level and type of classification to be adopted.

In general, the activity pattern is a dependent variable, dependent on the predisposing and preconditioning attributes of the sub-population which it represents. The activities that make up an activity pattern require classification. The factors which influence activity or upon which it is dependent (the predisposing or preconditioning factors of everyday life) also require classification, as does time which has to be parcelled up into units of a scale appropriate to the study environment. Following Chapin three types of activity system are generally recognized — the activity system of firms, the activity system of institutions and the activity system of households. It is with the latter activity type that we are concerned because most recent time use studies have been concened with households.

Firms Institutions Households

Frequency, duration, and sequence position relative to some marker can all be used as a basis for activity classification. The result is that we can group together all activities that consume a quantity of time within some *a priori* period of time, or according to the frequency with which they occur in a population or within a specified period. Again, activities can be classified in terms of their tendency or probability to be predecessors or successors (Parkes and Wallis, 1976, 1978, and Table 4.10 following) or according to the human company that is kept while they are being undertaken (as was done in the multinational time budget study) or according to the purpose they have; 'are they obligatory or discretionary?'

Activity predecessors and successors

Interviewer's Name Schedule Number

A RESEARCH STUDY OF
LEISURE-TIME ACTIVITIES
OF URBAN RESIDENTS

(Third Modification)

A Project of the
Center for Urban and Regional Studies
University of North Carolina
at Chapel Hill

INSTRUCTIONS

In this game we are borrowing the trade stamp idea - - the green stamps that merchants give you when you are buying something. However, we will use black stamps. You will paste in your supply of stamps in the boxes shown according to the way you would like to spend your spare time. Assume that during the next 7-day week, after subtracting out the time taken up in sleep, work, meals, grocery shopping, doctors' visits and similar necessities, you have a total of hours (to be filled in by interviewer) to use as you see fit for any of the activities you wish to choose on the other side of this sheet.

Your stamp supply corresponds to days of the week distributed, say, as follows:

Sat. Sun.

morning

M T W Th F afternoon

evening

There are a few rules we ask you to observe. You have to spend the full amount of time shown for any choice shown. Some choices call for four hours (four stamps), some three, some two, some single hours. We ask you to spend the full amount indicated for each choice you make. Thus, it is against the rules to put down one stamp in a place where a four-hour, three-hour, or a two-hour strip is called for.

The asterisk or star (*) by some choices indicates that these are things which you may want to do regularly, but not every week. In these cases, you may average the time you would spend on them, as long as you do not divide whole hours. For example, going to church is a two hour choice, but if you would only go to church every second or third week, you may put a single hour on this choice.

You will notice that the sheet has eight panels formed by the folds to the game sheet, each containing one or two major groups of activity. You may find it helpful to look these over before you start making choices. You should make your choices considering the present age levels of family members and your present income circumstances. Lay your stamps on the boxes in appropriate numbers, and after you are satisfied with your choices, paste them in.

(Choose as many single hours as you wish)

1. Bringing work home or working overtime on the job

(Choose by four-hour strips)

2. A second job (moonlighting)

(Use as many strips of four hours as you wish, but you must fill an entire strip with stamps)

(Choose as many single hours as you wish)

3. Other activities to increase income such as

taking care of rental property
watching stock market prices
earning income from some hobby
military reserve, national guard, etc.

(Choose by two-hour strips)

4. Job-related activities (attending union or professional meetings, job training, night courses to learn new occupation, etc.

5. _____

(specify other)

Table 4.3 An extract from Chapin's trading stamp game (Chapin, 1974. Reproduced by permission of John Wiley and Sons, Inc)

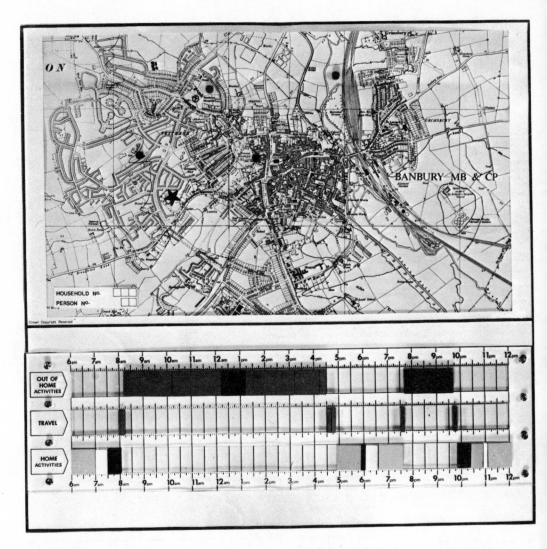

Figure 4.2. The HATS game board. (Jones and Oxford University Transport Study Unit, 1978. Reproduced by permission of the Transport Study Unit, University of Oxford)

Activity definition

Then again, what exactly do we mean by activity? According to Chapin (1974, 21) 'Activities are classifiable acts or behaviour of persons or households which, used as building blocks, permit us to study the living patterns or lifeways of socially cohesive segments of society'.

The first of the activity classifications to be considered here is the Chapin-North Carolina scheme. An important aspect of the classification issue is the distinction between the approach adopted in 'basic research which seeks the meaning of an activity to a subject and so seeks to define a classification system that grows out of the subject's own characterization of the activity (and) the applied approach (which) seeks a designation of an activity in terms

Table 4.4. Chapin's classification code for Household activities. An extract to illustrate detail (after Chapin, 1974. Reproduced by permission of John Wiley and Sons Inc.)

---

12. *Shopping and household business*
   120. Shopping or household business, n.e.c. or unknown
      — window shopping (concurrent with 640)
   121. Retail goods
      — convenience goods (food, drug, hardware, variety, optician, etc.)
      — shopper goods (clothing, shoes, fabrics, etc.)
      — consumer durables (home furnishings, sporting goods, automobiles, etc.)
   122. Retail services
      — caterers, etc.
      — home cleaning services, domestic employment services
      — laundry, dry cleaning, etc., laundromat
      — tailoring, dressmakers for custom tailoring
   123. Personal services
      — barber, beauty, pedicure, sauna, masseur, etc.
   124. Repair services
      — repair of automobiles, appliances, and other consumer durables
      — contractors, home repair services
      — tailors, dressmakers for repair, mending
   129. Household business and related errands (not travel itself)
      — routine use of governmental services and agencies (for non-routine, see 031) — e.g. post office, applying for and taking examinations for permits and licenses, welfare services, etc.
      — routine use of private services (banks, ticket offices, telegraph offices, etc.)
      — reading and writing letters for household business, writing checks, paying bills by mail, keeping records, etc.
      — talking over household business matters with spouse
      — telephoning for purposes of household business
      — paying bills at company or agency offices
      — seeing salesmen, peddlers who arrive unsolicited, agents, interviewers
      — showing repair men, inspectors, etc., around
      — writing invitations to parties
      — paying hotel bills; checking into hotels, airports, etc.
      — picking up or dropping off items, misc. errands
      — picking up paychecks when not done on company time
      — moving a car, finding a parking space, feeding the meter, etc.

---

of its meaning for suppliers of specific services or facilities, such as suppliers of medical care, adult education and transportation' (Chapin, 1974, 68). At a very elemental level, human activity consists of physiologically regulated behaviour and learned behaviour and by considering the activities which an individual engages in during the course of a single day it becomes clear that in 'a very fundamental way properties of physiologically-regulated

Physiologically regulated behaviour

Learned behaviour

activity set the temporal rhythm of the individual's activity routine and influence the scheduling of all other activity' (Chapin, 1974, 26, 26).

Operant conditioning

Motivated behaviour

Motivation choice and outcome components

There are two conceptualizations of human behaviour which Chapin adopts from psychology: Skinner's operant conditioning theory and Maslow's theory of motivated behaviour. This means that an activity consists of a motivation component, a choice component and an outcome component. The outcome component is the observable action that time budgets and activity studies in general record in empirical situations. But it is the motivation component, in conjunction with certain factors related to operant conditioning, (Skinner includes the notion of wants and these may be generated from physiological needs or self-actualizing needs (Maslow, 1970)) which provides the basis for Chapin's preliminary categorization of activities into *obligatory* type and *discretionary* type. In effect this is a continuum but any discrete grouping which is devised may be guided by this principle. This means that 'the ordering of activities in this fashion connotes that some are more postponable than others ... thus food and sleep intake would be classified as obligatory activities, and watching TV and rest and relaxation would normally be classified as discretionary activities' (Chapin, 1974, 37). (This idea of an obligatory-discretionary range to activity types can also be tied to the notion of activity elasticity which we have already mentioned briefly (4.2)).

The classification scheme which Chapin and the North Carolina group have developed is 'based on a "dictionary" of about 225 activity codes' which are grouped into activity classes on two levels of aggregation, one containing forty classes the other twelve. The complete activity classification code is reproduced in their Appendix Table A-1 (Chapin, 1964, 220ff) and an extract from the scheme is shown in Table 4.4 above. The listing is comprehensive and has been found to be valuable as a basis for comparative study (Parkes and Wallis, 1976, 1978).

The multinational time budget study classification system is perhaps the best alternative to the North Carolina scheme. In some ways, however, it is less easy to adopt because the published classification does not provide as much detail as the North Carolina scheme does. The essence of the scheme is as follows. 'The full 24 hours of the day are broken down either into the original category code of 96 categories, or into the more manageable 37 activity code that was obtained by collapsing certain of the detailed codes. The 37 activities are reducible to nine subtotals, such as work, etc. An example of the scheme and the towns which were involved in the multinational study follows in Table 4.5. (The boxed categories are blank codes — this explains why the classification seems to have 100 categories).

Primary and secondary activities

The first classification which is adopted in the multinational study system is into primary and secondary activities. This is based

Table 4.5. Multinational time budget classification — an extract. Places sampled, sample sizes. (Szalai *et al.*, 1972, 576-77. Reproduced by permission of Mouton Publishers, The Hague)

Sample size	
Belgium	2077
Kazanlik, Bulgaria	2096
Olomouc, Czechoslovakia	2193
Six cities, France	2805
100 electoral districts Fed. Rep. Germany	1500
Osnabrück, Fed. Rep. Germany	978
Hoyerswerda, German Dem. Rep	1650
Györ Hungary	1994
Lima-Callao, Peru	782
Toruń, Poland	2759
Forty-four cities, USA	1243
Jackson, USA	778
Pskov, USSR	2891
Kragujevac, Yugoslavia	2125
Maribor, Yugoslavia	1995

00 regular work	25 outdoor playing	50 attend school	76 party, meals
01 work at home	26 child health	51 other classes	77 cafe, pubs
02 overtime	27 other, babysit	52 special lecture	78 other social
03 travel for job	28 blank	53 political courses	79 travel, social
04 waiting delays	29 travel with child	54 homework	
05 second job		55 read to learn	80 active sports
06 meals at work	30 marketing	56 other study	81 fishing, hiking
07 at work other	31 shopping	57 blank	82 taking a walk
08 travel to break	32 personal care	58 blank	83 hobbies
09 travel to job	33 medical care	59 travel, study	84 ladies hobbies
	34 administrative service		85 art work
10 prepare food	35 repair service	60 union, politics	86 making music
11 meal cleanup	36 waiting in line	61 work as officer	87 parlour games
12 clean house	37 other service	62 other participation	88 other pastime
13 outdoor chores	38 blank	63 civic activities	89 travel, pastime
14 laundry, ironing	39 travel, service	64 religious organization	
15 clothes upkeep		65 religious practice	90 radio
16 other upkeep	40 personal hygiene	66 factory council	91 tv
17 garden, animal care	41 personal medical	67 misc. organization	92 play records
18 heat, water	42 care to adults	68 other organization	93 read book
19 other duties	43 meals, snacks	69 travel organization	94 read magazine
	44 restaurant meals		95 read paper
20 baby care	45 night sleep	70 sports event	96 conversation
21 child care	46 daytime sleep	71 mass culture	97 letters, private
22 help on homework	47 resting	72 movies	98 relax, think
23 talk to children	48 private, other	73 theater	99 travel, leisure
24 indoor playing	49 travel, personal	74 museums	
		75 visiting with friends	

on the respondent's description in the diary. However, this means that a particular activity may on some occasions be described as primary and on others as secondary. Whether the two activities occupying a particular period of time were concurrent or alternating is one of the difficulties posed by this kind of distinction and another is the degree of consciousness attached to a secondary activity; it may be so habitual when carried out in conjunction with a primary activity that in some degree one might argue that it was not an activity at all (e.g. having the radio on while holding a conversation with friends).

The main classification scheme is applied to all activities, primary or secondary. There are 96 activity categories (Szalai, 1972, 561), and a two digit code is used. The first digit divides activities into ten main groups:

(0) work (1) housework (2) child care (3) shopping (4) personal needs (5) education (6) organizational activity (7) entertainment (8) active leisure (9) passive leisure

whilst the second elaborates on these main groups. The complete two digit activity code is illustrated in Table 4.5. The complete two digit classification was also reduced to a 37 activity code. This amounted to grouping those activities which were included in the 96 activity code if they had the same general meaning, for instance: in the reduced categorization the following activities in the 96 category code, given here with their appropriate code number, became category 1 in the 37 activity classification

1. Main job 00 Regular work
               01 Work at home
               02 Overtime
               03 Travel for job
               04 Waiting, delays

Location data

Contract data

In addition to such descriptions of each activity, information was also collected and coded about the company, with whom, or in the presence of whom, the activity was carried out and about the location where the activity took place. The former data were called contact data and a further classification coding was then applied; an extract is shown in Table 4.6.

For the locational data a functional basis (e.g. home) was adopted for coding rather than a geo-grid system or place name; Table 4.7 shows an extract from these location tables and makes the point that a functional category-space may be associated with each time location rather than territorial space. Graphs were produced for the location data, for instance to show the proportions of people from different social or demographic groups at home during different times of the day for each of the countries studied and Figure 4.3 illustrates this.

The classifications we have mentioned so far are really only a preliminary classification method. Other techniques may be used for classifying activities into clusters of one sort or another. Chapin (1974) reports use of a factor analytic approach and Parkes (1974) has used principal components analysis to try and describe the activities of a sample population in terms of a smaller number of descriptive factors than the original forty or so activity descriptions used in the survey, but the results, in both cases, were not encouraging. Boh and Sakisda (1972) use a cluster analysis method, also with only limited success (this method is discussed in the Appendix). In using any of these methods the results obtained are entirely dependent on the original data input. Their utility is probably restricted to identifying activity structure within the particular sample rather than as a basis for a general classification scheme.[5]

For the time being it seems likely that activity classification, following a time-budget, will be restricted to schemes like those outlined above from Chapin and the multinational time budget study, in spite of the fact that Stone (1972) and Kranz (1970) have been developing computer-based semantic dictionaries which allow the classification of individual descriptions of activity into predetermined categories which try to minimize the aggregation of activity descriptions. These methods demand vast quantities of computer storage but have value in that a large number of categories can be handled rather than the small number which most taxonomies usually aim for. Such dictionary approaches were first proposed by Foote and Mayersohn (1959). In Kranz's approach every word in a subject's diary can be allocated a coded number and be stored for later matching with a dictionary of key words stored on file.

Semantic dictionary classification

(iv) Measures and models

This is a very considerable topic and in order to limit the discussion we only refer to some fundamental measures and models which can be readily applied to the small scale projects which might be undertaken by an individual researcher with limited time and funds. A good review of measures can be found in Hedges (1974). As we have already pointed out, the three most pervasive aspects of any time budget are the description of activities in terms of duration, frequency, and sequence.

(a) Duration

Any activity can be evaluated simply in terms of the amount of time spent on it. It may then be considered as a ratio either to the total

Table 4.6. Distribution of daily time according to different social contacts[1] (in average hours per day)[2] (Szalai et al., 1972, 778. Reproduced by permission of Mouton Publishers, The Hague)

	Belgium	Kazanlik, Bulgaria	Olomoue, Czechoslovakia	Six cities, France	100 electoral districts, Fed. Rep. Germany	Osnabrück, Fed. Rep. Germany	Hoyerswerda, German Dem. Rep.	Györ, Hungary	Lima-Callao, Peru	Toruń, Poland	Forty-four cities, USA	Jackson, USA	Pskov, USSR	Kragujevac, Yugoslavia	Maribor Yugoslavia
Employed men, all days															
all alone	11·2	12·4	9·5	11·0	12·8	12·8	11·9	9·7	10·1	9·8	12·8	13·4	11·0	9·5	11·1
alone in a crowd	0·6	0·8	1·1	0·7	0·8	0·0	1·1	1·0	0·9	0·6	0·0	0·0	1·0	1·4	0·6
alone with children	0·5	1·9	1·0	0·4	0·8	0·5	0·8	0·7	0·7	0·6	0·6	0·6	0·8	0·5	0·8
with children and spouse	1·6	0·2	1·6	2·1	1·2	1·5	1·0	1·3	0·9	1·3	1·3	1·6	1·5	1·4	1·2
with spouse only	2·9	0·8	3·1	2·5	3·5	3·7	2·3	1·8	1·6	2·0	2·2	2·1	2·1	2·1	2·3
with other adults in household	0·6	0·0	0·5	0·3	0·7	0·5	0·1	0·6	1·3	0·4	0·1	0·2	0·5	0·7	0·5
with friends and relatives	0·8	0·7	0·5	0·6	0·9	1·0	0·3	1·1	1·4	0·5	1·0	1·0	0·5	1·1	1·0
with neighbours	0·1	0·1	0·1	0·0	0·2	0·0	0·0	0·1	0·0	0·1	0·3	0·2	0·2	0·3	0·3
with work colleagues	3·8	5·9	5·4	5·7	4·9	4·8	6·3	7·4	5·4	7·8	3·5	3·8	6·5	6·8	6·3
with organization members	0·1	0·0	0·1	0·0	0·2	0·1	0·1	0·0	0·0	0·0	0·2	0·2	0·1	0·1	0·0
with administrative personnel	1·1	0·4	0·0	0·0	0·5	0·2	0·0	0·4	0·2	0·3	0·0	0·0	0·0	0·4	0·4
with other persons	1·4	0·8	1·1	1·5	0·6	0·1	0·2	0·2	3·1	1·2	2·1	1·6	0·2	0·0	0·0
Employed women, all days															
all alone	11·8	12·9	10·3	11·5	13·3	13·4	12·7	10·3	10·8	10·7	14·2	13·6	11·4	10·3	11·5
alone in a crowd	0·6	0·7	1·0	0·6	0·4	0·0	1·2	0·8	0·8	0·7	0·0	0·0	1·3	1·1	0·6
alone with children	0·9	2·1	1·9	1·3	1·5	1·4	1·9	1·4	2·0	1·4	1·3	1·7	1·6	2·0	1·7
with children and spouse	1·0	0·6	1·7	1·1	0·8	0·8	0·8	1·3	0·7	1·2	0·6	1·1	1·1	1·1	1·1

with spouse only	2·5	0·6	2·4	2·5	2·6	2·6	2·0	1·6	0·7	1·6	1·6	1·7	1·2	1·9	1·7
with other adults in household	1·3	0·0	0·8	1·3	1·0	1·3	0·2	1·0	3·3	1·2	0·8	0·2	0·8	1·2	0·7
with friends and relatives	0·9	0·7	0·4	0·8	1·2	1·6	0·5	0·6	0·9	0·6	1·5	1·5	0·5	0·8	0·7
with neighbours	0·1	0·1	0·1	0·1	0·3	0·1	0·1	0·2	0·2	0·2	0·7	0·4	0·5	0·3	0·1
with work colleagues	2·6	5·0	4·4	4·4	2·8	3·2	4·1	6·5	3·4	5·7	2·8	2·9	5·7	5·9	6·0
with organization members	0·0	0·0	0·1	0·0	0·0	0·1	0·1	0·1	0·0	0·0	0·1	0·1	0·0	0·0	0·0
with administrative personnel	1·3	0·4	0·1	0·2	0·7	0·3	0·0	0·5	0·1	0·3	0·0	0·0	0·0	0·4	0·8
with other persons	1·6	0·9	1·0	1·3	0·8	0·3	0·4	0·3	0·5	1·5	2·4	2·5	0·3	0·0	0·0

Housewives, all days (married only)

all alone	13·7	16·4	11·7	12·4	13·8	13·6	14·3	14·0	12·2	13·2	14·4	14·1	14·0	11·9	14·4
alone in a crowd	0·5	0·5	0·8	0·5	0·2	0·0	1·2	0·6	0·5	0·4	0·0	0·0	1·1	1·2	0·4
alone with children	2·7	3·0	4·1	3·8	2·9	2·7	4·4	3·0	4·3	3·1	3·2	3·6	3·2	3·4	2·8
with children and spouse	2·2	2·0	2·5	3·1	1·6	2·0	1·1	2·4	1·7	2·2	1·5	1·8	1·6	2·1	1·5
with spouse only	3·2	0·9	3·2	2·9	4·2	4·4	2·1	2·5	2·1	2·9	2·5	2·3	1·7	2·9	2·7
with other adults in household	0·4	0·2	0·5	0·2	0·3	0·3	0·1	0·5	2·6	0·4	0·2	0·2	1·4	1·1	1·0
with friends and relatives	0·8	0·3	0·5	0·9	0·8	0·7	0·4	0·6	0·7	0·4	1·3	1·2	0·6	0·9	0·7
with neighbours	0·2	0·3	0·2	0·1	0·3	0·2	0·2	0·4	0·3	0·4	0·5	0·5	0·8	0·7	0·4
with organization members	0·0	0·1	0·0	0·0	0·0	0·1	0·0	0·0	0·0	0·0	0·1	0·1	0·0	0·0	0·0
with administrative personnel	0·3	0·0	0·1	0·0	0·5	0·0	0·0	0·4	0·0	0·3	0·0	0·0	0·0	0·4	0·5
with other persons	0·5	0·3	0·5	0·5	0·4	0·1	0·2	0·1	0·8	1·2	1·2	0·9	0·4	0·0	0·0

1 Data are weighted to ensure equality of days of the week and number of eligible respondents per household.
2 The data may total more than 24 hours because of overlapping categories in case of simultaneous presences

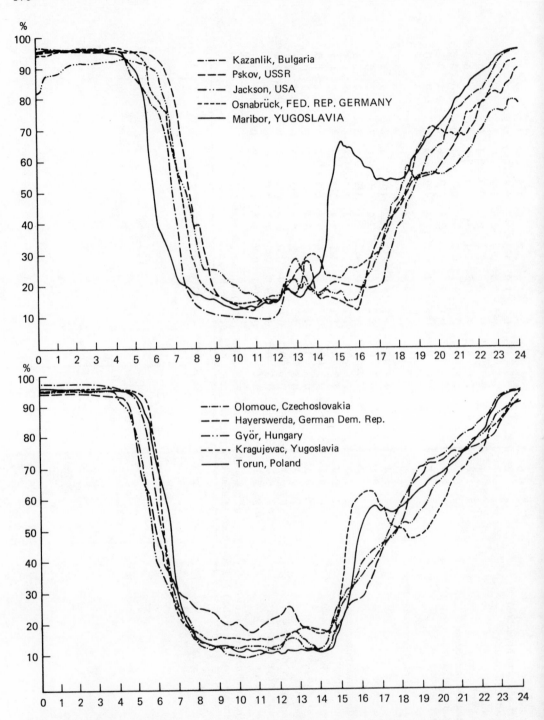

Figure 4.3. The time when people are at home — multi-national time budget study. (Szalai *et al.*, 1972, 803. Reproduced by permission of Mouton Publishers, The Hague)

Table 4.7. Distribution of daily time according to different locations (in average hours per day) (Szalai et al., 1972, 795. Reproduced by permission of Mouton Publishers, The Hague)

	Belgium	Kazanlik, Bulgaria	Olomouc, Czechoslovakia	Six cities, France	100 electoral districts, Fed. rep. Germany	Osnabrück, Fed. Rep. Germany	Hoyerswerda, German Dem. Rep.	Győr, Hungary	Lima-Callao, Peru	Toruń, Poland	Forty-four cities, USA	Jackson, USA	Pskov, USSR	Kragujevac, Yugoslavia	Maribor, Yugoslavia
Employed men, all days															
inside one's home	15·2	12·5	14·3	13·6	13·6	14·2	13·8	12·0	12·9	14·0	13·4	13·6	13·4	12·9	13·0
just outside one's home	0·5	0·7	0·3	0·3	0·7	0·5	0·4	1·0	0·1	0·2	0·2	0·3	0·3	0·3	1·4
at one's workplace	5·0	7·7	5·9	7·2	5·4	5·1	6·8	7·5	6·4	7·0	6·7	6·5	6·8	7·1	6·1
in transit	1·5	2·1	1·6	1·5	1·7	2·2	1·7	2·0	2·5	1·7	1·6	1·5	2·0	1·8	2·2
in other people's home	0·5	0·3	0·3	0·5	0·5	0·4	0·6	0·3	0·3	0·5	0·5	0·6	0·2	0·7	0·5
in places of business	0·7	0·6	0·6	0·5	0·4	0·4	0·6	0·4	0·7	0·4	0·7	0·7	0·4	0·5	0·5
in restaurants and bars	0·2	0·0	0·1	0·2	0·5	0·4	0·1	0·2	0·3	0·0	0·4	0·4	0·2	0·2	0·0
in all other locations	0·4	0·2	0·9	0·2	0·9	0·6	0·3	0·6	0·6	0·2	0·5	0·4	0·7	0·3	0·3
total	24·0	24·0	24·0	24·0	24·0	24·0	24·0	24·0	24·0	24·0	24·0	24·0	24·0	24·0	24·0
Employed women, all days															
inside one's home	17·1	14·6	16·0	15·3	17·0	16·7	16·7	14·5	16·1	15·0	15·4	15·3	14·0	15·0	15·0
just outside one's home	0·1	0·3	0·2	0·0	0·7	0·4	0·3	0·1	0·4	0·1	0·0	0·1	0·1	0·3	0·4
at one's workplace	3·6	6·5	5·1	6·3	3·6	3·6	4·9	6·8	4·4	5·8	5·2	5·0	6·7	6·1	6·4
in transit	1·2	1·6	1·3	1·1	1·1	1·3	1·1	1·4	1·8	1·5	1·3	1·3	1·7	1·4	1·5
in other people's home	0·4	0·2	0·2	0·5	0·4	0·9	0·2	0·3	0·3	0·6	0·7	0·6	0·2	0·6	0·2
in places of business	1·0	0·6	0·8	0·6	0·6	0·8	0·7	0·5	0·7	0·8	0·9	1·1	0·6	0·4	0·4
in restaurants and bars	0·2	0·0	0·0	0·1	0·3	0·3	0·1	0·0	0·1	0·0	0·2	0·2	0·2	0·0	0·0
in all other locations	0·4	0·2	0·4	0·1	0·4	0·2	0·1	0·2	0·2	0·2	0·3	0·4	0·5	0·2	0·1
total	24·0	24·0	24·0	24·0	24·0	24·0	24·0	24·0	24·0	24·0	24·0	24·0	24·0	24·0	24·0
Housewives, all days (married only)															
inside one's home	21·6	20·4	20·9	21·7	20·4	20·5	21·3	19·7	21·0	20·9	20·5	20·9	19·6	20·5	19·7
just outside one's home	0·2	1·4	0·3	0·1	0·8	0·4	0·3	2·1	0·5	0·1	0·1	0·1	0·4	0·8	2·3
in transit	1·0	0·9	1·2	1·0	1·0	1·0	1·0	0·9	1·2	1·2	1·0	0·9	1·9	1·5	1·1
in other people's home	0·4	0·4	0·3	0·6	0·6	0·6	0·3	0·2	0·4	0·5	0·8	0·7	0·7	0·7	0·7
in places of business	0·5	0·7	1·1	0·6	0·7	1·1	0·9	0·9	0·7	1·2	1·2	1·1	0·7	0·8	0·7
in restaurants and bars	0·1	0·1	0·0	0·1	0·1	0·1	0·0	0·0	0·0	0·0	0·1	0·1	0·0	0·0	0·0
in all other locations	0·2	0·1	0·2	0·1	0·3	0·3	0·2	0·2	0·2	0·1	0·3	0·2	0·3	0·1	0·1
total	24·0	24·0	24·0	24·0	24·0	24·0	24·0	24·0	24·0	24·0	24·0	24·0	24·0	24·0	24·0

1 Data are weighted to ensure equality of the week and number of eligible respondents per household

amount of time in the period or to time spent on some other activity. A wide range of elementary statistical measures can be applied to these values; mean, variance, standard deviation, coefficient of variation and correlation and regression methods can all be used for considering the co-variation of two or more activities within a sample population. Multiple discriminant measures can also be applied to duration data in order to classify individuals into activity profile types (Parkes, 1974; cf. Chapter 8 following). If some measure of the average duration of an activity within a group is required, then the total amount of time spent on an activity can be divided either by the total population number (or sample N) or by the number of individuals that actually participated in the activity (cf. Appendix). The latter measure is clearly preferable.

If the ratio of the standard deviation to the mean of an activity duration $(sd/\bar{x})$ is calculated and expressed as a percentage then a measure of group elasticity of the activity is obtained (Chapter 8). This approach was adopted by the Hungarian Central Statistical Office (1965) and has also been used in an urban social geographic context (Parkes, 1972, 1974) for comparing the activity elasticity between social status groups. The higher the coefficient the more elastic the activity for the group. Table 4.8 shows the activity elasticities calculated in the Hungarian study. It is also possible to calculate an index of structural difference, T. This and other methods are described in the Appendix.

Group elasticity of activity

(b) Frequency

The frequency with which an activity occurs may be considered in two ways. First of all there is the frequency of occurrence of an activity for an individual within some defined period, say a day. On the other hand we may calculate the frequency of occurrence of an activity across individuals in a population or sample, within some defined period. The frequency with which an activity occurs may be an indicator of the importance of an activity, just as duration is sometimes seen to be: it may also be the basis for investigating recurrences and cycles. Some examples will be given later and also considered in the Appendix.

(c) Sequence

This is possibly the most difficult aspect of time budget analysis. 'The sequential distribution of activities may be interpreted as (a) manifestation of values and habits, or life styles ...' (Chapin and Hightower, 1966, 63). However, there are a number of methods for the analysis of sequence and which one is used depends on 'deciding exactly which is the most meaningful way to examine sets

Table 4.8. Activity elasticities for different groups

STATUS	Total time devoted to earning and producing activity	House-work	Actively spent leisure	Sleeping time
	Coefficient of Variation as Dispersion/Mean × 100			
Workers and employees				
Men	47·8	154·7	95·6	26·0
Women	52·9	74·3	114·8	24·2
Members of coopera-tives in winter				
Men	51·2	162·1	120·0	21·8
Women	108·8	50·0	109·5	19·8
Non-earning				
Women		38·8	83·2	20·6

of behavioural sequences ... (for) any stream of events can be studied either in relation to some objective standard or in relation to its own unique properties' (Cullen and Godson, 1975, 46). The most awkward part of the study of sequence lies not in the description of the order of activities against the clock but in determining 'which types of transitional behaviour are relatively independent of absolute time, how independent they are, and what other objective clocks might act on transitional behaviour in the form of constraints experienced by most people and accepted as such' (Cullen and Godson, 1975, 46). Time series analyses (Appendix following) provides a way of describing the sequence of activities over a period such as a day using autocorrelation methods and by determining the fixity of activities in terms of space and time location it is possible to gain an idea of the structure of a period in terms of a single activity; some examples follow in Chapter 5. Then again, by examining cross-lag correlations the extent to which certain activities tend to lead or lag behind others can be determined, revealing the structure of the day, week or other period. However, when these autocorrelation and cross-correlation methods re used sequence is being investigated against clock time rather than in terms of its own internalized metric, but it is also desirable to try and understand the sequential structure of activities in a framework which is 'independent of absolute (universe) time' (Cullen and Godson, 1975, 73, *et seq.*) by calculating transition probabilities. A matrix of transition probabilities 'gives the probability that an episode of a certain type will be immediately followed by one of another type in any person's sequence of behaviour ... thus what any transition probability index measures is in fact the number of occasions on which people move from activity type i to activity type j, M_{ij}, expressed as a proportion of

Transitional behaviour

Autocorrelation

Activity fixity

Cross-lag correlations

Transition probabilities

the total number of occasions on which activity i occurs (so long as it is not the last episode in a sequence),

$$P_{ij} = M_{ij}/_j M_{ij}. \quad [6]$$

(Cullen and Godson, 1975, 73).

There are some restrictions on this approach, perhaps the most severe is the fact that a transition from one activity to another is assumed to be dependent only on the immediately preceding activity. However, a powering of the matrix does allow the determination of the probability of a sequence of episodes after an intervening period, depending on the weight of the power used (Appendix).

Another approach to the study of activity structure and to the identification of sequence has been suggested by Parkes and Wallis (1978). Here the interest was entirely in activity structure independent of clock time. The theory and methods of directed graphs and the idea of hypergraphs were used. To illustrate the approach consider the four directed graphs or digraphs in Figure 4.4 below.

Directed graphs

Each digraph shows the structure of a Sunday in Newcastle, Australia, for an individual adult. The digraphs G_s^1 and G_s^2 are the digraphs for the two most similar individuals in the sample. The digraphs G_d^1 and G_d^2 are the digraphs for the two most dissimilar individuals. The activity classification from the original time budget diaries, which allowed a free format description of the activity occurring at predetermined half hour intervals was based on the North Carolina classification. You will note that the digraph has been ordered according to Chapin's obligatory-discretionary continuum. The identification of the most similar and dissimilar individuals was based on an index,

Similar individuals

Dissimilar individuals

Ordered digraph

$$D_{ij} = \frac{|d_{ij}^1 - d_{ij}^2|}{d_{ij}^1 - d_{ij}^2}$$

where $|x|$ denotes the absolute value of x (d^k is a descriptive superscript and not an exponent.) Thus D_{ij} is a measure of the difference between the two graphs in the (ij) connection. If they are identical D_{ij} will be zero for all i and j. Various weighting factors for the numerator were also tried (Parkes and Wallis, 1976) but results to date have been based on the unweighted simple index as shown above. The output of the analysis is a matrix of dissimilarity coefficients and is shown in Table 4.9 following. The matrix of values in Table 4.9 shows the similarity of structure in the activities over a single Sunday, for ten males and females. The most similar pair found were in fact a husband and wife. In the next section we will report briefly on the results of applying the method to a sample of 70 subjects in a NSW country town when a diary was kept for

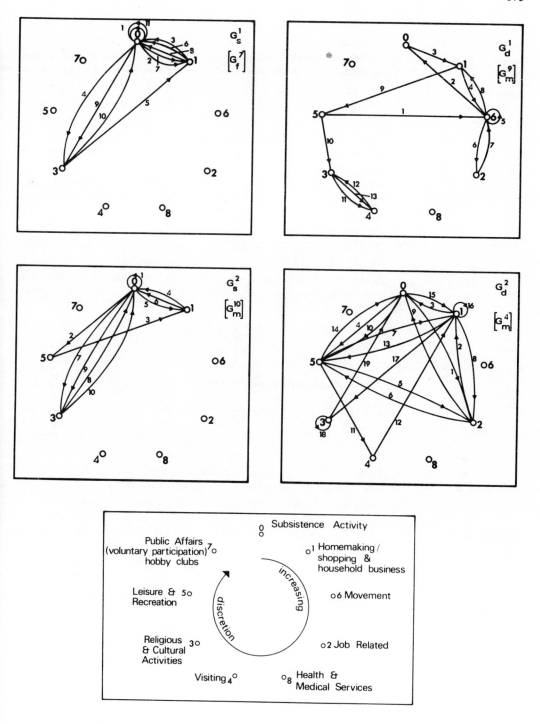

Figure 4.4. Digraphs of individual activity structure and sequence. (Parkes and Wallis, 1978. Reproduced by permission of T. Carlstein *et al.* and Edward Arnold (Publishers) Ltd)

176

Incidence matrices

**Activity
predecessors and
successors**

fourteen days (Robb, 1977). Apart from the digraphs method an incidence matrix analysis was also used and the matrices in Table 4.10 show activity incidence for each pair of most similar and dissimilar individuals shown in Figure 4.4 in terms of predecessors and successors among the activities. In the first incidence matrix between the most similar pair (top left), activity A0 precedes activity A1 twice, but A0 is also followed by A0 twice. Comparison of the subset female and the subset male by the incidence matrix reveals some interesting aspects of 'lifestyle' difference between males and females!

In Table 4.11 the activities are represented as sequences of triples and the frequency with which the triples occur is also recorded. Higher order sequences may also be considered. You will note that the two 'most similar' individuals, $G_s^1\ G_s^2$, (M10, F7) in Table 4.9 have rather similar triple sequences but the most dissimilar individuals, $G_d^1\ G_d^2$, identified as Male 9 and Male 4 (M9, M4) in Table 4.9 of relative isomorphisms are still very different. The activity sequence for the four subjects over the day was as follows:

G_s^1: 0 0 1 0 3 1 0 1 0 3 0 0　　　　　　　(Female 7)

G_s^2: 0 0 5 1 0 1 0 3 0 3 0　　　　　　　　(Male 10)

The most dissimilar were like this:

G_d^1: 5 6 0 1 6 6 2 6 1 5 3 4 3 4　　　　　Male 9

G_d^2: 0 2 1 0 5 2 5 1 2 0 5 4 1 5 0 1 1 3 3 0　(Male 4)

Hypergraphs

**Pivotal role of
activities**

A number of other measures related to analysis of activity structure and sequences were also developed, for instance, the linkage coefficient 'lambda' which aims to measure the pivotal role of activities, 'which activities appear to "precede" the greatest range of other activities?' Parkes and Wallis (1978) gives a more complete discussion (see also Appendix). However, whichever approach is taken, or whichever combination of approaches, 'the ultimate framework must be a compromise between methods that yield statistical reliability and those that yield findings relevant to major issues' (Meier, 1959, 33).

4.4　Examples

In this concluding section some results from the application of time budget studies to communities in different parts of the world will be briefly illustrated. First of all, the multinational study will be considered, then an Australian study based on the multinational time budget approach, undertaken by a Federal Government

4.9. The matrix of D_{ij} values of similarity. The D_{ij} values are a measure of relative isomorphism and the method is called the relative isomorphism measurement procedure: RIMP (Parkes and Wallis, 1978. Reproduced by permission of T. Carlstein *et al.* and Edward Arnold (Publishers) Ltd)

	F1	F2	F3	F4	F5	F6	F7	F8	F9	F10	M1	M2	M3	M4	M5	M6	M7	M8	M9	M10
F1	0·0																			
F2	12·5	0·0																		
F3	20·7	17·5	0·0																	
F4	17·7	13·8	13·8	0·0																
F5	15·1	13·0	15·3	9·6	0·0															
F6	12·3	14·1	16·5	15·9	16·1	0·0														
F7	10·5	6·4	13·5	11·1	6·7	10·3	0·0													
F8	14·3	17·7	17·2	13·8	16·2	12·7	13·8	0·0												
F9	18·2	14·6	15·3	12·6	12·1	19·8	12·2	20·0	0·0											
F10	18·5	12·8	15·0	14·8	12·0	16·2	10·8	15·5	12·1	0·0										
M1	16·7	11·4	14·1	14·3	11·3	14·3	11·2	12·8	15·3	11·2	0·0									
M2	21·7	16·5	5·7	13·2	14·6	17·5	12·5	16·7	17·7	15·3	14·3	0·0								
M3	17·3	17·0	21·9	18·2	20·5	15·7	14·7	19·0	19·2	20·0	20·7	17·5	0·0							
M4	19·7	24·8	21·7	14·8	15·3	19·3	18·8	15·3	18·8	18·3	18·2	17·7	22·3	0·0						
M5	13·5	16·2	14·7	13·8	13·9	14·5	10·1	13·2	14·5	15·3	13·4	12·8	13·7	16·4	0·0					
M6	15·5	8·5	14·1	15·0	16·3	17·5	10·3	15·5	15·5	11·0	13·0	13·5	14·8	14·9	14·0	0·0				
M7	18·3	14·3	18·5	17·9	14·1	16·0	12·1	24·7	16·8	17·8	19·0	17·0	17·8	21·7	25·0	15·3	0·0			
M8	16·9	14·0	15·3	14·6	17·3	18·3	11·5	16·8	16·8	15·8	16·5	14·7	12·4	18·0	21·8	16·1	7·7	0·0		
M9	21·0	18·5	13·3	22·0	20·3	21·3	16·3	17·3	17·3	18·7	19·0	19·3	14·3	24·0	26·0	17·0	14·0	21·0	0·0	
M10	10·7	10·2	12·9	8·7	8·2	11·7	4·2	11·0	11·0	14·0	10·5	11·2	12·3	15·7	16·0	9·3	9·8	15·4	11·4	0·0

Table 4.10. The incidence matrices for activities, predecessors, and successors (Parkes and Wallis, 1978. Reproduced by permission of T. Carlstein *et al.* and Edward Arnold (Publishers) Ltd

$G_s^1:(G_f^7)$

Activity			Successors							
	A0	A1	A2	A3	A4	A5	A6	A7	A8	σ
A0	2	2		2						4
A1		3								3
A2										
A3		1	1							2
A4										
A5										
A6										
A7										
A8										
σ	4	3		2						

(Predecessors)

$G_s^2:(G_m^{10})$

Activity			Successors							
	A0	A1	A2	A3	A4	A5	A6	A7	A8	σ
A0	1	1		2		1				4
A1	2									2
A2										
A3	2									2
A4										
A5	1									1
A6										
A7										
A8										
σ	4	2		2		1				

(Predecessors)

$G_d^1:(G_m^9)$

Activity			Successors							
	A0	A1	A2	A3	A4	A5	A6	A7	A8	σ
A0		1								1
A1					1	1				2
A2						1				1
A3				2						2
A4					1					1
A5							1	1		2
A6	1	1	1							3
A7										
A8										
σ	1	2	1	2	2	1	3			

(Predecessors)

$G_d^2:(G_m^4)$

Activity			Successors							
	A0	A1	A2	A3	A4	A5	A6	A7	A8	σ
A0		1	1			2				4
A1	1		1	1		1				4
A2										1
A3		1		1						1
A4	1									1
A5	1	1	1		1					4
A6										
A7										
A8										
σ	3	3	1	4						

(Predecessors)

Subset female total

Activity	Successors									σ
	A0	A1	A2	A3	A4	A5	A6	A7	A8	
A0	23	17	2	11	3	5	2			40
A1	21	3	2	8	1	3	2			37
A2	2	2					1			5
A3	9	5		3	3	7				24
A4	4	2					4			10
A5	3	9		2	1	4	1			16
A6	3	2	1	3	1					10
A7										
A8										
σ	42	37	5	24	9	15	10			

Subset male total

Activity	Successors									σ
	A0	A1	A2	A3	A4	A5	A6	A7	A8	
A0	22	11	3	4	4	10	3	1		36
A1	5	1	3	5	1	4	4			22
A2	5	1				2	2			10
A3	9	1	1	2	2	1	1			15
A4	1	3		3		2				9
A5	10	5	1	2	1	3	3			22
A6	7	1	2	1	3	2	3	1		17
A7	2									2
A8										
σ	39	22	10	15	11	20	4			

(Set) Grand Total

Activity	Successors									σ
	A0	A1	A2	A3	A4	A5	A6	A7	A8	
A0	45	28	5	15	7	15	5	1		76
A1	26	4	5	13	2	7	6			59
A2	7	3				2	3			15
A3	18	6	1	5	5	8	1			39
A4	5	5		3		2	4			19
A5	13	14	1	4	2	7	4			38
A6	10	3	3	4	4	2	4	1		27
A7	2									2
A8										
σ	81	59	15	39	20	34	25	2		

(σ is the row-sum or column-sum minus the value in the principal diagonal)

Department, the Cities Commission, as well as some results from the study of a small New South Wales country town, Scone. The latter shows the kind of study an individual student or researcher can carry out over a short period.

(i) The multinational time budget study

Twelve countries and fifteen sites were involved in the multinational study. We cannot cover such an ambitious scheme, which occupied 830 pages, in a comprehensive way (Szalai *et al.*, 1972). Therefore we have selected one rather unusual approach from the essays, first because it has a particularly geographical flavour, and second because it introduces a methodology which urban and behavioural geographers have been attracted to in recent years, multidimensional scaling (Kruskal, 1964). The Guttman-Lingoes method of smallest space analysis was also used (Lingoes, 1965).

Multidimensional scaling and smallest space analysis

Converse (1972) aimed at the identification of main dimensions of site differentiation in time use. The 37-activity classification, outlined in the previous section, was the starting point for the analysis and Table 4.12 summarizes the total sample time use for primary activities.

The analysis began with a space of 37 dimensions, i.e. equivalent to the number of activities that were being analysed. The absolute distance between each pair of sites (i.e. their difference without regard to sign), was then calculated using the index of structural difference (cf. Appendix A), T. This produced a matrix of dissimilarities as shown in Table 4.13 and it was this matrix which was the input to the smallest space and multidimensional scaling analysis. This matrix of dissimilarities was reduced to two dimensions which together accounted for most of the variance. In addition, little difference was apparent between solutions from the Guttman-Lingoes or Kruskal scaling methods. Converse's surprise at the two-dimensional solution is quite understandable: 'we have retrieved from these time use profiles a picture that bears a substantial resemblance to a map of the western world' (Converse, 1972, 150). The first dimension or principal axis is an East-West differentiation. The second dimension a North-South axis. Figure 4.5 illustrates this intriguing solution. The contribution of the various individual activities to the solution is dependent not so much on the absolute amounts of time allocated but on the elasticity or variance about the mean. The magnitude of the standard deviation and the degree to which an activity's inter-site variation is correlated with the dimension allowed the role of specific activities to be considered in terms of their contribution to the solution. Figure 4.6 displays this for both the east-west and north-south dimensions.

Index of structural difference, T.

Table 4.11. Triples or sequences of length 3 and their frequency (Parkes and Wallis, 1978. Reproduced by permission of T. Carlstein *et al.* and Edward Arnold (Publishers) Ltd)

G_s^1	G_s^2	G_d^1	G_d^2
001	005	560	021
010	051	601	210
103	510	016	105
031	101	166	052
310	010	662	525
101	103	626	251
010	030	261	512
103	303	615	120
030	030	153	205
300		534	054
		343	541
		434	415
		150	
		501	
		011	
		113	
		133	
		330	

Triple	G_s^1	G_s^2	G_d^1	G_d^2
		Frequencies		
001	1			
005		1		
010	2	1		
011				1
016		1		
021				1
030	1	2		
031	1			
051		1		
052				1
054				1
101	1	1		
103	2	1		
105			1	
113				1
120				1
133				1
150				1
153			1	
166			1	
205			1	
210			1	
251			1	
261			1	
300	1			
303		1		
310	1			
330			1	
343			1	
415			1	
434			1	
501				1
510		1		
512			1	
525				1
534			1	
541				1
560			1	
601			1	
615			1	
626			1	
662		1		

Table 4.12. Average time spent in 37 primary activities in twelve countries (Robinson, Converse and Szalai, 1972, 114. Reproduced by permission of Mouton (Publishers) The Hague)

	Belgium	Kazanlik, Bulgaria	Olomouc, Czechoslovakia	Six cities, France	100 electoral districts, Fed. Rep. Germany	Osnabrück, Fed. Rep. Germany	Hoyerswerda, German. Dem. Rep.	Gyor, Hungary	Lima-Callao, Peru	Toruń, Poland	Forty-four cities, USA	Jackson, USA	Pskov, USSR	Kragujevac, Yugoslavia	Maribor, Yugoslavia
total N	2077	2096	2192	2805	1500	978	1650	1994	782	2754	1243	778	2891	2125	1995
total minutes*	1440	1440	1440	1440	1440	1440	1440	1440	1440	1440	1440	1440	1440	1440	1440
1. main job	255	338	297	242	225	210	254	315	200	287	225	225	324	230	254
2. second job	4	0	1	5	2	4	2	3	10	3	5	5	2	1	11
3. at work other	4	25	6	8	6	4	22	15	4	8	12	11	13	9	17
4. travel to job	24	41	33	22	18	16	32	41	37	37	25	19	33	27	29
total work	287	404	337	277	250	234	310	374	251	334	266	259	371	267	311
5. cooking	46	39	64	45	59	49	65	60	71	59	44	45	55	70	76
6. home chores	64	36	51	70	71	73	78	55	40	51	58	57	38	49	57
7. laundry	22	12	31	26	25	20	40	35	45	34	26	24	28	28	41
8. marketing	13	14	27	20	22	26	23	14	16	16	14	16	10	22	14
total housework	145	100	172	162	177	167	206	164	172	160	142	141	131	168	188
9. care to garden/pets	8	23	8	11	31	18	11	33	2	3	3	3	8	6	49
10. shopping	6	4	6	6	3	4	5	4	9	12	18	17	14	5	5
11. other household care	15	18	27	22	19	21	16	21	6	19	24	25	17	26	27
household care	29	45	41	39	53	42	32	58	17	33	45	45	39	37	81
12. basic child care	12	9	16	32	16	14	30	12	18	16	22	23	18	14	16
13. other child care	5	8	15	9	11	11	15	17	5	18	10	8	17	9	13
total child care	17	17	31	40	27	25	45	30	23	34	32	31	35	23	29
14. personal care	44	55	71	57	54	59	49	53	47	56	69	61	49	58	47
15. eating	104	86	65	106	102	103	76	73	100	72	81	78	72	79	69
16. sleep	501	418	168	498	510	503	474	473	497	467	470	480	462	472	477
personal needs	649	618	604	661	665	665	600	599	643	595	620	619	583	609	592

17. personal travel	17	24	15	16	4	7	14	15	25	21	31	31	34	24	19
18. leisure travel	14	18	12	15	13	19	11	14	28	17	19	23	21	24	18
non-work travel	30	42	27	31	17	25	26	30	52	38	50	54	55	48	36
19. study	16	11	16	13	6	12	11	16	36	21	12	9	38	14	20
20. religion	5	0	1	4	5	6	0	1	4	5	10	11	0	0	1
21. organizations	4	7	7	2	2	4	12	3	2	4	6	6	8	5	4
study-participation	25	18	24	19	13	22	23	20	42	31	28	26	46	19	24
22. radio	8	20	11	5	7	4	4	11	8	10	4	3	10	16	6
23. TV (home)	81	14	64	55	61	72	80	39	52	64	91	99	33	34	41
24. TV (away)	3	2	2	3	2	2	1	4	2	6	1	2	5	3	0
25. read newspaper	16	14	13	14	12	13	13	12	10	16	24	25	15	20	19
26. read magazine	5	1	3	4	12	13	2	1	6	3	6	5	5	1	1
27. read books	14	21	20	7	5	6	7	14	2	17	5	4	29	7	8
28. movies	4	10	4	3	3	3	1	5	6	4	3	2	15	7	6
total mass media	131	79	116	91	98	112	108	85	87	120	134	140	113	87	81
29. social (home)	15	5	7	12	13	18	10	7	10	25	25	27	4	29	13
30. social (away)	25	8	15	20	32	32	16	16	19	22	38	39	9	42	20
31. conversation	15	9	11	17	17	18	11	13	27	13	18	16	8	28	13
32. active sports	2	2	2	1	5	4	1	2	2	1	6	5	4	0	2
33. outdoors	10	24	12	11	39	32	18	17	13	10	2	5	14	13	19
34. entertainment	5	14	2	3	4	3	2	3	4	2	5	5	3	2	2
35. cultural events	3	1	3	1	1	1	1	1	1	1	1	3	1	1	0
36. resting	27	41	17	33	17	19	10	15	63	24	9	12	11	31	20
37. other leisure	27	13	18	23	14	20	20	9	14	11	20	18	13	36	10
total leisure	128	116	86	121	140	147	91	81	152	95	123	126	67	181	99
total free time	297	231	239	245	264	300	233	200	309	262	301	310	247	311	222
total travel	56	89	62	58	39	58	60	74	90	78	78	76	88	77	78

* Because of rounding, subtotals do not sum to exactly 1440 minutes

184

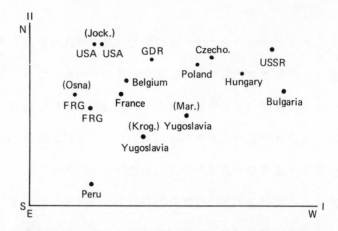

Figure 4.5. Two dimensional solution to the dissimilarity matrix of time use. (Converse, 1972, 151. Reproduced by permission of Mouton Publishers, The Hague)

The East-West axis (Dimension I) summarizes a work-free time axis. Activities which occur at the top of the graph in Figure 4.6(a) (for instance, Read books, Trip to work, Movies and radio, At work) imply that 'time spent on the activity is systematically increasing as one moves from West to East' (Converse, 1972, 155). In the lower half of the graph various socializing activities combine to provide a further discriminant factor between East and West in that it seems likely that as one moves from East to West there is a significant trade-off between time released from work and gains in time spent in informal social life.

Studies in the United States have indicated that people of higher status and means tend to engage more frequently not only in formal organizations, but also in informal socializing. (Converse, 1972, 157; Blum, 1964). Thus the East-West axis arises 'primarily from the variation in work and its social organization on one hand, and a factor involving technological development and consumer affluence on the other' (Converse, 1972, 159).

The North-South axis (Dimension II) is structured by activities to which a climatic explanation can be attached but there are many other factors which must also be taken into account. Among the most important of these is the time of year at which sampling was undertaken. There is also a hint of a North-South development axis, similar to the East-West dimension display. Mass media, especially TV at home, is also prominent in defining the northern pole of the axis.

A third dimension was also identified by the two techniques but there was now some disparity between them. This third dimension was not easily interpreted from the Guttman-Lingoes smallest space analysis, although it was independent of the other two. The Kruskal solution also reproduced the first East-West dimension.

Table 4.13. Matrix of dissimilarities in overall time use between each pair of fifteen sites (Converse, 1972, 148. Reproduced by permission of Mouton Publishers, The Hague)

	Olomouc, Czechoslovakia	Hoyerswerda, German Dem. Rep.	Lima-Callao, Peru	Kazanlik, Bulgaria	Győr, Hungary	Kragujevac, Yugoslavia	Belgium	Osnabrück, Fed. Rep. Germany	USA (nat.)	Jackson, USA	Maribor, Yugoslavia	Six cities, France	FRG (nat.)	Toruń, Poland	Pskov, USSR
Olomouc, Czechoslov.	0·0														
Hoyerswerda, GDR	4·68	0·0													
Lima-Callao, Peru	8·98	7·40	00·0												
Kazanlik, Bulgaria	5·56	8·15	10·37	0·0											
Győr, Hungary	3·52	6·00	9·84	3·73	0·0										
Kragujevac, Yugoslavia	6·39	5·77	5·61	8·16	7·42	0·0									
Belgium	5·78	4·41	6·44	7·88	6·79	5·69	0·0								
Osnabrück, FRG	7·97	5·46	6·08	10·31	9·05	5·66	4·31	0·0							
USA (nat.)	6·61	4·70	7·10	10·13	8·47	5·36	4·76	4·95	0·0						
Jackson, USA	6·96	4·89	7·18	10·48	8·81	5·75	4·61	4·82	1·35	0·0					
Maribor, Yugoslavia	5·32	4·73	7·06	6·90	4·84	5·07	6·02	6·54	6·64	6·90	0·0				
Six cities, France	6·13	4·41	5·62	7·79	6·96	4·85	2·96	3·72	4·94	5·08	5·45	0·0			
FRG (nat.)	7·51	5·23	6·47	9·33	7·98	5·77	4·44	2·15	5·91	5·80	5·51	3·75	0·0		
Toruń, Poland	2·16	4·15	7·77	5·66	3·75	5·66	4·86	7·35	5·71	6·06	4·88	5·38	7·01	0·0	
Pskov, USSR	4·48	7·52	10·52	4·28	3·74	8·29	7·88	10·52	9·21	9·62	6·83	8·26	9·87	4·38	0·0

Cell entries are values of T (see text) for the 37-activity time-budget profile, as compared between each designated pair of sites. A small volume of T means that the time use profiles from two different profiles are relatively similar. If two sites showed exactly the same profiles, T would be zero.

186

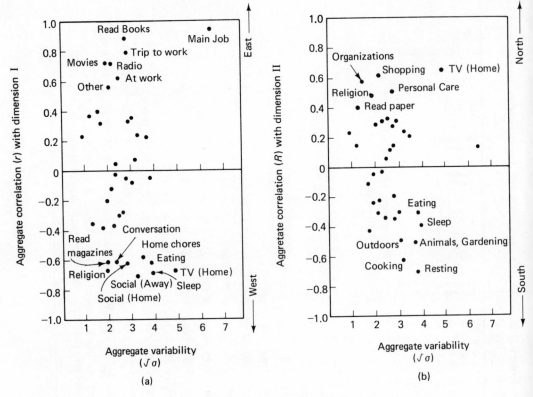

Figure 4.6. The contribution of specific activities to each dimension. (Converse, 1972, 154, 161. Reproduced by permission of Mouton Publishers, The Hague)

However, if the smallest space dimension suggests anything of substance about the time use structure differences among the twelve countries and fifteen sites, it is that there are personal travel and leisure travel differences which contribute heavily to the differentiation. However, no simple explanation can account for this. Apart from the particularly geographical characteristics of these results, they do imply that analytical designs of this kind might also be applied to different time locations rather than to different sites.

(ii) Time budgets of Australian places

There were no Australian settlements included in the multinational time budget study. But the Cities Commission of the Federal Government, based in Canberra, did undertake a time budget study of two Australian places in 1974, based on the multinational time budget approach. Their preliminary results were published in October 1975 (Cities Commission, 1975). The two settlements which were studied were Melbourne, with a population of nearly 3 million, and Albury-Wodonga with a population of about 50,000.

Table 4.14. Individual selection procedure (Cities Commission, 1975)

One eligible respondent was selected from each listed household. To select
the respondent the interviewer listed all the members of the household aged
between 18 and 69 years, and listed their respective age, starting from the
oldest. An example of such a table is given below:

No.	Household members		Household number													
	Name	Age	1	2	3	4	5	6	7	8	9	10	11	12	13	14
1	John Citizen	56	1	1	1	1	1	1	1	1	1	1	1	1	1	1
2	Mary Citizen	54	1	2	1	1	2	2	1	1	2	1	2	2	1	2
3	Alan Citizen	27	3	3	2	2	1	3	2	3	1	2	1	1	3	3
4	Ruth Citizen	25	4	1	3	4	3	1	2	2	2	1	3	4	4	1
5	....	...	1	1	5	3	2	2	4	5	4	1	3	5	1	1
6	....	...	6	4	1	5	4	1	2	6	3	5	2	3	6	4
7	....	...	5	2	3	1	7	7	3	2	6	4	4	6	5	2
8	....	...	2	5	4	1	1	3	5	4	8	7	6	3	2	5
9	....	...	3	4	6	7	5	8	1	9	2	6	7	2	3	4
10	....	...	7	8	3	2	4	1	6	1	5	9	10	7	8	8
11	....	...	11	10	9	6	8	5	3	3	7	2	1	4	11	10
12	....	...	1	3	7	5	6	4	8	10	12	9	11	2	1	3

So as to keep a proper balance of people of all ages, 18 to 69, the
respondent was selected in the following way: the interviewer (1) read down
the appropriate column using the random household number and (2) read
across from the youngest person listed 18 years of age or older. The number
at the intersection of the column and row determined the person to be
interviewed. A convenient time was arranged for the interview to be
conducted if the selected respondent was not home. In this example, the
household number is '8', the youngest eligible respondent is Ruth Citizen,
and the intersection of the appropriate column and row is '2'. The person
to be interviewed is therefore Mary Citizen.

For each selected household a Dwelling Form was completed (provided
someone was at home), and one person within the age group 18 to 69 years
was selected for interview by this method already outlined. No substitution
of persons or dwellings was permitted.

Albury-Wodonga is a government sponsored growth centre
spanning part of the border of Victoria and New South Wales.

The focus of the study was 'the way people in two Australian
cities ... allocate their time'. While the potential which lies in the
Lund time-geographic approach (to be discussed in Chapter 6) was
acknowledged, especially as a means to answering 'policy'
questions about 'how society should be organized and how
settlement patterns can be structured to ensure "liveable" day-to-
day existence for individuals' (p.2), the methodology of the time-
geographic approach was not followed. Instead a 'classical' time
budget with activities classified according to the multinational time
budget scheme was adopted, with some minor modifications. The
theoretical framework, however, owed some of its structure to
some rather early notions of Chapin's (Chapin, 1968). This early

paper however, does contain many of the essential components of his more recently developed choice model (outlined in Chapter 5.2 following).

The report argues that 'the analysis of activity pattern is the key to improved policy-oriented models, but the fundamental need is for a clearer theoretical understanding of the inter-relationship between human values, social systems, spatial structure, and market forces. Chapin's descriptive model of household activity patterns is a first step which may lead to a predictive planning model' (Cities Commission, 1975, 3). There is seen to be need for 'much more work to be done in the Australian context on the inter-relationship between use of time and use of space, on the one hand, and on the inter-relationship between social organization and the use of space on the other' (p.4).

The Cities Commission, given the responsibility for developing growth centres in Australia as an inducement to decentralization away from the coastally located capital cities, 'sought from this study some insight into how life styles of individuals and social groups differ in a large metropolis compared with a provincial city' (p.5). The authors were of the opinion that:

> 'differences in the amount of time people spend on different activities have implications for social planning. For example, the incorporation of married women into the paid workforce[7] has implications for the provision of childminding services, domestic help, and substitute facilities such as restaurants, take-away food centres, and pre-prepared meals. Shorter working hours mean more discretionary time to be spent on an almost unlimited range of activities, and depending on what they are (walking, organized sports, pleasure driving, swimming, reading, watching television, doing community welfare work, and so on), the land-use implications may be vastly different. Time-budgets therefore are a crucial input for evaluating alternative plans for community development, since they provide knowledge of the existing pattern of time allocation in the community, how it differs among groups, and how it might change, or be changed in the future'
>
> (p.6)

A brief summary of some of the preliminary findings is now offered. There were 776 representative people aged 18-69 sampled in Albury-Wodonga and 717 in Melbourne. Table 4.14 outlines the sampling procedure adopted. The study covered activities over a 24-hour period. Final response rates were 67 per cent in Albury-Wodonga and 58 per cent in Melbourne. Most of the reasons given for the loss of response are familiar enough, for example, 'too busy', 'not enough time' but in Melbourne the high proportion of 'New Australians', mainly migrants from Southern Europe (Italy, Greece, and Yugoslavia) meant that too much difficulty was encountered in handling the English language. A pre-diary interview was followed by the diary-day and then by a post-diary interview.

Amongst the many results, journey to work patterns were found to be significantly different as was flexibility of mode choice which was much greater in the smaller settlement. Women in Albury-Wodonga felt that they had more spare time than did women in Melbourne. Face-to-face contacts with relatives and friends were a 'prominent feature of life in the smaller city', but telephone ownership rates were also much lower — 54 per cent in Albury-Wodonga and 79 per cent in Melbourne. Frequency of attendance at or participation in picnics, drive-in cinemas, spectator sports, participant sports, driving for pleasure and church attendance, were all more highly represented in Albury-Wodonga than in Melbourne. A summary of some of the time allocation values obtained from the study is illustrated in Table 4.15.[8] Preceding Table 4.15 are two simple bar graphs (Figure 4.7) which give further summary information on comparative 'life styles' with regard to the use of time in Melbourne and Albury-Wodonga.

Our second example follows a rather different approach, taken by Robb (1977). Scone is a small country town in the Hunter Valley of New South Wales with a population in 1976 of 3,234 people. Seventy residents of Scone volunteered their co-operation in the survey, following the appearance of a short feature in the local newspaper (Figure 4.8). Every subject kept a diary for each of the fourteen days of the study, which took place in the summer of 1977 (January). Pre-diary interviews were held with each subject and randomly selected subjects were visited during the period the diary was being kept. 'Problems' were followed up after the diary period. Figure 4.9 illustrates the situation and relative location of Scone and Figure 4.10 shows the distribution of the volunteer residents within the settlement.

Clearly with a volunteer population it is not possible to make inferential statements about the population as a whole, based on the usual probability models, but on the other hand, there is much to be gained from the enthusiasm and commitment of individuals who participate. This is especially so when a rather long period such as 14 days is used in a diary study. Entries were made in the diary using a free recording format, but times for the activity record were pre-set at half hourly intervals. This was done simply because processing time required was thereby considerably reduced. An example of the diary sheet was illustrated in Table 4.2(c).

Chapin's choice model (Chapin, 1974, 1978) permitted some preliminary structuring, as did the routine and deliberated choice model to be outlined in Chapter 5.3 (Cullen and Godson, 1975). In this latter approach interest centres on the notion of fixity of **Activity fixity**

190

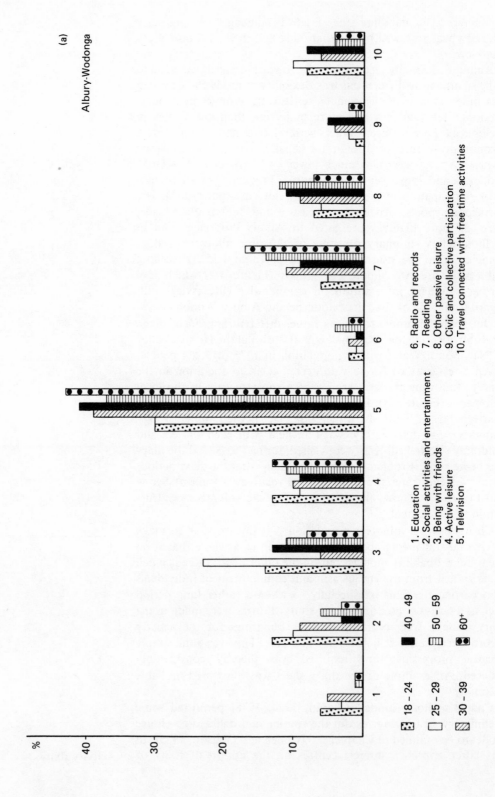

(a)

Albury-Wodonga

1. Education
2. Social activities and entertainment
3. Being with friends
4. Active leisure
5. Television

6. Radio and records
7. Reading
8. Other passive leisure
9. Civic and collective participation
10. Travel connected with free time activities

18 – 24   40 – 49
25 – 29   50 – 59
30 – 39   60+

%
40
30
20
10

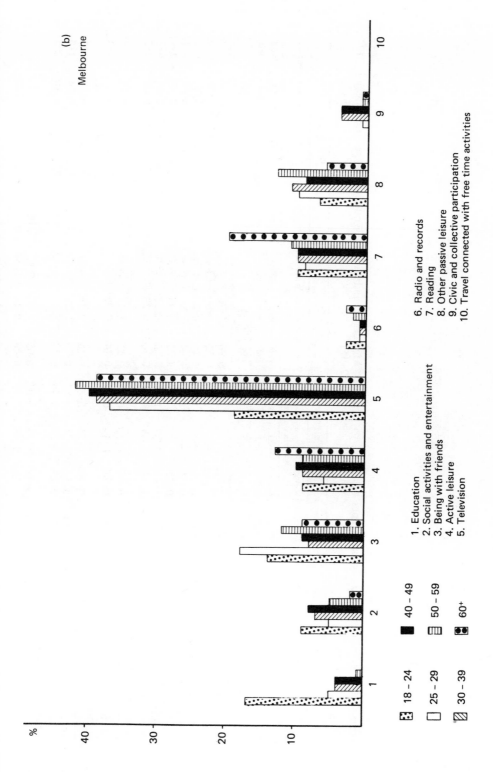

Figure 4.7. Percentage distribution of free time among various activities by age. (Cities Commission, 1975)

Table 4.15. A summary of average times allocated to activities Melbourne and Albury-Wodonga (Cities Commission, 1975; Appendix 2, 15)

Activities	Melbourne						Albury-Wodonga					
	Wage-earning women		Non-wage-earning women (housewives)		Wage-earning men		Wage-earning women		Non-wage-earning women (housewives)		Wage-earning men	
	Week end	Week end	Week end	Week end	Week end	Week end	Week end	Week end	Week end	Week end	Week end	Week end
Number in each cell	130	50	140	67	236	69	127	77	154	87	239	65
Normal work	4·97	0·59	—	—	6·80	1·76	4·50	1·55	—	—	6·69	1·50
Overtime	—	—	—	—	0·20	0·06	—	0·08	—	—	0·06	0·08
Breaks during work	0·64	0·06	—	—	1·18	0·25	0·58	0·10	—	—	1·38	0·37
Trips to and from work	0·80	0·07	—	—	1·04	0·16	0·43	0·13	—	—	0·60	0·16
Work related	6·41	0·72	—	—	9·22	2·23	5·51	7·86	—	—	8·73	2·11
Cooking and washing up	1·19	1·61	2·09	1·93	0·20	0·22	1·30	1·32	2·23	1·93	0·18	0·30
Housework	1·21	1·68	2·49	1·54	0·10	0·22	1·55	1·73	2·57	1·65	0·09	0·33
Other repairs and house maintenance	—	0·01	0·10	0·12	0·12	1·18	—	0·05	0·11	0·08	0·18	0·55
Gardening	0·03	0·08	0·17	0·17	0·14	0·48	—	0·17	0·28	0·28	0·11	0·42
Animal care, other domestic activities	0·18	0·19	0·29	0·26	0·16	0·16	0·26	0·25	0·39	0·20	0·18	0·35
Housework	2·61	3·57	5·31	4·02	0·76	2·26	3·11	3·52	5·58	4·14	0·74	1·95
Child care	0·37	0·27	1·23	1·46	0·21	0·16	0·43	0·31	0·92	0·75	0·18	0·21
Travel connected with above activities	0·08	0·05	0·23	0·05	0·05	0·02	0·07	0·05	0·10	0·03	0·02	0·05
Care of children	0·45	0·32	1·46	1·51	0·26	0·18	0·50	0·36	1·02	0·78	0·20	0·26

Purchase of goods	0·42	0·31	0·69	0·30	0·13	0·26	0·47	0·30	0·68	0·18	0·21	0·36
Purchase of services	0·08	—	0·02	—	0·01	0·01	0·09	0·06	0·09	—	0·02	—
Personal care outside home	0·05	—	0·04	0·04	0·04	0·07	0·04	0·06	0·05	0·02	0·06	0·03
Travel connected with above activities	0·22	0·15	0·38	0·14	0·14	0·24	0·19	0·13	0·27	0·10	0·10	0·19
Purchase of goods and services	0·77	0·46	1·13	0·48	0·32	0·58	0·79	0·55	1·09	0·30	0·39	0·58
Personal care at home	0·85	0·99	0·76	0·70	0·68	0·69	0·81	0·89	0·73	0·70	0·65	0·68
Eating	1·09	1·48	1·67	1·66	1·04	1·48	1·12	1·36	1·61	1·62	0·97	1·27
Sleeping	8·12	9·00	8·64	8·87	7·82	8·66	8·30	9·24	8·69	9·40	8·19	8·76
Physiological needs	10·06	11·47	11·07	11·23	9·54	10·83	10·23	11·49	11·03	11·72	9·81	10·71
Education	0·30	0·04	0·13	0·09	0·15	0·16	0·05	0·05	0·12	0·08	0·13	0·07
Social activities and entertainment	0·24	0·83	0·13	0·23	0·41	0·82	0·30	0·86	0·07	0·43	0·48	1·63
Being with friends	0·47	1·90	0·64	1·45	0·46	1·65	0·55	1·33	0·73	1·21	0·37	1·11
Active leisure	0·33	0·85	0·71	0·83	0·23	0·79	0·38	0·88	0·63	0·81	0·36	1·02
Television	1·08	1·21	1·94	1·69	1·51	2·35	1·37	1·50	2·02	2·18	1·56	1·86
Radio and records	0·05	0·20	0·08	0·28	0·07	0·24	0·07	0·06	0·08	0·13	0·08	0·19
Reading	0·40	0·56	0·50	0·50	0·40	0·41	0·37	0·33	0·51	0·40	0·53	0·41
Other passive leisure	0·47	0·57	0·48	0·57	0·32	0·45	0·38	0·29	0·51	0·64	0·28	0·43
Civic and Collective participation	0·07	0·25	0·14	0·28	0·05	0·03	0·11	0·18	0·18	0·37	0·08	0·22
Travel connected with above activities	0·28	0·98	0·28	0·82	0·29	0·85	0·21	0·68	0·32	0·68	0·31	1·13
Free time	3·69	7·39	5·03	6·74	3·89	7·75	3·79	6·16	5·17	6·93	4·18	8·07

194

75-100 PEOPLE NEEDED TO KEEP DIARY

Local girl Jennifer Robb urgently requires the assistance of local residents to help her in her honours work for her university studies.

Jennifer is the daughter of Mr and Mrs Bob Robb, of Scone, and is in her fourth year at the Newcastle University.

The Honours work to be undertaken by Jennifer in Scone; has two aspects to it.

The first aspect involves gathering data on how people use their time.

The second aspect of the study is at a more practical, planning level as it deals with Scone's role at the town level, and its relationship to the district that surrounds it, and the Hunter Valley Region, of which Newcastle is the capital.

It is considered that this sec-tion will point out many of the needs of Scone in terms of transport, retailing, business and professional services.

In order to undertake the study Jennifer requires some 75 to 100 people to participate by keeping a diary for two weeks, these being the last week in January and the first week in February.

During these weeks the par-ticipant will have to note the time and place of various ac-tivities they undertake during the course of a day, particular-ly those activities outside of the work and home environ-ment. Naturally nothing of a

personal or business nature will have to be noted.

The basic aim of the study is to find out how the people use their time and the facilities available to them in Scone, the district and the region. During the period of diary keeping Jennifer will keep in contact with all those participating in case any problems arise. After this period of diary keeping a final interview with each par-ticipant will be necessary and then the field work section of the research will be finalised.

Jennifer expects to have the work completed and bound by the beginning of October.

Anyone who will be resident in the district the last week of January, and the first week of February and would be willing to assist in the study are asked either to contact Jennifer Robb on Scone 45-1731 or leave their name and address at the Advocate Office.

Of the people that are to be participating Jennifer hopes that about half will be under the age of 60 years, and the other half over this age.

They may be married or single and live in Scone itself or the immediate surrounding area.

Figure 4.8. Scone NSW: a newspaper feature which produced 70 volunteers. (The Scone Advocate, 29th December 1976)

Activity elasticity

activities in space and time locations. Which activities in an individual's day are fixed or necessary at a specific time or at a specific location in space or are fixed as to where and to when they occur? From the diaries it was also possible to evaluate the similarity in the structure of a day between any two individuals (Parkes and Wallis, 1976, 1978), using the index of relative isomorphism outlined earlier. Apart from the calculation of durations, the elasticities of activities were also calculated. The graphs of individuals, describing the 'shape' of activities in time according to their fixity, are illustrated in the next chapter when the model with which they are related is discussed in more detail. In Table 4.16 the mean duration of the ten major activity categories is summarized. The Michelson-Reed (1975) classification scheme was used. (Essentially it is a 'hybrid' of the multinational classification and Chapin's scheme.) The mean durations are based on the number of participants in an activity, on the contingency 'N' and not the total 'N'.

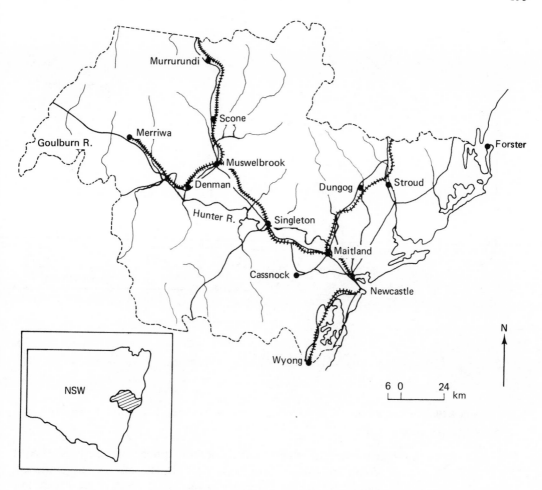

Figure 4.9. The situation of Scone NSW. (After Robb, 1977)

All activities are 'elastic' to some degree as was pointed out earlier in this chapter. In Table 4.17 and Table 4.18 some calculated elasticites for the Scone study are shown. These elasticity values are based on the coefficient of variation about the mean time for the classified activities. The first week of the study coincided with the last week of school holidays and the difference in the elasticities for the two weeks is reflected rather well in the elasticities for leisure activities, Table 4.18. From Table 4.17 you will notice that elasticities in week one are essentially unchanged in week two because they apply to those activities which can be considered obligatory. These are physiologically based in many cases (some operating rather strictly in biological time). On the other hand, in Table 4.18, there is quite a marked difference between the elasticities and this is in accord with their character as discretionary

Obligatory activities

Discretionary activities

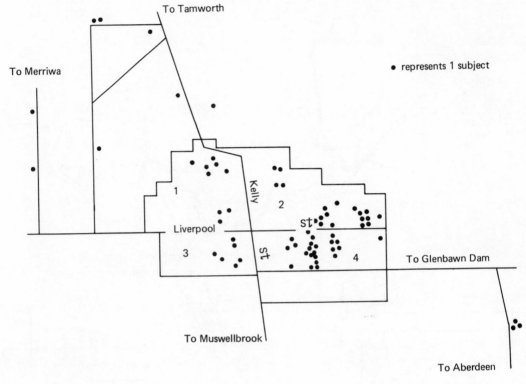

Figure 4.10. The distribution of individuals participating in the study. (After Robb, 1977)

activities. If an adequate sampling strategy had been adopted it would of course have been possible to test for the statistical significance of the difference as well as the research significance or 'importance' which seems to be revealed.

The fixity factor referred to above was used to further classify activity into 'degree of fixity' categories and autocorrelation and cross-lag correlation studies were then undertaken (see Appendix A.2). The 'day' began at 0600 and finished at 2400, for the purposes of these analyses. Table 4.19 summarizes some of the results.

Degree of fixity

To assist in the interpretation of this table first consider row 5 and column 2 of the upper table, A. Activities had been recorded every half hour from 0600 to 2400 and so there were 36 time points at which 'observations' were made. The value 0.8193 in row 5, column 2 (5,2) means that activities, classified in this way, tended to have the same sort of relationship with time of day. Activities fixed in space tended to occur at the same time as activities which were not fixed in any way. However, in the lower part of the table, (b), the correlation falls to 0.75 when the unfixed activities are lagged a half hour behind the space fixed activities. This implies that there is some tendency for space fixed activities to precede

Activities fixed in space

Activities not fixed in time or space

Table 4.16. Total of mean durations of each of the ten major activity codes for each day (in minutes) (Robb 1977)

Code	Mon	Tues	Wed	Week 1 Thurs	Fri	Sat	Sun
0. Work	815·29	727·98	817·02	995·12	850·42	480·54	459·86
1. Housework	630·27	842·77	745·67	665·78	664·27	726·95	722·83
2. Child care	223·34	198·75	357·29	400·50	489·39	216·11	457·08
3. Shopping	633·38	818·33	882·92	980·39	630·12	680·49	420·00
4. Personal needs	471·81	644·82	542·70	529·25	438·92	495·63	511·81
5. Education	650·00	340·00	510·00	—	350·00	187·50	307·50
6. Organizational activity	565·75	160·00	510·09	232·50	272·14	219·18	238·01
7. Entertainment	33·62	261·63	487·92	587·94	664·76	857·73	458·39
8. Active leisure	1842·09	1146·48	1514·58	1353·88	1136·41	1012·15	1227·18
9. Passive leisure	527·14	411·54	462·8	563·42	487·43	543·21	566·01

Code	Mon	Tues	Wed	Week 2 Thurs	Fri	Sat	Sun
0. Work	621·26	603·94	796·71	800·40	661·94	495·43	367·51
1. Housework	586·54	857·25	722·89	713·73	683·60	710·30	835·46
2. Child care	356·30	319·42	295·24	402·51	364·22	478·91	415·50
3. Shopping	235·00	343·82	356·48	430·13	342·11	1221·11	252·50
4. Personal needs	526·39	553·51	570·17	442·74	636·53	703·20	494·47
5. Education	315·00	727·50	735·84	763·79	657·90	509·17	438·00
6. Organizational activity	376·67	581·67	414·05	342·24	778·04	298·54	294·18
7. Entertainment	603·58	201·53	301·04	218·09	208·35	1037·06	380·35
8. Active leisure	1386·91	1718·78	535·25	657·80	1404·57	1581·80	390·78
9. Passive leisure	506·73	494·69	419·27	473·09	493·03	484·33	657·30

Table 4.17. Elasticity values for personal need activities (obligatory) (Robb 1977)

	Week one	Week two
Per week	0·13 (13%)	0·16 (16%)
Per weekdays	0·15 (15%)	0·15 (15%)
Per weekend days	0·02 (2%)	0·19 (19%)

Table 4.18. Elasticity values for leisure activities (discretionary) (Robb 1977)

	Week one	Week two
Per weekdays	1·24 (124%)	0·87 (87%)
Per weekend days	3·39 (339%)	1·19 (119%)

Table 4.19. Autocorrelations and cross correlation with lags (0 and 1) between 0600 and 2400 when activities are classified according to fixity level (after Robb, 1977)

Lag = 0 for rows	$N = 37$				(a)
	1	2	3	4	5
1	*1·000*	−0·0653	−0·5102	−0·4782	0·1125
2	−0·0653	*1·000*	0·5922	0·5652	0·8198
3	−0·5102	0·5922	*1·000*	0·9876	0·4522
4	−0·4782	0·5652	0·9876	*1·000*	0·4468
5	0·1125	0·8193	0·4522	0·4468	*1·000*

Lag = 1 for rows	$N = 36$				(b)
	1	2	3	4	5
1	1·000	−0·2251	−0·5967	−0·5568	0·0228
2	0·0770	0·9208	0·5268	0·5188	0·7870
3	−0·4790	0·5831	0·9537	0·9681	0·5284
4	−0·4469	0·5487	0·9379	0·9602	0·5148
5	0·2273	0·7460	0·2903	0·2949	0·9412

Where:
N	=	number of observations
1	=	time variable i.e. time intervals 1-37 (½ hourly from 6am to 12 midnight)
2	=	variable — activities fixed in space
3	=	variable — activities fixed in time
4	=	variable — activities fixed in both space and time
5	=	variable — in no way fixed

The principal diagonal (C_{ii}) with values italicized shows the autocorrelation at zero lag. The cells (C_{ij}) show the familiar product moment correlation at zero lag.

Cyclicity and the correlogram

unfixed activities, by half an hour. If you consider (1,3) in Table 4.19(a) and (b) you will see that there is an increase in the relationship following a half hour lag in the first variable. The first variable is simply the sequence of half hourly time units from 1 to 36; the relationship is negative. That is to say as the day progresses, the 'time unit' is increasing in size but the number of activities which are considered to be rigidly fixed in terms of time location are decreasing. This is much as one would expect because the evening time is available for discretionary activities. These results support the research reported by Cullen and Godson (1975) in England, which we discuss in the next chapter. A plot of the lags, which could be extended up to about ¼ of the number of observation points (in this case about 9) provides an indication of possible cyclicity in the relationship: such a plot is known as a correlogram, Appendix A.4.

Table 4.20. The incidence matrices for an individual (after Robb, 1977)

1. Housework	6. Organizational activity
2. Child care	7. Entertainment
3. Shopping	8. Active leisure
4. Personal needs	9. Passive leisure
5. Education	0. Paid work

		Activity category					Successors				
Week 1		1	2	3	4	5	6	7	8	9	0
	1	23	0	0	8	0	0	0	0	2	0
	2	0	0	0	0	0	0	0	0	0	0
	3	1	0	12	3	0	0	1	1	0	0
	4	8	0	5	171	0	1	0	5	6	0
	5	0	0	0	0	0	0	0	0	0	0
	6	0	0	0	0	0	1	1	0	0	0
Predecessors	7	0	0	0	2	0	0	2	0	0	0
	8	1	0	0	5	0	0	0	34	0	0
	9	0	0	1	7	0	0	0	0	28	0
	0	0	0	0	0	0	0	0	0	0	0
Week 2		1	2	3	4	5	6	7	8	9	0
	1	13	1	1	2	0	0	0	0	3	0
	2	0	1	0	1	0	0	0	0	0	0
	3	1	0	8	1	0	0	0	0	1	0
	4	6	0	1	164	0	0	2	4	11	0
	5	0	0	0	0	0	0	0	0	0	0
	6	0	0	0	1	0	6	1	0	1	0
	7	0	0	0	0	0	1	0	0	0	0
	8	0	0	0	5	0	0	0	37	0	0
	9	0	0	1	14	0	0	0	1	40	0
	0	0	0	0	0	0	0	0	0	0	0

Finally, as an example of the characteristics of time use and activity structure in Scone, consider the incidence matrices and graph theoretic results illustrated in Table 4.20. They show the incidence matrices for individual 16 in week one and again in week two. (An explanation of the meaning of an incidence matrix is given in the Appendix). In week one activity category 1 (housework) precedes itself 23 times and the same activity precedes activity 4 (personal needs) 8 times. However, in week two, with the school holidays over, the equivalent values are 13 and 2.

A very large number of matrices can easily be considered in this way and so it is possible to assess the changes in activity sequence structure for a large community thereby gaining an insight into the impact of 'environmental' changes on activity pattern and activity routine and hence of possible changes on space and time use. It is also possible, by powering the matrix of incidence relations

(Appendix A.2) to evaluate relations between activities which are two or more steps apart rather than adjacent as in the example shown.

For the same individual it is also possible to represent 'the day' as a digraph (Appendix A.2). Figure 4.10 illustrates this facility. The vertices of the graph, each representing an activity class, were ordered following the principles in Chapin's obligatory-discretionary dimension (Chapter 5). The digraph is a useful illustrative device in itself, but quantitative operations required in order to compare the similarity between pairs of digraphs cannot be achieved directly from looking at them when they are as complex as these are, and this is where the index of relative isomorphism comes in. It allows a single number to be computed which summarizes digraph similarities. (A worked example is given in the Appendix, and so the details need not concern you immediately.)

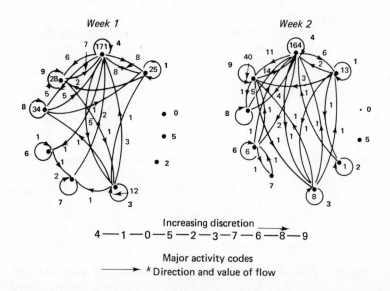

Week 1   Week 2

Increasing discretion ⟶
4 — 1 — 0 — 5 — 2 — 3 — 7 — 6 — 8 — 9

Major activity codes
⟶ ᵏ Direction and value of flow

Figure 4.11. The digraph of an individual for each of two weeks. (Robb, 1977)

As illustrated in Figure 4.10, the digraph does prove to be a handy way of representing activity structure, and can of course be programmed for machine drawing, but there is clearly a limit to the number of lines which the eye can take in, and a limit to the availability of computer funds for drawing pictures! If you note the activity loop at the top of the digraph for Week 1 you will see the number 171 against activity 4. Refer back to the incidence matrix in Table 4.20 for Week 1 and locate cell (4,4). The number you find there is the number of times on which the activity preceded itself during the week.

Table 4.21. Relative isomorphism values for Scone (matrix section only) (after Robb 1977)

Week Two

Individuals

	11	12	13	14	15	16	17	18	19	20	21	22	23	24	25	26	27
11	0·00																
12	23·56	0·00															
13	32·42	30·21	0·00														
14	27·26	26·67	28·21	0·00													
15	10·26	21·17	29·58	21·48	0·00												
16	31·99	29·48	29·27	27·45	27·98	0·00											
17	17·55	23·72	29·87	22·03	16·43	33·23	0·00										
18	15·98	22·04	26·69	22·41	12·87	27·25	17·77	0·00									
19	22·95	24·64	31·92	30·11	27·77	24·34	32·61	26·34	0·00								
20	22·89	32·09	40·12	33·09	23·39	32·45	34·92	25·02	27·66	0·00							
21	35·04	34·09	35·57	31·28	27·03	29·04	38·21	29·50	30·21	27·98	0·00						
22	25·13	29·83	31·60	30·43	27·03	29·79	34·75	24·93	25·62	22·13	26·58	0·00					
23	32·88	34·91	39·04	44·06	36·17	40·24	41·67	31·42	28·33	25·85	29·95	25·35	0·00				
24	29·62	35·19	40·15	39·32	30·49	36·39	41·75	32·28	24·96	22·85	29·19	29·74	26·17	0·00			
25	31·96	37·13	35·08	37·77	30·91	31·09	37·55	30·99	28·65	24·96	25·31	19·74	27·25	23·06	0·00		
26	28·07	34·15	37·12	32·99	29·88	29·97	28·95	28·65	24·96	35·53	30·24	21·01	23·34	25·31	28·56	0·00	
27	34·86	40·26	40·34	38·84	37·04	41·69	38·79	37·28	35·91	35·53	30·24	25·87	33·38	33·06	24·13	23·06	0·00
28	31·36	29·47	39·49	31·58	29·49	29·84	39·25	20·22	29·81	20·22	28·31	28·21	29·89	33·38	30·85	30·05	33·64
29	28·19	33·51	42·11	29·38	30·24	28·98	34·12	22·12	29·81	22·27	26·59	23·84	35·35	25·79	24·82	21·52	34·25
30	30·71	27·73	31·19	24·95	24·24	23·06	34·55	22·53	32·12	26·59	34·12	22·73	28·88	30·25	30·25	25·79	37·04
31	19·12	34·71	24·14	24·53	21·15	23·78	31·04	18·87	26·98	34·12	29·58	35·23	40·29	39·08	37·65	39·08	29·81
32	25·64	32·49	39·34	35·21	25·03	25·28	22·87	26·52	20·59	32·18	32·18	22·14	28·02	29·64	29·85	29·64	40·99
33	26·49	39·41	34·75	29·78	26·11	31·17	32·81	25·59	27·03	22·99	22·99	26·60	33·16	29·69	34·94	33·16	38·60
34	34·06	31·48	34·02	32·15	31·17	31·36	30·24	30·29	27·54	34·59	34·59	28·96	36·68	30·81	30·22	36·68	42·58
35	29·43	26·69	39·39	36·27	31·36	35·47	30·62	30·73	26·88	40·89	40·89	34·65	42·96	39·85	39·33	42·96	44·19
36	32·47	34·67	32·58	33·17	33·67	22·79	31·18	32·03	30·29	35·38	35·38	34·93	47·88	42·91	42·96	37·07	42·81
37	19·86	26·49	35·93	26·87	38·38	33·36	33·67	21·84	30·73	39·24	39·24	32·82	34·91	37·07	32·04	26·91	32·09
38	31·73	32·14	32·94	31·59	22·79	26·61	27·58	31·41	22·27	30·23	30·23	25·81	44·50	34·23	38·69	44·50	44·79
39	40·59	32·87	37·12	37·31	33·36	35·49	29·83	37·99	26·66	39·46	39·46	33·39	39·75	44·50	41·56	41·56	44·46
40	31·71	31·47	31·65	34·32	40·69	27·35	36·07	27·97	31·51	44·57	44·57	39·75	50·92	48·22	38·69	43·12	42·01
41	21·43	30·23	34·24	31·19	33·17	27·45	33·51	19·72	26·88	38·74	38·74	32·87	43·12	38·49	47·59	31·74	28·09
42	30·06	29·62	32·92	26·46	22·35	27·81	25·45	30·25	27·54	30·10	30·10	33·59	31·74	37·17	37·76	41·10	41·81
43	26·93	20·28	37·46	35·64	32·31	26·44	31·05	31·43	20·15	30·03	30·03	28·71	41·10	34·93	34·46	32·91	41·68
44	38·58	38·69	42·25	43·19	40·78	37·49	39·59	39·33	37·34	44·37	45·41	44·56	50·81	46·62	46·65	49·03	51·40

Recall that there were 70 subjects in the Scone study. The lower triangle of the relative isomorphism index matrix for all pairs of individuals has 2520 cells in it. In other words, 2520 diagraph pairs are summarised in terms of their relative similarity, including 70 values representing the isomorphism (or identity) of an individual with itself. Table 4.21 is a section of the full matrix but shows only the lower triangle because, of course, the matrix is symmetric. The two boxed values show the most similar and most dissimilar pairs of individuals in the studied community[9].

Individuals 11 and 15 were the most similar and individuals 27 and 44 the least similar (Figure 4.11). The demographic and other locational characteristics of these individuals were readily retrieved and it is therefore possible to map the location of similar pairs. A second stage, and up to an nth stage analysis can be undertaken by combining similar pairs of individuals and rerunning the analysis.

You might like to refer to Table 4.12 and find out for yourself what the value for the new pairs (11 and 27), (11 and 44), (15 and 27) and (15 and 44) was. The same individuals have been used but from looking at the digraphs of these additional pairs (Figure 4.11) you might reasonably expect values to be nearer to the maximum dissimilarity value of 51.40 than to the maximum similarity value of 10.26, as the digraphs do look rather different. Of course, the fact that the number of edges of lines joining adjacent activities is much greater in the dissimilar pair is somewhat misleading to the eye. Activities 8, 6, 5, and 2, however, appear rather frequently in digraphs 27 and 44, but not at all in digraphs 11 and 15.

These examples show just some of the ways that time budgets can be analysed in a geographical context. There are, of course, many more and some of these can be found in Carlstein, Parkes, and Thrift (1978).

Your attention has been drawn to the distinction between time use and time allocation studies. This distinction is not always adhered to and this arises through the use of the time budget as a method of recording how people use their time or allocate it to activities. When allocation per se is being considered, the mechanisms for allocating time as well as the consequences are of interest whereas in time use studies such objectives are not necessarily involved although, as we have seen, they can be incorporated. We now turn to the study of human activities and the underlying theoretical basis which has been developed.

Notes

1. The social notion of time has been considered in recent empirical work by Rudolf Rezsohazy (1972) as part of the multi-national time budget study under the direction of Szalai *et al*. The development of a 'new

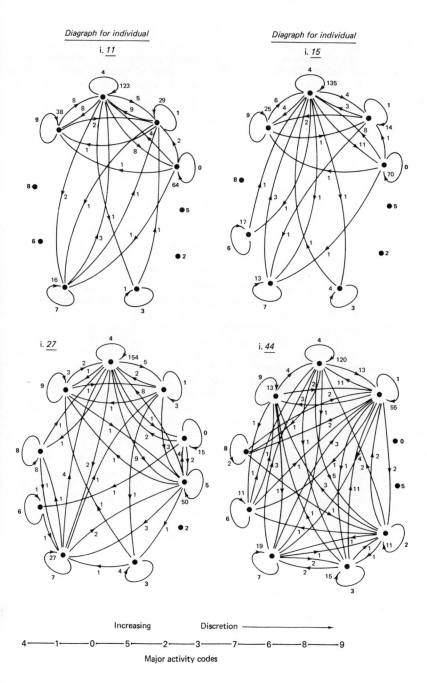

Figure 4.12. Diagrams for the most similar and the most dissimilar pairs.
(Robb 1977)

value placed upon time' is a necessary requirement before 'take off' for traditional societies in transition to industrialized societies.

The precision of meetings, spacing of activities, foresight (time perspectives), sense of progress and time as value in itself are five relations in social time, reflected in time use in the process of development.

2. *cf.* Carlstein and Thrift 1978 for a more comprehensive summary. Only a very brief overview of the contribution to the study of time allocation by economic theorists can be given here. A knowledge of the principles of micro-economic theory is essential to proper understanding of this work, especially of utility theory.

3. In macro-economic theory four time types are usually identified, these being:
   (i) the timeless world of equilibrium models,
   (ii) mechanical time of dynamic economic models,
   (iii) the expectation times of Keynesian economics (Shackle, 1972),
   (iv) the evolutionary time of, for instance, Marshall's long term supply curve and of entropy models.

4. Sorokin and Merton (1937, 617-8) when they wrote that 'astronomical time is only one of several concepts of time' and 'in the course of our daily activities we often make use of this (social time) as a means of indicating points of time'.

5. Boh and Sakisda (1972) use Cattell's Q-technique of cluster analysis based on an index of profile similarity. Here the typology produced relates to clustering individuals rather than activities, into clusters with similar time use profiles. They describe their attempt as 'largely unsuccessful' (Boh and Sakisda, 1972, 229), see Appendix A following.

6. In Chapter 10 when we outline briefly a number of analytical methods appropriate to time budget and time series analysis the transition probability approach and the graph theoretic approach to be introduced in a moment will be illustrated more fully.

7. In Chapter 6, following the introduction of the principles of the Swedish time-geographic approach, a study by two American geographers, Palm and Pred, is given as an example of the application of the method to just such issues as are raised here by a public service authority with power to influence government policy development: similar topics have been tackled by Hägerstrand on behalf of various government departments in Sweden.

8. Unfortunately it is not clear what the basis for the computation of the mean times was. If the total sample has been used, rather than the participant 'N', then some devaluation in the time quantity actually allocated to the activities will take place. A more precise summary of the central tendency is obtained if the number of participants is used, i.e. if the contingency 'N' is used as the divisor. Please bear this in mind when comparing these results with those for the study of Scone, a small country town in New South Wales, which is to be discussed in a moment, and when comparing the results with a study of activity structure in Newcastle NSW undertaken in 1970, to be discussed in Chapter 8.3.

9. The measure has not yet been developed sufficiently for us to be able to assume interval scale relations between the two outlier values but a crude ordering of individuals in terms of these limits is possible (Parkes and Wallis, in preparation).

5
Human activity

'Most of our time is devoted to living out a fairly well-ordered and neatly integrated routine.'

(Cullen, 1978)

5.1 Preliminaries

The time budget is a way of assessing, in a systematic manner, the 'record of a person's use of time over a given period' (Anderson, 1971, 353). It is a descriptive technique. When spread over a number of time periods the individual's activity routine may be estimated. Individual records, collected over a period of time and following suitable methods of aggregation of individuals and classification of activities, indicate an activity pattern. But the more difficult task of explaining activity routine and pattern requires a theoretical basis to guide survey design and analysis.

Activity routine

Activity pattern

Human activity analysis, guided by theory and using time budget records, 'offers promise of supplying some conceptual guidelines for relating ... behaviour patterns to the spatial organization of the city' (Chapin and Logan, 1969, 306). Such theoretically oriented studies should be distinguished from simple empiricism which is unable of itself to advance understanding of activity routine and pattern.

Human activity analysis

To 'understand human spatial behaviour, rather than just monitor it, we must treat time explicitly, ... as the path which orders events as a sequence, which separates cause from effect, which synchronises and integrates ... (In fact) the lack of an explicit treatment of time in behavioural studies, apart from the odd desultory venture into the field of predictive analysis ... has been truly remarkable'. One very real problem has been the way that 'research boundaries compartmentalize activity sequence (leaving) little to say about ... temporal structure' (Cullen 1978, 28-31).

Sequence cause and effect

206

In this chapter three theoretical approaches and some results of their application to the design of empirical studies will be discussed. The chapter also briefly introduces a number of other studies. Recognition of the dominance of routine in daily conduct is central to each approach. The view taken is that the individual operates within a 'framework which is fundamentally structured by physical patterns and needs' (Cullen, Godson, and Major 1972, 284). The object of human activity study 'must be to understand and explain action' (Cullen, 1976, 407), through the development of theoretical frameworks which give explicit status to the role of time, space, choice, and constraint.

**Time-space choice
constraint**

The first approach, developed at the University of North Carolina, under the direction of Chapin, is the *transductive* approach, following Chapin's own use of the term in the Washington DC human activity study (Chapin, 1974, 40. Cf. also Hammer and Chapin, 1972). In the transductive approach the focus of study is the solution of urban planning problems through a better understanding of 'the living patterns of city residents — the way they allocate their time to different activities in the course of a day, the rhythm of these activities around the clock, and the locus of these pursuits in city space ... Variations in activity patterns among subsocietal segments in the population are studied' and 'the postulated antecedent ties which these patterns may have with felt needs and preferences' are explored (Chapin, 1974, vii).

**Transductive
approach**

**Felt needs and
preferences**

The second approach, developed at the Joint Unit for Planning Research, London University, over the past decade or so, is what we have called the *routine and deliberated choice* theory (Cullen and Godson, 1975). For any population it is the recurrent routine activities which structure the day. The day tends to be structured as the result of a relatively small number of deliberated choices made rather infrequently; such as the decision to marry, to have children, to take a certain job, to buy a car, to live in a particular place. There is little demand, on a day to day basis, for deliberated choice about the majority of activities which are engaged in; in fact, there is usually neither time nor opportunity. In effect a few key times and activities act as pegs around which the day is organized.

**Routine and
deliberated choices**

Activity pegs

The third approach is based on what we have called the *routine and culturally transmitted structure* theory, developed at the Martin Centre for Architectural and Urban Studies at Cambridge University. However, unlike the other two approaches, here there is no explicit attempt to explain behaviour (activity) in terms of motivations or other aspects of 'the psychology of decision-making' (Shapcott and Steadman, 1978). Instead the postulate is that time rhythms have a culturally transmitted structure. The 'pattern of constraints which ... restrict and limit social and spatial behaviour do so in fact, in just such a way as to confer freedom and possibilities for a variable choice of patterns[1] of personal activity, onto the individual' (Shapcott and Steadman, 1978). There is no recourse to the role of motivations in this

**Routine and
culturally transmitted
structure**

scheme, instead a structuralist method is adopted which implicitly includes motivations as part of a determined and determinate social structure.

5.2 The transductive approach

(i) Model

An aggregated, survey-based approach is adopted with sufficient sociodemographic categorization to interface with the processes that govern urban spatial organization in the city. The model is sensitive to predisposing and preconditioning factors in the motivation of individuals but also aims to be sensitive to the economic system and to other systems in the city, as distinct from the household which usually forms the primary unit for the collection of time budgets. The household and the individual's behaviour, which is centred on residential activities, define the subsystem which has to date been given most attention. However, two other subsystems are also recognized: a productive activities subsystem and a general welfare activities subsystem, centred around institutions.

Predisposing and preconditioning factors

Household activities

Productive activities

General welfare activities

The model is described as transductive (Chapin, 1974, 40) because it is first conceived of in terms of the individual and the primacy of physiologically regulated behaviour (biological time) coupled to learned behaviour (social time). This initial concept gains its generality by aggregation of individuals, for instance in terms of ethnic or racial constituencies, socio-economic groups, etc. However, the approach aims at explanation as well as description of the patterns of activity which result from population aggregates.

Explanation is attempted by assessing the relative importance of the role and personal background factors which precondition people from different population categories to adopt particular activity patterns as well as the relative significance of motivations and other attitudinal factors which affect predisposition to act in one way or another, the predisposing factors being most closely associated with activity routine.

The study of human activities, based on the pattern of time allocation by households, can be approached 'in the traditions of economic theory' but when the emphasis is 'on the social construction of human activity, it is work from 'other behavioural sciences which is more relevant, particularly psychology, social psychology, sociology and anthropology' (Chapin, 1974, ix). The juxtaposition of behavioural phenomena with processes of spatial organization and development demands a recognition of the need for an aggregative and holistic level of study. This suggests the possibility of using a systems approach in the analysis of the interface between activities and the physical organization of space in the city (Chapin, 1974, x) and this squares with the move in urban theory and planning away from a conception of the city as a static, timeless arrangement of people and things in space and toward an approach that views the city as a system of activities (Ottensmann, 1972, 1). People come together in cities for a wide variety of purposes and human activity analysis should seek to understand this behaviour: this coming together, or meeting.[2]

Meeting

Definition of activities

Lifeways

Activity system

'What then are activities?' They are 'classifiable acts or behaviour of persons or households which, used as building blocks, permit us to study the living patterns or lifeways of socially cohesive segments of metropolitan area society' (Chapin, 1974, 21). The urban activity system becomes an 'umbrella kind of term for the patterned ways in which individuals, households, institutions, and firms pursue their day in day out affairs in a metropolitan community and interact with one another in time and space' (Chapin 1974, 23).[3]

Physiologically regulated behaviour

Learned behaviour

Motivated behaviour

Operant conditioning

Chapin draws a distinction between a micro-analytic view of human activity, a disaggregated approach, and an aggregative model of human activity. In the former, micro-analytic view, human activity can be distinguished as either physiologically regulated behaviour or learned behaviour. The former tends to be inelastic and 'in a very fundamental way ... sets the temporal rhythm of the individual's activity routine and influences the scheduling of all activity' (Chapin, 1974; 26. Cf. also Parkes, 1974). Learned behaviour or acts of an individual has been distinguished in the micro-analytic approach as motivated behaviour, the 'ongoing behaviour of an individual initiated by a want and directed to the end of satisfying that want through the use of suitable elements in the environment' (Lee, 1945, 12; cited in Chapin, 1974) or as operant conditioning, in the Skinnerian tradition (Skinner, 1971; also cited in Chapin, 1974). In the operant conditioning scheme the 'environment not only provides the opportunity for the act but also conditions the act through its consequences' (Chapin, 1974, 27).

Of these two schools of thought the motivational approach is preferred and is adopted in principle as the basis for the development of the transductive model. It is chosen because it brings 'into urban planning more direct consideration of the way in which people in a metropolitan community go about their round of activities on a day-to-day basis and how variations in these activity patterns call for differing emphases in planning for facilities and services' (Chapin, 1974, 29).

An act consists of:

(i) a motivational component (felt need or want directed toward a goal)
(ii) a choice component (alternatives are considered)
(iii) an outcome component (observable action).

'The want aspect of motivation is sometimes characterized in terms of levels that are graded from physiological needs to what are called self-actualising needs, (Chapin, 1974, 80; on Maslow, 1970). Such ordering is related to an activity's potential for postponment;[4] its elasticity in the terms used in previous chapters.

But various 'situational contingencies' in the individual's environment may impinge on ability to satisfy a motivation to act in a particular way, for instance as a member of particular subsocietal categories (social status, ethnic status, etc.) and 'the status hierarchy of all segments ... in the community' must also influence behaviour, including as it does the 'roles assigned to individual members of families or households by the society at large and by the subsociety. Still another factor is the spatial aspect of an activity' (Chapin, 1974, 31, 32). If the notion of a behaviour or action sequence is to be applied to the study of population activity patterns then it will be through the combinations of the factors illustrated in Figure 5.1 below that the activity pattern will be structured. In its simplest form the choice model used in the transductive approach has four principal components:

Situational contingencies

   I Motivations
  II Roles
 III Availability
 IV Quality

Motivations and roles combine to produce a 'propensity' whilst 'availability' and 'quality', as perceived by the individual, combine to produce an 'opportunity'.

Propensity and opportunity

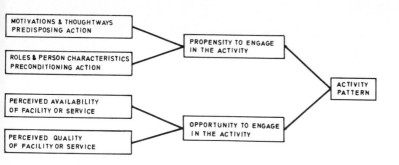

Figure 5.1. Basis for a general model to explain activity patterns. (Chapin, 1974, 33. Reproduced by permission of John Wiley and Sons, Inc.)

The simplified scheme shown in Figure 5.1 has recently been expanded by Chapin to illustrate his preferred systems-oriented approach. This expanded scheme is shown in Figure 5.2. We shall now discuss its content and meaning.

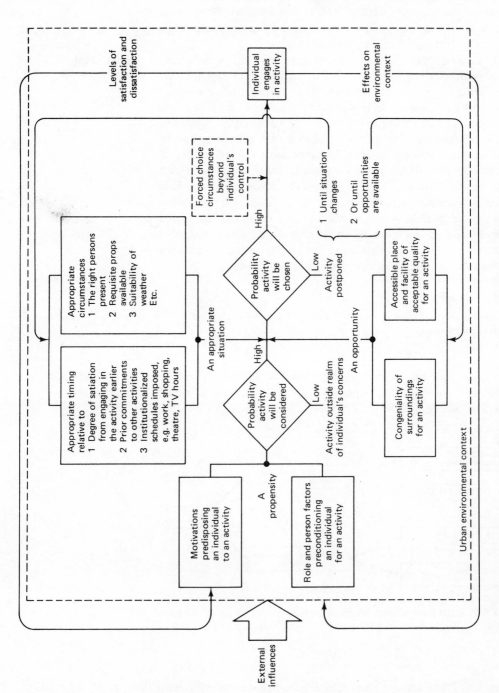

Figure 5.2. Choice model of time allocation to daily activity. (Chapin, 1978. Reproduced by permission of D. Parkes *et al.* and Edward Arnold (Publishers) Ltd

Once again four factors are identified as the most influential sources in the generation of an activity pattern. Two of them have been explicitly defined in Figure 5.1, propensity and opportunity.

i. Propensity: This is the motivational basis for an activity but it is conditioned by person-specific constraints. Propensity determines what activities are likely to fall into a person's realm of concerns, thus defining the scope of choice.

ii. Opportunity: The availability of a physical place facility suited to the activity and the suitability (congeniality) of surroundings.

iii. Situation: This incorporates the appropriateness of timing and other circumstances for the activity.

iv. Urban environmental context: This is the milieu within which choices are made: it is everything of a non-physiological nature influencing a person's behaviour, including his/her own previous behaviour, thus establishing a feedback loop.

In Figure 5.2 the choice process is represented by the flows on the diagram. The urban environmental context (dashed outermost line) envelops and influences the whole process of behaviour choice. Motivations, which predispose an individual to an activity, combine with certain role and person characteristics to precondition an individual for an activity. Both predisposing and preconditioning factors are susceptible[5] inputs to the activity system: they are influenced by the very activities which they give rise to. Together they initiate a propensity to consider an activity. If the probability of engaging in the activity is high, then two additional factors must still come into play before the activity is selected: an appropriate situation must exist and so must an opportunity to engage in the activity.

First, consider the part played by situational factors in the probability that an individual will engage in an activity. There are really two distinct criteria. First of all, timing must be appropriate relative to a number of known or anticipated factors. (Remember the experiential aspects of time introduced in Chapter 1 and discussed as psychological time in Chapter 2). Some of these are itemized in the box at centre left of the diagram, Figure 5.2. Secondly, the circumstances (or context) of the act must be acceptable. For instance, having the right persons present, the necessary 'props' to support the action, suitable weather, etc. When each of these characteristics of a situation are considered appropriate, the information is hardened; the activity probability is raised, possibly even to the level of forced choice, as in, 'This is too good an opportunity to miss ... '.

Choice process

Precondition

Susceptible

Situation and opportunity

Timing

Circumstances

Hardening of information

212

Opportunity

Congeniality and accessibility

Quality

Obligatory discretionary continuum

The second factor, coincident with situation, is the opportunity factor. Two principal elements are again involved. They are the congeniality of surroundings and the accessibility, in space and in time, to facilities which have a perceived acceptable quality. Thus access is not independent of other less tangible elements included in predispositions and preconditions: it is not simply a function of opportunity.

When situation and opportunity do combine to induce a high probability of participation it will usually only be as the result of unexpected events that an activity will not be engaged. This may result in postponment of the activity, but whatever the outcome, the urban environmental context will have been brought into the choice process.

Once an activity is engaged the environmental context impinges on the future by reinforcing existing predispositions and possibly even adjusting existing preconditions. Taking and passing a particular examination, for instance, adjusts the preconditions of entry to alternative lifeways, introducing new opportunities, situations and probabilities of specific activity routines.

Variations in the degree of postponability of an activity (its elasticity or information source type as hard or soft — Chapter 1) lead Chapin to introduce an ordering of activities along a continuum from obligatory activities at one pole to discretionary activities at the other. The elements of the transductive choice model are particularly pertinent to the discretionary activity mode. 'The ordering suggests that on the average people have more latitude for choice at the discretionary end of the continuum and little or no latitude for choice at the obligatory end' (Chapin 1974, 37). However, it is important not to treat this notion of a continuum too rigidly. 'What is discretionary and what is obligatory are relative concepts. An activity is discretionary if there is a greater degree of choice than constraint' (*Ibid.* 1974, 38). The basic physiological needs of every-day living like eating or sleeping offer limited choice possibilities, are low in elasticity and reveal little or no difference between populations drawn from different sociodemographic sets (excluding age of course). (cf. Chapter 8 following). On the other hand, self-actualizing or discretionary needs are continuously confronting the sort of variabilities incorporated in the transductive choice model.

(ii) Results

The Washington DC survey of 1968 and further participant observation and survey in 1971, also in Washington, allow some illustration of the scheme discussed above. Two communities were sampled, an inner-city black community and a fairly close-in transitional white community.

High and low income groups were distinguished, as were male and female and whether employment was full time or part time.

Table 5.1. Activity patterns, all persons according to race and income Washington DC 1968 (Chapin, 1978. Reproduced by permission of D. Parkes *et al.* and Edward Arnold (Publishers) Ltd)

Activity Measure and population segment	Work	Eating	Shopping	Home-making	Family activities	Church and organizations	Recreation and hobbies	Social activities	Watching television	Resting and relaxation	Miscellaneous activities	All forms of discriminatory activities	All out-of-home activities
Percent of persons in each segment engaging in activity[a]													
All persons (n = 1,667)	58	96	36	76	31	6	29	36	67	57	96	99	87
Black (n = 358)	59	93	19	71	13	5	13	22	69	41	90	99	78
Nonblack (n = 1,309)	54	97	40	78	35	7	34	40	66	62	97	99	90
Low income (n = 592)	44	92	27	73	25	5	21	31	69	50	89	97	75
Middle income (n = 863)	64	97	40	76	33	7	31	37	65	59	97	99	51
High income (n = 212)	64	97	42	76	33	8	40	44	58	69	97	98	54
Mean hours allocated per participant[b]													
All persons	8·7	1·7	1·6	3·6	1·7	2·2	2·3	2·0	2·5	1·6	2·9	5·9	10·4
Black	9·1	1·5	1·7	3·7	1·6	2·2	2·7	2·2	3·5	2·4	2·9	5·8	10·6
Nonblack	7·6	1·8	1·6	3·6	1·7	2·2	1·8	2·0	2·2	1·5	2·9	5·9	10·3
Low income	8·8	1·6	1·6	4·3	2·0	2·4	2·2	2·3	3·1	2·1	2·9	6·3	9·9
Middle income	8·8	1·8	1·5	3·3	1·6	2·2	1·8	1·9	2·1	1·5	2·9	5·7	10·6
High income	8·7	1·9	1·8	3·1	1·6	1·7	1·7	1·9	1·7	1·4	3·0	5·9	10·8

[a] Income figured on a per-member-of-household basis. For the derivation of income groups, see Chapin (1974, Figure 111-3, p.64).

[b] Includes time spent in travel to and from places where activity took place out-of-home.

Note: Table 5.2 examines the activity patterns enclosed in the boxes using results from supplemental investigations of low income persons.

214

The activity patterns for an average spring weekday, classified by colour, low income, middle income, and high income, according to eleven activity categories from the Chapin scheme (Chapter 4, above), are illustrated in Table 5.1.

The table shows the proportion of all people engaged in each of the activity categories, based on an initial grouping of about 225 possible activities. The lower half of the table shows the average number of hours allocated to the activity by those who engaged in it.

The 'boxes' in the table refer to population and activity categories which were given more detailed attention in further survey work in Washington in 1971. In Table 5.2 the results from this 'narrower' study are shown (Chapin, 1978) with emphasis on three discretionary pursuits which were particularly amenable to explanation by the transductive choice model.

Table 5.2. Patterns of discretionary activity by sex and employment status (Chapin, 1978. Reproduced by permission of D. Parkes *et al.* and Edward Arnold (Publishers) Ltd)

| | Inner city black community low income only[a] (1969) | | | | Close-in white community Low income only[a] (1971) | | | |
	(n)	Social activities	Watching television	Resting & relaxing	(n)	Social activities	Watching television	Resting & relaxing
Percent persons in each segment engaging in activity								
All persons	(223)	34	71	56	(241)	56	70	48
Women working FT	(45)	18	67	49	(31)	81	71	39
Women not WFT	(120)	40	71	56	(110)	59	75	50
Men working FT	(32)	28	75	47	(50)	32	54	38
Men not WFT	(26)	42	73	81	(41)	61	73	59
Mean hours allocated per participant[b]								
All persons		2·0	4·2	3·2		2·4	2·9	2·3
Women working FT		1·4	3·6	2·2		1·6	2·0	1·2
Women not WFT		2·1	4·0	3·2		2·6	3·4	2·0
Men working FT		1·1	3·9	2·4		2·2	2·4	2·2
Men not WFT		2·7	6·1	4·9		2·6	2·9	3·5

[a] Income figured on a per-member-of-household basis. For the derivation of the 'low-income' category, see Chapin (1974, Figure 111-3, p.64).

[b] Includes time spent in travel to and from places where activity took place.

Note: Table (following) applies the choice model on low-income male heads of households for activity patterns enclosed in the box shown here. WFT means 'working full time.'

The distinction made between male and female working full time or working part time means that the sample can be related to the causal factors of precondition, and predisposition, which the transductive choice model suggests are activity pattern determining factors. Because the results refer to activities which have actually been engaged in, it can be assumed that situational and opportunity components have been satisfied. In addition, because only a single 24 hour period is involved, the urban environmental context may be taken as constant, having little or no influence on the activity choices which were made during the course of the day for which the budgets and diaries were made. This means that there is no feed-back of new information to alter predisposition or precondition, and nor are there any changes in the timing and spacing of activities: there is persistence (Chapter 3). Satisfaction of needs is based on what has been learned from the previous day and on the experience of persistence over a longer period of time.

Timing and spacing

Consider only the men working full time (WFT) and men working part time (Not WFT) categories. Chapin is able to summarize the influence of the motivational factors scored according to a rank derived from a step-wise multiple regression analysis, in Table 5.3 below.

(Amplification of the various scales used to provide the so-called 'proxies' for the motivational elements listed in the table is available in Chapin (1978) but from the results shown above interpretation should be quite straightforward).

As an example of the kind of interpretation that the transductive model allows, consider the status component of esteem (need for status) represented by the declared importance of having neighbours of similar socio-economic background. For men in the black community this was (apparently) less important than for men in the white community, especially as a predisposing factor before engaging in social activities. The transitional white community was in fact about three miles further from the city centre than the black community. During the summer in which the survey was taken the black community had experienced considerable violence on the streets in their neighbourhoods. The concern about violence on the streets became a very significant predisposing factor, generated by the unsettled urban environmental context in the days previous to the day under study. Social activities of all kinds which involved leaving home were affected, perhaps especially in terms of their time location. In-home activities were adjusted in consequence. Reference to preconditioning factors like stage in the life cycle or health status suggested that both groups had their activity pattern determined to similar degree, but that health status factors perhaps had marginally more significance in the determination of activity choice in the white community than in the black community.

From this brief discussion of some of the results of Chapin's activity analysis, the value of having a theoretical structure upon

Table 5.3. Application of the choice model to some survey results from Washington DC (Chapin, 1978. Reproduced by permission of D. Parkes et al. and Edward Arnold (Publishers) Ltd)

	Rank order importance of factors for male heads of households[a]											
	Inner city black community (1969)						Close-in white community (1971)					
	Social activities		Watching television		Resting and relaxing		Social activites		Watching television		Resting and relaxing	
Proxies of motivational factors	WFT	Not WFT	WFT	Not WFT	WFT	Not WFT	WFT	Not WFT	WFT	Not WFT	WFT	Not WFT
Predisposing factors												
Need for security: concern for violence in neighbourhood	—	1	2	4	—	2	—	—	1	2	—	—
Need for achievement: evaluation of chances of 'getting ahead'	1	—	—	—	—	—	1	2	1	2	3	—
Need for status: desire for neighbours of same SES level	—	2	3	—	3	3	3	—	—	—	—	3
General social adjustment: degree of alienation	4	4	1	2	2	—	2	3	—	1	1	4
Preconditioning factors												
Stage in life cycle	2	—	4	3	1	1	—	—	3	4	2	2
Health status	3	3	—	1	—	4	4	4	4	3	4	1
R^2	0·20	0·18	0·08	0·17	0·08	0·22	0·12	0·56	0·24	0·63	0·10	0·15

[a] Rank order established from step-wise regression analysis; beyond the ranks shown, the value of the multiple determination coefficient R^2 changes very little.

which to base interpretations is clearly illustrated. The theory guides the selection of empirically derived relations. It should be noted that by emphasizing choice factors in activity participation this becomes an ideal tool for planners. When making decisions they are better able to accommodate aggregated choices assessed through a model which is initially sensitive to the motivations which predispose action; to the role and person characteristics which precondition action and to the perceived availability and quality characteristics of facilities in time and over space. Jones (1978) has suggested that this approach contrasts with the time-geographic approach of Hägerstrand (Chapter 6), where the concern is to 'understand the operation of constraints so that planning can relax them'. But the differences in goals are less obvious than the differences in method which exist between the two approaches. Although Chapin's approach is avowedly instrumentalist a careful reading of his work suggests that he is certainly aware of the issues Jones raises, although, at one and the same time, the role of constraints is perhaps underemphasized.

Model sensitivity

5.3 The routine and deliberated choice approach

(i) Model

How do we 'find out more about the way in which the individual's time-space (space-time) dimensions are structured?' (Cullen and Godson, 1975, 7). The role of motivational and psychological factors of various types assume a central position in the scheme which is now outlined. The structure of activity patterns during a day is highly routinized and largely beyond the control of the individual except in terms of a small number of significant decisions made during a lifetime. However, motivational and other psychological factors like stress levels of expectation and degree of commitment combine to influence these significant, if infrequent, decisions or 'deliberated choices'. In general, 'activities to which an individual is strongly committed and which are both space and time fixed (that is to say they must occur at a particular space location and at a particular time), tend to act as pegs around which the ordering of other activity is arranged and shuffled according to their flexibility ratings' (Cullen and Godson, 1972, 8). The result is that 'most of our time is devoted to living out neatly integrated routines' (Cullen, 1978), punctuated by purposeful behaviour.

Motivational and psychological factors

Stress levels of expectation and degree of commitment

Deliberated choices

Pegs

While behaviour may not be consistently rational it does contain highly organized episodes, structured by physical patterns and needs with priorities based on deliberations like the financial importance of the activity, the presence of certain other participants, the order in which activities are planned to occur and the likes and dislikes which the individual may have for certain

Episodes

Priorities

218

Constraint and choice
Fixity rating
Schedules

Time budget as accounting medium

Time budget as linking medium

Deliberated premeditation

activities. Constraints operate on choice and, as a result, when related to priorities, any single activity choice is subject to a fixity rating according to degree of commitment. In accordance with fixity ratings scheduling is carried out, 'in order to facilitate synchronization of activities and movements and to consume time' (Cullen, Godson, and Major, 1972, 286).

The relations among activities should be considered outside their relations in clock (universe) time, as well as in relation to clock (universe) time, although the latter approach is more typical in time budget and diary studies at our present level of understanding of paratimes. The time budget approach, according to Cullen and Godson, interprets activities in an accounting medium; the amount of time allocated is a measure of utility. However, treated as a linking medium, time becomes a path which orders events as a sequence, separating cause from effect, always synchronizing and integrating.[6]

Of course, classification of activities undertaken purely in terms of functional characteristics (as in the transductive approach, but see Chapin, 1974, 41, footnote 7) loses some of the power to explain why certain activities are adopted, because it is harder to identify the motivational factor involved. A partial answer to this problem, is to classify activities directly according to motivation or degree of commitment or of deliberated premeditation. In these terms the functional features of human conduct which are usually used in activity classification have enhanced meaning. For example, the activity 'shopping' may be classified into various subcategories according to whether it was premeditated or impulsive, planned at some specified time in advance or forced by circumstances, etc. In general all actions may be considered in terms of whether they are planned or unplanned, deliberated or spontaneous, expected or unexpected, fixed in time or not, etc.

However, although some daily activities are deliberated quite carefully, the profile of activities in a 'normal' day seems to be 'swamped by a dominant pattern of repetition and routine' (Cullen, 1978). Everyday behaviour seems to be the manifestation of a few important deliberated choices concerning a relatively small number of key events in life which then act as pegs around which other activities are scheduled. In a sense the year is father and mother to the day. Key events such as marriage, the job one has, the school one is sent to, whether one has children or not — these are the primary structuring agents for the daily activity round. When the day begins, its programme is already printed, and the stage is set for another performance; much like yesterday's.

'Action is a response to the meanings which are attached to and define the social and spatial institutions against which action is set' (Cullen, 1976, 407). The process of adaptation, 'whereby long term (deliberated) choices about where to live, what job to take and which clubs to join are translated into a pattern of daily and weekly routines' (Cullen, 1978) is illustrated in Figure 5.3.

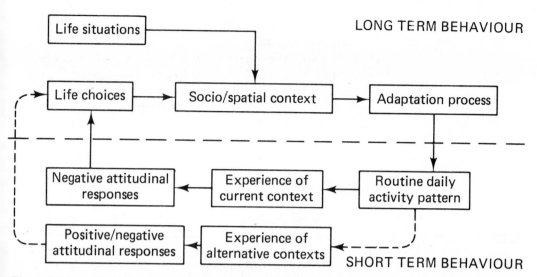

Figure 5.3. Routine activity and long term choice. (Cullen, 1978, 32. Reproduced by permission of D. Parkes *et al.* and Edward Arnold (Publishers) Ltd)

Here 'the context in which we operate everyday is the product of a mixture of choice and constraint. In the long term we may seriously deliberate over certain aspects of our situation, and accept others that are beyond our control. Either way, a context is established in the long term which has definite implications — time-consuming implications — for the way we operate day to day' (Cullen, 1978). Time enters the cycle (dotted line shown in the diagram) (Figure 5.3) and helps to overcome the paradox whereby actions which are dependent on motives and attitudes are in turn the initiators of motive and attitude through experience gained 'in action'! This paradoxical cycle is broken by the explicit reference to time which allows us to differentiate 'the long-term attitude-behaviour relationship from that which operates day to day' (Cullen, 1978).

Cullen's paradoxical cycle

To a greater or lesser degree every activity is fixed in space or time and sometimes in both. These characteristics of relative fixity are related to four categories of activity which in turn affect the subjective rating of fixity. They operate *in addition* to existing objective fixity factors — like opening and closing times of shops, or the location of a football ground, etc. These four subjective factors of fixity are associated with:

i. *Arranged* activities, involving joint activity with others, may be more difficult or less difficult to arrange at some other time or some other space location depending on the status of those involved; (in Chapin's terms on their preconditioning and predisposing factors).

ii. *Routine* activities undertaken with sufficient regularity and frequency may acquire the status of being highly fixed, especially in time.

iii. *Planned* activites need not involve others but will be assigned a level of subjective fixity according to commitment. Generally speaking the nearer in time future an activity is, the harder it is to adjust its planned space-time location.

iv. *Unexpected* activities may impose fixity ratings onto existing arranged, routine or planned activities.

(ii) Results[7]

Frequency counts

Time allocation

The first stage of analysis, following classification of activities from a diary survey, involves frequency counts using activity episodes as a unit. This allows cross tabulations to be undertaken, for instance of prearrangement and subjective fixity. Time allocation analysis, involving statistical description of the time budget, is based on computed mean durations and other statistical moments about the mean, with activities further classified according to fixity, deliberation, commitment, etc. Table 5.4 illustrates some results.

Having computed the means and variances for the duration of time allocated to classified activities, classified in terms of function, premeditated choice, etc., it is possible to calculate the intercorrelations among all activity classes and then to use a method such as factor or component analysis to indicate any underlying structure in the day. Twenty-one indices of time allocation were treated in this way by Cullen and Godson (1975) and five factors were able to be interpreted. The results obtained could not be generalized to the population of a city because they were derived from a rather special sample from a university population. However, the opportunity for undertaking analyses of this sort has been provided, and the interpretation of factors or components is assisted by the theory of routine and deliberated choice.

Relative time location, that is when something happens in relation to some other event, may be considered with the aid of various methods of time series analysis. Lagged autocovariance and cross lagged autocovariance describe the changing pattern of a single activity (autocovariance) and changes in relative incidence between two or more activities may be indicated by cross-covariance (Appendix).

Such analyses help to answer questions such as 'Which activities tend to lead and which tend to lag other activities?' 'How does one activity tend to lead (i.e. come before) or lag (i.e. come after) another activity?' The answer to the second question, of course, demands additional information, perhaps theoretical.

Table 5.4. Allocation of time by degree of pre-arrangement and fixity (mean duration of episode figures are given in minutes) (Cullen and Godson, 1975, 21. Reproduced by permission of Pergamon Press Ltd)

	Mean number of episodes per day	Mean % of episodes	Mean time in min/ day	Mean % of time	Mean duration of episodes
Pre-arrangement					
Arranged	2·8	8·8	133	9·2	47·0
Planned	5·4	16·8	225	15·6	41·7
Routine	9·8	30·6	301	20·9	30·6
Unexpected	6·8	21·1	196	13·6	29·0
Time-space fixity					
Not able to do anything else at that time	7·1	24·7	324	22·6	45·6
Not able to do activity at any other time	9·2	28·5	375	26·1	40·8
Not able to do activity else-where	10·1	49·8	534	37·2	52·9
Could not have been else-where at that time	9·1	29·7	351	24·4	38·6
Generalized fixity	3·6	11·2	203	14·1	56·3

An integral part of this approach is the study of the sequence of activities, independent of clock (universe) time. Two methods suggest themselves — computing transition probabilities and time series analysis. Using the method of transition probabilities some progress has been made toward understanding the internalized structure of activity patterns, independent of time allocation and durations. Rather than operating in the time domain, as with autocorrelation methods discussed in the previous paragraph, consideration is now given to activities in the frequency domain. The object is to investigate the likelihood or probability of occurrence of given sequences of activity. This is done by computing the frequencies of changes between one activity and another. These values are converted into transition probabilities, probabilities of transition (change) from the existing activity to some other specified activity. It is possible, by powering up this matrix of transition probabilities (Appendix) to estimate what activity is likely to occur two, three, or four stages ahead. For example squaring the matrix will estimate the probabilities for the

Transition probabilities

Time domain

Frequency domain

'next-but-one' episode. The probability of transition from one activity to another is conditional upon certain assumptions in the transition probability model used. The most usual assumption made is that an episode is dependent on the episode immediately preceding, but there are ways of overcoming this constraint so that one may move further back into the day, to two or three episodes. It may also be desirable to remove or filter unwanted activities and re-estimate the probability of activity change from a new starting point. Table 5.5 shows one example of a set of computed transition probabilities of various types of activity. Consider Table 5.5(a). Here travel, which is activity 8, will most probably be followed by a social activity.[3] The probability is 0.297. The next most probable episode to follow travel is shopping[7] with a probability of 0.213, but the probability that shopping will be followed by travel is much higher than either of these sequences, as we might expect, at 0.794. In Table 5.5(b) note that sleep[5] will most probably be followed by a routine activity[3], 0.774; but a routine activity will be followed by another routine episode with greater probability than any other type, 0.551. In Table 5.5(c) the fixity notion is considered in terms of transition probabilities and we find that the probability that sleep[5] will be followed by an activity[13] fixed in space is 42%. But the probability that sleep might be followed by an activity which is fixed in both space and time[4] is almost as high, at 40%.

Time series analyses allow activities and various attributes of activities to be considered right through the day, or any other period. We shall take an example of a study by Cullen and Godson (1975). The period selected extended over a 17-hour period, from 0730 to midnight, the wakeful period, broken down further into 100 equal ten minute intervals. At each interval the number of people engaging in a particular activity or, more precisely, the number who described the activity in which they were engaged as being of a certain fixity or deliberated choice type, such as routine, planned alone, fixed in space, fixed in time, etc., was used as an indicator of the changing shape or structure of the activity and its perceived constraints during the day. Nineteen activities, or attributes of activities, were considered in some detail. We have selected three of these for illustration here. We can compare these results with those from a recent Australian study in which some aspects of the deliberated choice model were used for a time budget and diary based survey of a sample population in a small Australian country town, Scone (Robb, 1977).

The various autocorrelation and cross-correlation matrices resulting from these studies cannot be illustrated here but an idea of the changing pattern through the day, for the selected activities, can be illustrated by a number of graphs. In Figure 5.4 the graphs from Cullen and Godson's study are shown (excluding the various parameters derived from the harmonic regression method which

Routine
Planned alone
Fixed in space
Fixed in time

Table 5.5. Activity transition matrices (Cullen and Godson, 1975, 73, 76, 77. Reproduced by permission of Pergamon Press Ltd)

(a)

Transition variables	Eight-way activity classification: transition matrix							
	1	2	3	4	5	6	7	8
Meals	0·030	0·258	0·119	0·108	0·246	0·097	0·006	0·136
Domestic, personal	0·206	0·346	0·123	0·086	0·079	0·010	0·002	0·149
Social	0·062	0·160	0·252	0·049	0·110	0·063	0·009	0·294
Leisure	0·107	0·295	0·177	0·107	0·162	0·049	0·003	0·100
Non-formal work	0·124	0·133	0·164	0·062	0·245	0·098	0·006	0·168
Formal work	0·226	0·024	0·125	0·033	0·181	0·183	0·010	0·218
Shopping	0·013	0·012	0·063	0·0	0·012	0·012	0·094	0·794
Travel	0·080	0·117	0·297	0·067	0·111	0·076	0·213	0·040

(b)

Transition variables	Activity premeditation: transition matrix				
	1	2	3	4	5
Arranged (A)	0·306	0·191	0·216	0·280	0·008
Planned Alone (P)	0·113	0·367	0·240	0·263	0·016
Routine (R)	0·068	0·144	0·551	0·179	0·058
Unexpected, etc. (U)	0·101	0·222	0·250	0·389	0·039
Sleep	0·029	0·055	0·774	0·127	0·016

(c)

Transition variables	Levels of fixity: transition matrix				
	1	2	3	4	5
Free in space and time	0·422	0·043	0·266	0·232	0·037
Fixed in time only	0·279	0·119	0·252	0·320	0·030
Fixed in space only	0·274	0·033	0·406	0·231	0·057
Fixed in space and time	0·217	0·044	0·247	0·471	0·021
Sleep	0·099	0·030	0·423	0·400	0·048

they used). The continuous line shows the best fit curve for the observed values. We may treat this curve as a first sighting of the possible shape of time of the day for the particular activity types which are considered.

For the Australian study orthogonal polynomial functions were used (Appendix A.2) to find the best fit line allowing a clearer picture of the structure of a day in terms of fixity levels. The results are shown in Figure 5.5. For the Australian study the sample population was 70 people. The study was based on a household diary covering fourteen days during January 1977 (summer time). The day commenced at 0600 and terminated at midnight, but 30 minute intervals, rather than 10 minute intervals, were used for analysis. The curve fitting methods were different to those used in the Cullen and Godson study and so in comparing the graphed structures you should note the pattern of observed values carefully.

Fixed in space and time
Fixed in space

The similarity of the 'fixed in space and time' graphs is most noticeable. Some oscillation in the 'fixed in space' graph for the Australian study is discernible, but it is more marked in the English study. This can be explained by the fact that the Australian study is dealing with a 'normal' community rather than an institutional community. It also deals with a diary kept over 14 days. In addition, the curve fitting systems used are different. The orthogonal polynomial method does not fit a curve which follows the low amplitude, high frequency oscillations as closely as the harmonic regression.

In contrast to the 'fixed in space and time' graphs, the structures for the activities 'fixed in time' differ markedly. The peculiar order and timing of key activities in a university environment might have a lot to do with such a difference! Reference to Cullen and Godson's results for the 'no way fixed' activities also shows greater similarity than difference, bearing in mind the scale differences between the studies as well as the functional spaces occupied by both samples. (In the Australian study the values on the vertical axis are the number of times, over 14 days, that activities at the given times are reported as fixed, etc. In the English study actual numbers involved on a single day are calibrated).

As was noted earlier, stress is an explicit feature of the routine and deliberated choice approach. In a study of 50 married couples in Hackney, Cullen and Phelps (1975) report that while the structure of the normal day was highly routinized, subjects did note the stresses of everyday life and were more prepared to talk about them than they were to talk about the 'unstressed' highlights of the day. Such stresses have to be coped with, and they form part of the adaptation processes which operate in long term behaviour (Figure 5.3). They have been graphed in a manner similar to that used in Figure 5.4; in Figure 5.6. (In order to allow ready comparison with the graphs of Figures 5.4 and 5.5, Cullen's original diagrams have been adjusted so that the time scale of the graph runs from 0600 to 0600, rather than from 1800 to 1800 as in Cullen (1978)).

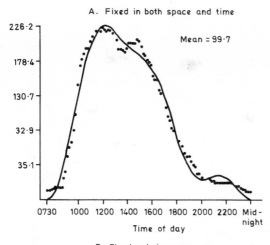

A. Fixed in both space and time

Mean = 99·7

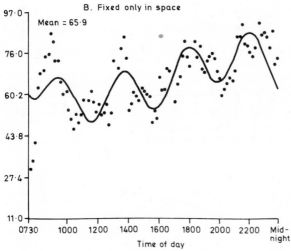

B. Fixed only in space

Mean = 65·9

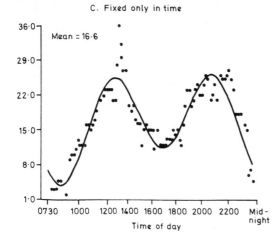

C. Fixed only in time

Mean = 16·6

Figure 5.4. Daily activity structure with time and space fixity. (After Cullen and Godson, 1975. Reproduced by permission of Pergamon Press Ltd)

226

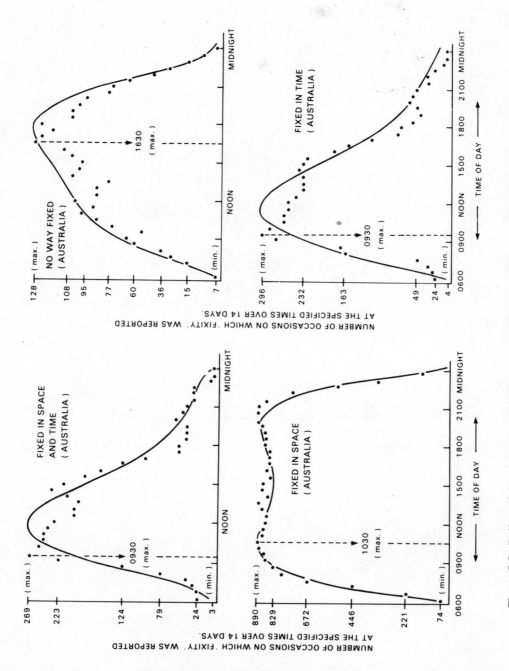

Figure 5.5. Daily activity structures with time and space fixity: Scone, NSW, Australia. (After, Robb 1977)

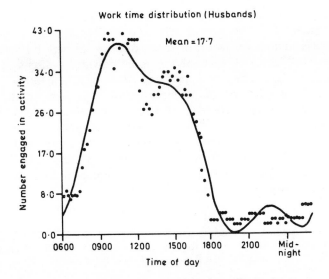

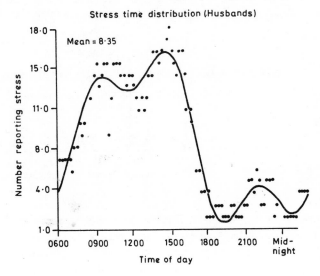

Figure 5.6. Work and stress time distribution, Hackney. (After Cullen, 1978. Reproduced by permission of D. Parkes *et al.* and Edward Arnold (Publishers) Ltd)

The Hackney survey sample and the environment in which it is situated is functionally more like the Australian country town study than the university population referred to in Figure 5.4. It is interesting to compare the work distribution of husbands with the shape of time for activities fixed in time and space, Figures 5.4 and 5.5. Looking more closely at the distribution of observed values than at the fitted curve reveals the number of husbands who reported feeling under stress at various times of the day and, as might be expected, the stress peak lags the work distribution peak by about four hours.

There are other ways of assessing activity structure independently of clock time in order to get some idea of the relation between activities for which there are different perceived levels of choice. In Table 5.6 each of the four individuals discussed in chapter 4.3 (iv) (when we considered simple graph theoretic and matrix representations of activity structure) is now compared according to the characteristics of their activity sequence structure. Activities are now defined first in terms of maximum and minimum choice (akin to the obligatory-discretionary continuum) and secondly in terms of level of expectancy (Parkes and Wallis, 1976, 1978). Recall that G_s^1 and G_s^2 were the most similar individuals and G_d^1, and G_d^2 were most dissimilar individuals.

Table 5.6. Activity sequence structure by choice level and by expectancy level (Parkes and Wallis, 1978. Reproduced by permission of D. Parkes *et al.*, and Edward Arnold (Publishers) London)

G_s^1 choice 1 0			G_s^2 choice 1 0			G_d^1 choice 1 0			G_d^2 choice 1 0		
1	0 3		1	1 4		1	4 3		1	11 3	
0	3 5	(11)	0	4 1	(10)	0	3 3	(13)	0	3 2	(19)
Expectancy 1 0			Expectancy 1 0			Expectancy 1 0			Expectancy 1 0		
1	4 1		1	4 1		1	4 3		1	5 6	
0	1 5	(11)	0	1 4	(10)	0	3 3	(13)	0	6 2	(19)

(The values in brackets represent the sum of different activities undertaken)

Consider the top left 2 × 2 table. The bits 1,0 represent[1] for 'maximum choice felt to be associated with a particular activity' (i.e. discretionary) and (0) 'minimum choice' (i.e. obligatory). On no occasion was a discretionary activity followed by a different type of discretionary activity for this individual, out of the 11 major activity categories engaged in during a single Sunday. The individual G_s^1, derived from the measure of relative isomorphism reported in 4.3 was most similar in the activity structure of his day with G_s^2, a female whose activity structure sequence is shown in the 2 × 2 table immediately to the right. The most dissimilar activity structures were for graphs G_d^1 and G_d^2. They are shown in the two tables to the right. The difference in structure is quite clearly revealed. In the lower part of Table 5.6 a similar approach is used for the identification of the expectancy level for the same four individuals. An expected activity is coded (0) and an unexpected activity is coded (1). Once again the most similar pair, derived from the relative isomorphism level, reveal very similar motivational and anticipation patterns for their daily activity structure but the dissimilar pair show quite distinct structures. More detailed reference to these results can be found in Parkes and Wallis (1976, 1978).

So it seems that from the routine and deliberated choice approach it is possible to structure the analysis of human activity study in a manner which allows a very positive contribution to be made towards the explanation of some of the basic characteristics of human activity. The possibility for a coupling of the transductive and deliberated choice approaches exists and seems likely to be fruitful, for instance in transport choice studies, see Kutter (1975) or Brög (1977, 1978).

5.4 The routine and culturally transmitted behaviour approach

(i) Model

In the strictest sense this approach is a simulation approach. The estimation of the distribution of aggregates of people, in the three dimensional frame of location, time, and activity, is the principal object. Various environmental constraints are built into an elementary entropy maximising method. Sets of equations which allow the prediction of the population aggregates' distribution in relation to the three principal dimensions of location, time, and activity are estimated (Tomlinson *et al.*, 1974).

Simulation aggregates Location time activity Entropy maximizing

'A rhythmic pattern of behaviour is established, and is transmitted culturally in the form of a known "timetable" of activities' (Shapcott and Steadman, 1978). Once again it is routine rather than erratic conduct which is taken to be the primary factor in the structuring of an activity pattern for a population aggregate. The research programme at the Martin Centre in Cambridge has been directed towards the development of 'a computable model to predict patterns of activities' (Bullock *et al.*, 1974, 46).

Timetable as culture transmission Routine

What has been the underlying rationale in the theoretical stance adopted? Emphasis is placed on the value of an aggregate statistical approach; 'no attempt is made to examine patterns of activity from the standpoint of the individual as decision maker ... nor has there been concern ... with the reasons for which individual people choose to do this or that' (Bullock *et al.*, 1974, 46). In this approach the time budget is taken as the starting point for the modelling of activity patterns in time and space and the concern is 'with the overall pattern of behaviour of groups of people who are identified by similar easily measured characteristics' (*Ibid.*, 46). Thus we have an approach which has similarities to the transductive approach of Chapin because of its explicit interest in the activity patterns of aggregates as well as in the identification of the differences between sociodemographic groups. However, there is no underlying individual behaviour model which attempts to

Overall pattern of behaviour

230

Motivation

accommodate motivation, as in the transductive scheme, which used a theory of individual behaviour to assist the interpretation of the analysis. Similarities with the deliberated choice model also exist, especially in the significance assigned to the structuring powers of routinized behaviour, and in the hypothesised stability of activity patterns over long periods of time. However, we find differences once again because of the absence of a motivational, premeditative component in the model.

Premeditation

The first of the assumptions 'made about the behaviour of people in aggregate for the purposes of the model', is that the proportion of time spent in various activities will remain the same for a given group of people. Over a repeated period of time such as a day their time budgets will remain broadly the same. Secondly, the behaviour of people is subject to a number of limiting restrictions which determine the time and space locations of the activities which they engage in. Thirdly, 'timing and location of some activities is determined or affected by cultural conventions and physiological necessities'. Finally, for the purposes of this approach, 'it is critical that the overall proportions of time spent in different activities ... be independent of the location of the facilities in which these activities take place in a particular instance' (Bullock *et al.*, 1974, 48).

Limiting restrictions
Cultural conventions
Physiological necessities

Behaviour time location and activity

A simple diagrammatic scheme is used to illustrate the various normal constraints which operate on daily behaviour. These constraints operate through the combination of the three dimensions time, location, and activity; as shown in Figure 5.7. The sum of the cells formed by the intersection of the planes of time, location, and activity, accommodate all the people who will be engaged in various activities. The modelling problem involves the allocation of 'population groups to activities in time and space' (Bullock *et al.*, 1974, 47), in accordance with the times allocated to similarly classified activities derived from time budgets. Thus the parameters of the dimensions have to be estimated, subject to various restrictions on the availability of locations. In Figure 5.7(d) an 'interrupted matrix' is shown, this indicates that certain cells are not available for occupation because time, locational or activity restrictions of one sort of another are operating. Any permutation of these restrictions is possible. The problem is to find the most probable permutation. Additionally demographic characteristics of some aggregates will make some cells unavailable. For instance juveniles cannot use pubs and males cannot occupy female institutions. An entropy maximising method[8] is used to determine the most probable distribution of the population, subject to defined restrictions. The result is the distribution 'of the number of people engaged in each activity for each time period in each location, over the day' (Bullock *et al.*, 1974, 48).

Adapting the Hägerstrand time-geographic scheme (Chapter 6

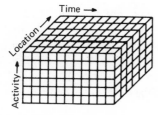

(a) A three-dimensional array of cells representing activity, time and location

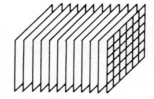

(b) The total population in cells in each activity/location plane must obey the population constraint

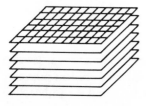

(c) The total population in cells in each location/time plane must obey the time budget constraint for that activity

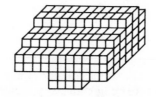

(d) Some activity/location combinations are not available, nor are some activity/time combinations

Figure 5.7. A three dimensional time, location and activity matrix. (Bullock *et al.,* 1974)

following) Shapcott and Steadman (1978) suggest a way of depicting 'the generally acting set of constraints (and opportunities) which apply in the society or culture as a whole'. Now the 'constraints (and the opportunities which they reveal) appear as rhythmic structures ... for example the fixed hours of work might be imagined as a series of clusters or "tubes" appearing regularly between the hours of 0900 and 1700, the location of the tubes in the horizontal plane corresponding to the relevant spatial positions of the work-places in the city or region in question ... An instantaneous "time slice" cut horizontally through this picture will reveal the effective map of the spatial area in question, as it is defined in activity terms at the chosen time of day' (Shapcott and Steadman, 1978, 52-54).[9]

The notion that activity patterns have a culturally transmitted structure owes part of its rationale to Popper's world three of objective knowledge (Popper, 1972; Shapcott and Steadman, 1978). This gives us a valuable insight, because it suggests that too often in human geography, space-time behaviour has been studied only in terms of the interaction between the material environment (world one) and the subjective environment of individuals (world two). An intermediary is needed — a third world which mediates between world one and world two. This is the world of

Popper's world three

World one
World two

objective knowledge, a cultural product. For the study of human activity it provides a perspective between that of a wholly subjective, psychological point of view and a wholly rigid cultural and environmental deterministic point of view. At any single point in time the social structure of temporal regularities 'is a "given" so far as the individual is concerned' (Shapcott and Steadman, 1978, 73).

We might think of social time as performing a similar mediatory role between universe time (a world one factor) and life time (a world two factor).

(ii) Results

Various parts of this approach have been applied, in at least three major studies to date. First of all in a study of student populations at Reading University, Leicester University, and the Polytechnic at Leicester. Secondly in a study of the population of the city of Reading, and recently in a study of the impact of the introduction of flextime working hours into a Civil Service Department, in Reading.

Flextime

For a student population, activity structures can be explained in terms of different restrictions which present different opportunities, for instance through the time-table of lectures and other classes. At Reading, for instance, lectures commenced on the hour, but at Leicester at half-past the hour. This meant that the markers for critical activities at the two institutions were 30 minutes apart and consequently such parameters must be included in any simulation model built to accommodate both institutions. Using the entropy maximizing method it was found possible to reproduce the results of the diary surveys rather closely using the entropy maximization scheme, and differences in the generated activity patterns were interpreted as being primarily a function of the culturally transmitted temporal regularities of the lecturing time-table (Tomlinson et al., 1973).

Time-tables

Markers

The second study considered 450 residents of the city of Reading. Table 5.7 summarizes the amount of time, averaged over the seven days of the week, spent on urban activities. The population is split into three categories. This survey was compared with a time budget study undertaken by the BBC in 1961. The reason for undertaking such a comparison was to assess the long-term stability of human activity structure, the argument being that the changes in culturally transmitted structure between 1961 and 1973 would have initiated significant differences in the structure of the daily routine. A comparison of the 1961 and 1973 figures for the population category, Work: men-not-at-home, is made in Figure 5.8. In spite of the 'large social changes which have occurred over this period, and which might well have been expected to have had a real effect on daily activities: (for instance) an average increase in real incomes

over the period of something like 30%, a change in the proportion
of women in the work force from 30% in 1961 to 42% in 1971'
(Central Statistical Office/Shapcott and Steadman, 1978, 59); it
seems 'that decisions about the timing of most of the major kinds
of activity are not made on a daily basis at all. Instead individuals,
and households collectively, commit themselves over a much longer
time scale, by a number of 'life decisions' — marriage, taking **Life decisions**
particular jobs, choosing a particular place to live, sending their
children to particular schools — to a series of rhythmic constraints
which then govern the greater part of their daily routine and

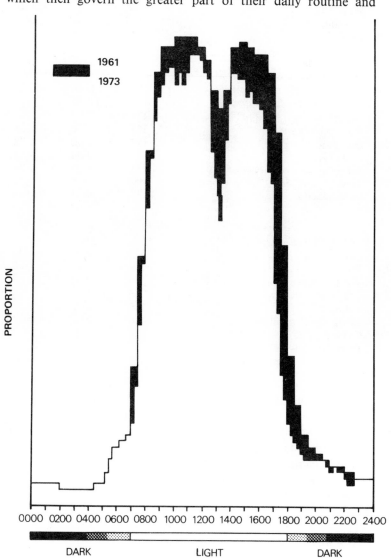

Figure 5.8. Comparison of work: men-not-at-home 1961 and 1973.
(Shapcott and Steadman, 1978. Reproduced by permission of D. Parkes *et
al.* and Edward Arnold (Publishers) Ltd)

Table 5.7. Reading survey of urban activities (Bullock *et al.,* 1974, 57. Reproduced with the permission of the Controller of Her Majesty's Stationery Office)

	Number engaging	Average daily hours spent by:	
		Those who engage	Whole group
Men			
Sleep	195	8·40	8·40
Work	163	5·13	4·29
Work (travel)	39	2·25	0·45
Full-time education	19	2·57	0·25
Eating	195	1·51	1·51
Drinking (alcoholic)	92	0·58	0·27
Causal social	188	1·13	1·09
Organized leisure, community activities, entertainment	86	0·66	0·29
Private leisure, study	184	1·34	1·26
Television	181	2·43	2·25
Personal hygiene	194	0·68	0·68
Domestic	172	1·00	0·88
Child care	50	0·44	0·11
Shopping, use of services	153	0·33	0·26
Travel	190	1·58	1·54
Miscellaneous	130	0·71	0·47
			24·00
Working women			
Sleep	144	8·53	8·55
Work	129	4·02	3·60
Work (travel)	6	0·36	0·01
Full-time education	13	1·58	0·13
Eating	143	1·37	1·37
Drinking (alcoholic)	46	0·40	0·13
Casual social	138	1·18	1·13
Organized leisure, community activities, entertainment	77	0·62	0·33

around which and within which those optional and variable activities which do occur at the day-to-day level must be fitted in' (Shapcott and Steadman, 1978, 67). To emphasize the links with the routine and deliberated choice approach outlined in the previous section, Shapcott and Steadman cite Cullen and Phelps (1975, 5) 'it is this which enables people to cope with the complexity of the urban environment they face by eliminating the necessity of considering anything but a small fraction of its true variety'.

Flex-time

Finally, the third study undertaken using this approach relates to the introduction of flex-time work hours. Flex-time is a work organization system aimed at giving the employee as much flexibility (discretion) as possible in the hours at which he/she reports for work (A flexibility which is deemed to be commensurate with the cost efficiency of the enterprise). Although the total

	Number engaging	Average daily hours spent by:	
		Those who engage	Whole group
Private leisure, study	125	1·15	1·00
Television	129	1·86	1·67
Personal hygiene	141	0·79	0·77
Domestic	143	2·92	2·92
Child care	40	0·58	0·16
Shopping, use of services	129	0·50	0·45
Travel	143	1·41	1·41
Miscellaneous	83	0·72	0·42
All activities			24·00
Non-working women			
Sleep	111	8·92	8·92
Work	5	1·31	0·06
Work (travel)	1	0·12	0·00
Full-time education	0	—	—
Eating	111	1·52	1·52
Drinking (alcoholic)	21	0·24	0·05
Casual social	107	1·54	1·48
Organized leisure, community activities, entertainment	35	0·89	0·28
Private leisure, study	101	1·50	1·36
Television	102	2·53	2·33
Personal hygiene	108	0·64	0·62
Domestic	111	4·62	4·62
Child care	65	1·38	0·81
Shopping, use of services	103	0·66	0·61
Travel	110	0·95	0·95
Miscellaneous	60	0·72	0·39
All activities			24·00

number of hours of work required of the individual is not altered, it may be allocated in a way which is suited to personalized 'attributes of predisposition' (cf. the transductive schema) but is constrained by a block of so-called core time which delimits a compulsory period of attendance. In the study of a Civil Service Department in Reading, a 'before' and an 'after' diary as well as a questionnaire were used. Comparing the results identified before the introduction of flex-time with the time allocation and activity pattern which was recorded six months after the introduction of flex-time, the central theoretical position of this approach seems to be generally supported: that is, routine dominates and quite dramatic changes in such aspects of social organization as work times, seem to be transmitted very slowly into significant changes in the structure of daily activity patterns (Shapcott and Steadman, 1978a).

Core time

5.5 Other approaches

Three other approaches to the study of human activity can be noted.

(i) A longitudinal diary survey has been completed by Michelson (1976) at the University of Toronto, Canada. A Space-time budget approach was used to complement a longitudinal study of residential choice behaviour. The theoretical basis lay in the motivational and life-style characteristics of a population of movers. This quite specific objective allowed the daily activity pattern to be studied in the light of longer term decisions: analogous to Cullen's deliberated choice theory of activity pattern. This indicates that although the outcome of choice is observable at the day to day level, this is no reason to believe that the day is the appropriate time scale for appreciation of the causal factors which produce the observed activity pattern for a population, or the observed routine for an individual.

Michelson's sample consisted of residence movers who were also married couples with children: who were not in their first home and who could afford 'reasonably high' monthly payments for their house. Such a sample of people, it was argued, would be able to exercise some choice over the selected environment because of financial advantages which they held in the market place. The purpose of the study was to examine movement between downtown high-rise apartments and suburban single family homes. A substantial questionnaire, as well as a diary, was completed over three phases — once before moving, two months after moving, and one year following the second interview. In other words there was opportunity for a longitudinal study to be undertaken, using the same population. There was only a 10% loss in the sample reported at each phase of the interviewing. The results, summarized in Michelson (1976), seem to support various aspects of the deliberated choice scheme. Studying a population group that had considerable advantage in the exercise of choice due to its financial 'security', Michelson was dealing with people for whom expectations of a good life were seen as attainable through the exercise of choice in the housing market. The reported results seem to indicate that on the whole expectations were met: the new life-styles and activity patterns were very much in line with wants.

(ii) A space-time budget study has been underway at the Institute of Public Affairs, Dalhousie University, Halifax, Canada (Elliott, Harvey and Procos, 1973, Harvey and Clark, 1975, and Elliott and Clark, 1975). The focus has been on disadvantaged sectors of the population and has been patterned on the multi-national time budget study. However spatial locations were given more consideration than in the multinational study (Chapter 4), and the study was tied to the Canadian Census year. This provided additional opportunity for making reliability tests, etc. The spatial data was analysed using nearest neighbour techniques. Recently a study of discretionary activities has also been reported (Harvey, 1978). 'Where, when, and with whom people spend discretionary time are matters of substantive interest'. This study investigates how discretionary time activities are influenced by certain contextual elements. There is some similarity to the notion of time spacing and space timing (Parkes and Thrift, 1975) outlined in Chapter 3 above. This approach, along with the deliberated choice approach of Cullen and Godson and certain aspects of the Swedish time-geographic approach, outlined in the next chapter, are seen as coming together in what Barker (1978) calls a 'behaviour setting'. The behaviour setting is defined by Barker as 'units of the environment that have relevance for behaviour ... a behaviour setting coerces people and things to conform to its temporal spatial pattern'.[10]

(iii) A simulation model for the study of daily activity sequences and space-time constraints has been developed by Stephens (1976). His research is designed for disaggregated data in the first place, with aggregation following at a later stage. The simulation is based on a Monte Carlo model with the 'principal aim of the research (being) to develop a methodology for investigating how individual's decisions about their space-time behaviour interrelate' (Stephens, 1976, 21). The time-geographic approach (Hägerstrand *et al.,* to be discussed in the next chapter) contributes to the formulation of a simulation model focussing attention 'on the paths, (or behaviour) of the individual through space-time' (Stephens, 1976, 28). Space-time budgets allow identification of 'sequence, linkage, timing, duration, and frequency of activities as well as the spatial and temporal coordinates of one's behaviour' and focus on both overt behaviour and 'the perceptions ... of physical and social environment' (*ibid,* 1976, 28). Cullen's fixity notion (discussed above) is prominent in the design as a four-part subjective fixity constraint.

Monte Carlo simulation

Paths space-time

Identification of routinized, recurrent patterns is Stephens' first task. It is followed by a grouping algorithm based on the amount of time which individuals spend on activities and the sequence of the activities. 'The algorithm focusses on those points in an individual's sequence where the activity performed is the same as that performed by another individual' (Stephens, 1976, 31). The resultant overlap time is used as the linking index for the grouping. The more overlap there is, the more alike are the activity sequences. Two distinct groups were identified in the university population which was used in the study. The results are similar to those found by Cullen in that once again there are two activity groups: (1) a rather loosely structured group and (2) a more highly structured activity group. Figure 5.9 compares some aspects of activity structure for these two activity groups when the fixity notion discussed in the deliberated choice schema is adopted.

Overlap time

238

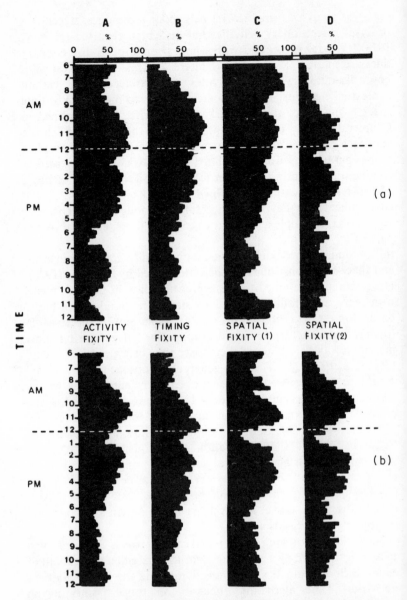

Figure 5.9. Temporal profiles of priorities and constraints as described by the subjects. (a) Loosely structured day. (b) More highly structured day. (After Stephens, 1976, 36 and 39)

The impact of the constraints imposed by the 'fixities' generated 'an oscillating pattern' and 'moreover, the extent of oscillation tends to narrow with increasing time' (Stephens, 1976, 40). This first generalization is illustrated in Figure 5.10. 'Amplifications of the wave represent the most highly constrained episodes while de-amplifications, or troughs, depict the periods of least constraint. Thus, the daily activity sequence (for the university population, at least) may be considered as a process which oscillates between constrained-high priority phases and relatively unconstrained-low priority ones' (Stephens, 1976, 40). This seems to be much as expected; the real value of the work will come when the more difficult task of identifying community activity structure is tackled.

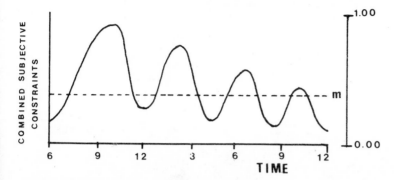

Figure 5.10. Generalized pattern of space-time constraints (damped oscillation form). (Stephens, 1976, 40)

The Monte-Carlo simulation exercise attempts to answer the question, 'Can activities be associated with this oscillating pattern of commitment and constraint in terms of their timing and location?' Some of the data requirements for the simulation are shown in Figure 5.11. Cumulative probability distributions are calculated for each of the activities and a Monte-Carlo method is then used to select activities, durations, and locations. Having isolated the pegs about which the day is assumed to be structured, based on the subject's ranking of their importance and level of commitment, further passes are made 'through the day' with different combinations of commitment and constraint and, by the use of random numbers, the probable activity duration and location for each type of fix is identified. The position of a fix or peg, in relation to an activity is then determined and the possibility that non-compatible locations have been assigned is checked.

Following the simulation a comparison of the results with the initial time budget was made. It showed that, for the majority of experiments which were conducted, it was possible to conclude that the actual (time budget-derived) and predicted (simulation-based) activity sequences were the same. From the experimental work to date Stephens makes the general conclusion that 'the subjective constraints acting on the choice, timing, and location of activities are modestly important in attempting to understand the formation of activity sequences and the paths people follow through space-time. The statistical results of the model's output suggest that they are not as significant as they were hypothesized to be. Hence, it must be concluded that subjective constraints are not the critical determinants which structure space-time paths at a daily scale' (Stephens, 1976, 55).

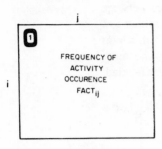

i = 1,2,...,n, the total number
   of time intervals (72)
j = 1,2,...,m, the total number
   of classes of activity (65)

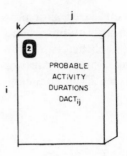

i = 1,2,...,n, the total number
   of classes of activity (65)
j = 1,2,...,m, the total number
   of activity duration categories
k = 1,2--average activity durations

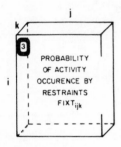

i = 1,2,...,n, the total number of
   of classes of activity (65)
j = 1,2,...,m, the total number of
   subjective constraint categories (16)
k = 1,2,...,l, the total number of
   degrees of commitment (4)

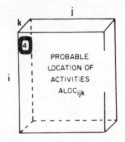

i = 1,2,...,n, the total number
   of classes of activity (65)
j = 1,2,...,m, the total number
   of activity locations (9)
k = 1,2,...,l, the total number
   of time intervals (6)

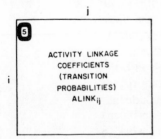

i = 1,2,...,n, the total number
   of classes of activity (65)
j = 1,2,...,m, the total number
   of classes of activity (65)

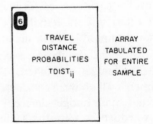

i = 1,2,...,n, the total number of
   time intervals (72)
j = 1,2,...,m, the total number of
   trip length categories (5)

Figure 5.11. Data requirements of the simulation model. (After Stephens, 1976, 43)

Stephens' conclusion suggests that some new strategy is needed in order to more fully understand the determinants of human activity. It seems that this would need to emphasise more fully the role of actual environmental constraints in 'non behaviour' (what cannot be done) rather than the psychological dimensions of choice, of what can be done. Such thoughts are enshrined in the approach known as time geography, the study of the physical limits to human action.

This chapter has discussed some of the theoretical underpinnings to human activity study. Three schemes were introduced in some detail, the transductive, deliberated choice and culturally transmitted activity structure approaches. Three other approaches were then discussed rather more tersely. The next chapter introduces the Swedish time-geographic approach, which we have already alluded to on a number of occasions. This approach has received considerable interest in human geography, especially since the early 1970's. However, it has tended to be the case that the Swedish time-geographic approach has been identified as *the* approach to the incorporation of a space-time framework in human geography. While at present it certainly exists as the most comprehensive approach, its real potential will only be achieved when it is considered in relation to the other theoretical and empirical studies of the role of time in social organization and individual conduct which we have been considering so far in chapters one to five.

Notes

1. The use of the term routines is generally to be preferred to the term patterns in relation to *individual* activity.
2. The concept of a 'meeting' is rather important in the Swedish time-geographic approach which we will discuss in Chapter 6.
3. Following the discussion in chapter 1, you will recognize that the coincidence of behaviour in time and space according to locational and experiential factors, defines place: a system.
4. You should recall at this point the concept of soft and hard information proposed in Chapter 1.
5. Susceptibility was introduced in Chapter 1 and again in Chapter 3 as a concept having general applicability in any space-time framework in human geography.
6. Treated in this way, once again we have the notion of time as social time (Chapter 2) in which the recurrent item or event relations which were relevant to social behaviour are themselves a basis for time reckoning.
7. The early stages in the development of this approach relied on surveys of institutional populations (universities and other tertiary education colleges).

8. Refer to Tomlinson *et al.*, (1973) and Wilson (1974) for details of the entropy maximizing method adopted.
9. For a more detailed account of this approach see Shapcott and Steadman, 1978.
10. A comprehensive account of behaviour settings will be found in Barker (1968).

6
Time-geography: The Lund approach

'In fact space and time always go together and we might as well, the sooner the better, try to get accustomed to seeing space and time as united into one compact four-dimensional entity.'

(Hägerstrand, 1974, 271)

6.1 Preliminaries

'Time-geography' is the result of attempts by geographers at Lund University, Sweden, to develop a model of society in which constraints on behaviour (activity) can be formulated in the physical terms, 'location in space, areal extension, and duration in time' (Hägerstrand, 1970, 11). The scheme has been a part of research at Lund, under Hägerstrand's direction, for about fifteen years. However, for human geographers outside Sweden it was with the publication of Hägerstrand's seminal paper, 'What about people in Regional Science?' (1970) that their work first became more widely known. Perhaps the most comprehensive treatment of their work, in English, will be found in Carlstein, Parkes and Thrift (1978, Volume 2). 'To see time-geography as a separate field would be inappropriate. It is rather an approach anchored in certain basic facts of life which one can possibly ignore or neglect, but hardly deny'. (Carlstein, Parkes, and Thrift, 1978, 121).

Most social scientists look at the world in a manner which might be called 'compositional'. That is to say they classify and then analyse 'objects' according to how alike or 'composed' they are. This is true of the studies discussed in the previous chapter. But there are some drawbacks to this compositional approach. In effect it means that new objects of study are put together and our understanding is then of this new composition and not necessarily of the initial context. What is needed to complement the

Physical

Facts of life

Compositional view of world

**Togetherness
likeness
indivisibility**

compositional approach is a 'contextual' approach. This treats items where and when they are; it treats them in context. Such an approach allows evaluation of items located in their appropriate space and time positions, and in this way their togetherness is given as much emphasis as their alikeness. In the time-geographic approach indivisibility is a central contention. A person is indivisible and without exception always occupies space and time.

Space-time flows

The time-geographic approach does not take a particular phenomenon and investigate it whenever or wherever it occurs. It does not set up a controlled experiment hoping thereby to fix relations, *in situ* as it were, as though time could be stopped and space could be sliced without distortions of any consequence. Instead the first step is always to try and define a bounded region of space and time and treat everything in it as a space-time flow of

**Organisms and
artefacts**

'organisms and artefacts' (Hägerstrand, 1970). The time-geographic framework therefore begins with the environmental structure that surrounds every individual and, whereas the compositional approach tries to simplify the complex traces of movement through space and time by classification, the time-geographic approach attempts to capture the complexity of interaction at the scale of the smallest indivisible unit which for human population is, of course, the individual. There is in effect no 'sacred lower scale limit' (Hägerstrand, 1973), but although aggregation is avoided it can be handled.

**Intended activity
programmes**

Time-geography supplies a notation. It permits the physical boundaries of space and time to be evaluated as they impinge on intended activity programmes. But there are no direct references to the motivational factors which are involved in intention. Why we do things is obviously an important factor in the understanding of human activity but it is also important to know what stops us from doing certain things.

Constraints

There are a relatively small number of primary factors in everyday life which impinge upon all individuals and constrain their freedom to occupy certain space and time locations. When these constraints are identified it becomes possible to deduce the reasons why a particular individual geography exists as it does rather than following some other path. Because the 'life of an individual is his foremost project' (Hägerstrand, 1970, 9) the logical place to begin a study of human geography is with the individual. A project in time geographic terms is a set of linked tasks which have to be undertaken somewhere at some time within a constraining environment.

Project

**Trajectories or
paths**

Any bounded region contains a population made up of individuals (organisms and artefacts), represented in the time-geographic approach as 'point-objects' (Lenntorp, 1976, 12). They describe continuous trajectories or paths through time and over space, from the point when and where they come into being (birth) to the point when and where they cease to be (death). For the purposes of time-geographic analysis, however, such paths are followed until they make a permanent exit from the bounded space-time region under study or until they are transformed into some other entity, as in the case of many materials. (Recall that organisms (individuals) and artefacts (tools and materials) are both essential elements in the time-geographic scheme). All point-object populations can be studied in two ways, either demographically, when concern is only with numbers and descriptive characteristics like age and sex, or geographically. Then the concern is to develop explanations of the ways in which different individuals and populations co-exist in the same space-time frame or region.

Demographic approach

Geographic approach

The time-geographic approach captures the spatial and temporal sequence and co-existence of events by using a 'dynamic map' to represent the path of an individual in motion over space and through time. It also represents the interactions among individuals. Figure 6.1 illustrates the basic elements of the dynamic maps which appear in time-geographic studies (see also Chapter 1).

Dynamic map

Paths and interactions

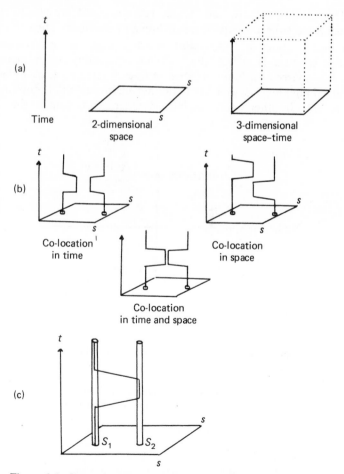

Figure 6.1. Some elements of the time-geographic dynamic map. (a) The time dimension and two-dimensional space: combined as 3-dimensional space-time. (b) The paths of two individuals with their space-time relations. (c) An individual moving from one station to another and back again, S_1 to S_2 to S_1

Each path shown in Figure 6.1(a, b and c) represents a particular trajectory of one individual. When the line of the path is vertical there is no movement over space; when the line is sloped velocity is registered. The shallower the slope the greater the speed or velocity of movement,[1] and so less time is consumed to cover a given distance than when the line is steep. If we were to keep a record of the movements of a number of individuals over a period, as in Figure 6.1(b) (and other figures in this chapter) it is easy to see that we build up a sort of web of the interacting trajectories. Repeated over the life time scale for each individual we construct the geographical biography of the population within a constrained environment. We can speak of a constrained environment because both space and time are limited in supply. The factors which cause this limitation of the supply of space-time will be outlined in a moment. Seamon (1979) has called these intricate webs 'place-ballets'.

Time and space are both scarce resources and are absolutely inseparable from the intricacies of human behaviour. Their use is contingent upon the constraints which operate on individuals. These constraints affect the 'individual's ability to influence his/her environment' and 'depend on a set of circumstances linked to the individual as well as to his/her environment' (Lenntorp, 1976a, 13). In other words, an individual's reach is limited, that is

Velocity

Webs

Geographical biography

Constraints

Reach

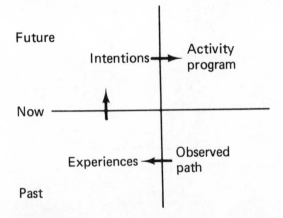

Figure 6.2. Observed paths and their relation to intentions. An illustration of the relation envisaged between an observed individual-path, experience intentions, and activity programme. An individual-path and programme can be portrayed in space-time and can be derived to a large extent by an observer. Individual experiences and intentions, on the other hand, are largely concealed from an observer in the environment. There exists no simple way of making direct projections of intentions into a space-time region other than in terms of intended activities. (after Lenntorp, 1976a, 14. Reproduced by permission of Bo Lenntorp)

the possibility of participating — 'physically' — in events is bounded. The physically accessible part of the environment constitutes a connected and continuous set of positions in space-time and is called a prism (Lenntorp, 1976a). **Prism**

The time-geographic scheme is sometimes described as physicalist because (at the present stage of its development) it is explicitly concerned only with the physical, concrete, observable realism of the location and movement of individuals and not with individual experiences and intentions. But like any label, this one must not be treated as a full and sufficient description of its ingredients. (Carlstein, Parkes, Thrift, 1978, 117). **Physicalist**

For just because there is no 'model' within the time-geographic approach for coping with the nature of motivation — this does not mean that there is no awareness of the significance of such factors in the generation of human conduct; rather intentions are seen as being elusive, and rather difficult to handle at the moment. (Hägerstrand, 1974; cited by Lenntorp, 1976). However, an activity programme does allow the basic features of intentions to be inferred from a suitably designed space-time projection as Figure 6.2 shows.

The foregoing discussion allows us to summarize the physical conditions of existence which must be considered in the development of a practical time-geography. They have been itemized by Hägerstrand (1975) and can be summarized as:

i. The indivisibility of the human being and of many other objects. It is always necessary to take into account the fact of the individual's corporeality, as van Paasen (1976) has termed it, i.e. the constraints that the particular physical limitations of the human body place upon action.

ii. The limited time-span of existence of all human and other physical entities. This condition alone is sufficient to account for many demographically based interactions among individuals, upon which all life ultimately depends.

iii. The limited ability to participate in more than one task at a time.

iv. All tasks are time demanding and commitment to a particular task diminishes the finite time resources of the individual and ultimately, of the population.

v. Movement uses time.

vi. Space has a limited capacity to accommodate events because (a) no two physical objects can occupy the same space at the same time and (b) any physical object has a limit to its outer size and so limits the number of objects that can occupy a particular space. Therefore every space has a packing capacity defined by the types of objects to be 'packed' into its area or volume. **Packing**

vii. Every physical object which has existence, has a history or biography. Most non-human objects are sufficiently defined by their past alone. Humans have the ability to plan or commit the future and are sufficiently defined only by considering the past and the future.

These 'time-geographic realities' (Pred, 1978) are *always* true regardless of any individual variations in the perception, conception and measurement of space and time. They are facts of life which spring from the nature of physical being and they are responsible for the 'local connectedness' of existence with which time-geography is concerned.

Local connectedness and time-geography

So we can now see that the web formed by the connectedness of individual trajectories (life-paths) is the outcome of collateral processes within bounded regions — 'processes which cannot unfold freely as in a laboratory but have to accommodate themselves under the pressures and opportunities which follow from their common existence in terrestrial space and time' (Hägerstrand, 1976, 332). There is still very little known about collateral processes or about the conflicts associated with them.

Collateral processes

6.2   Time-geographic notation

What are the components in an individual's movement and encounter with other individuals? First there is the factor that all individuals have goals. Success in a chosen career, physical fitness or suicide are all goals and to attain them projects must be formulated. Projects are composed of series of tasks and act as the vehicle for goal achievement. A project includes people and resources, space and time. If a project is to be completed it must be able to overcome the constraints which exist in the environment and to which we drew attention earlier. Three kinds of constraints are given explicit identification in the time-geographic design. These are:

Goals

Projects

Capability

i. *Capability constraints* limit the activities of individuals because of their biological construction and/or the tools which they can command. Some have a predominant time factor (e.g. of overwhelming importance are the needs for sleep and food), at rather strictly regular intervals. Others have a dominant space factor, limiting movement and communication.

Coupling

ii. *Coupling constraints* define 'where, when, and for how long, the individual has to join other individuals, tools, and materials in order to produce, consume, and transact and here the clock and the calendar are the supreme anti-disorder devices' (Hägerstrand, 1970, 8).

Coupling constraints largely determine the pattern of the paths which occur within an individual's daily prism. The volume of space and time which is within reach of an individual within a day is the daily prism, having not only a geographical boundary but 'time-space walls on all sides'. The principle of the prism is illustrated in Figure 6.3.

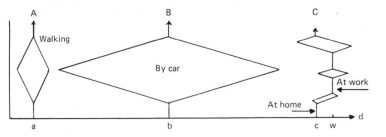

Figure 6.3. Cross-sections of prisms. (a) Walking; therefore prism sides are steep and spatial range available to the individual is narrow. (b) Motor car; therefore prism sides are gentle and spatial range available to the individual is wide. (c) Any time spent at a station reduces the range of the remaining prisms which can nest within the original. A fast journey to work (lower prism *c-w*) allows some room to manoeuvre in space-time so that a smaller prism could nestle inside the one shown. An a.m. period at work, located above *w* is followed by a lunch period prism and so on. None of the tasks which occur in these subprisms can overlap the prism boundary thus breaking the constraint rules

iii. *Authority constraints* impose limited access to either space locations or time locations. Every environmental context is replete with control areas or domains of authority. Their purpose 'seems to be to protect resources' (Hägerstrand, 1970, 9). On dynamic maps the domain of authority appears as a cylinder. The inside is accessible according to certain conditions. For instance, the authority domain of the alcohol licencing laws puts public houses 'out of bounds' to certain age groups at all times; and out of bounds to all age groups at certain times. There is a hierarchy of authority domains which range from near absolute regardless of individual attributes to subordinate domains which can be entered given social power of one sort or another. The notion of a hierarchy of domains is illustrated in Figure 6.4

Authority

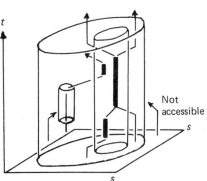

Figure 6.4. The hierarchy of domains. (Hägerstrand, 1970, 17. Reproduced by permission of the Regional Science Association)

250

Stations

Bundles

Aquarium

Everyone and everything has a history or biography, consisting of all movements made between sets of fixed space-time locations or stations.

Figure 6.5 illustrates the biographies of three individuals using a dynamic map and also introduces to the notation the idea of a space-time bundle. When the paths of two or more individuals meet together at a station for some period of time, they form a bundle. The paths, stations, and bundles, when represented in this fashion, are sometimes described as occurring in an aquarium; in Figure 6.5 this is the outer cube.

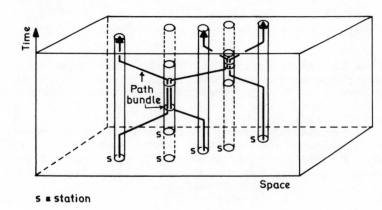

s = station

Figure 6.5. Paths, stations, and bundles

Life-spans

Meetings

Paths will have different life-spans or durations. They meet at various stations as the result of different tasks and projects. Because stations have physical extent in time and space they are represented as tubes of varying size according to the length of time for which they are in operation. The length of time a station is open may result directly from an authority constraint such as a licencing law.

The time-geographic model of society adds to our power to describe the organization of society by opening up new avenues for analysis of social processes. Description is not the object at all; the time-geographic approach is about development of deductive methods which will lead to explanation and intervention at any political level; local, regional, national, international.

6.3 Three scales of analysis

At least three quite distinct scales of analysis can be discerned in the time-geographic approach — those of the individual (upon whom the conceptual basis rests), the station (with bundle), and finally the population or activity system scale. We shall deal with each in turn.

i. *The individual scale:* at this level the simple graphical device of the dynamic map can reveal the joint interplay of path, location and constraint. In particular, it illustrates the fact that the choice of one task by an individual implies that there will be less time available to devote to alternative tasks. Such a choice may lead to a blocking effect, constraining other individuals from interacting or coupling with that individual. This can lead to a displacement or knock on effect of these blocked activities into other activities, or possibly into a queue.

The interrelatedness among individuals is well exemplified in the family. Decisions on the part of one member have quite substantial effects on the others. In Figure 6.6 these dependencies amongst members of a family are graphically represented

ii. *The station scale:* At this level, which may include segments of the paths of individuals, the method of time-geography involves mapping out the reach of each individual, that is to say 'all the events in which an individual can participate in one way or another' (Lenntorp, 1976(a), 13). These possible events (which must occur at stations) are specified in time-geographic terms by means of a prism, or activity area, the volume of space-time in which it is possible to carry out a particular activity or set of activities given a point at which the activity must start and finish. The prism is therefore, less evocatively, but more accurately, known as the potential path area (PPA), a term more often used in simulating activity programmes using the PESASP model, aspects of which are outlined below.

Reach

Potential path area
PPA

PESASP

The shape of a prism[2] is described by the individual's speed of travel and whether the station of origin (x) is the same as the station of destination (y) (Figure 6.7). If it is the same the prism is symmetrical, otherwise it will be asymmetrical. (In fact prisms are three dimensional as shown in the top right portion of the Figure, but for convenience they are usually represented in simple cross section). Given an origin and a destination several factors act as limits to the size and shape of the prism. Most notably these are:

(a) The set of constraints operating at the origin, for instance poor public transport which will limit coupling opportunities.
(b) The maximum speed at which the individual can move from origin to destination.
(c) The distance in space and the interval in time between destination and origin — in general the greater the distance over which movement is possible, the larger the prism, but the effect of different velocities must be taken into account.

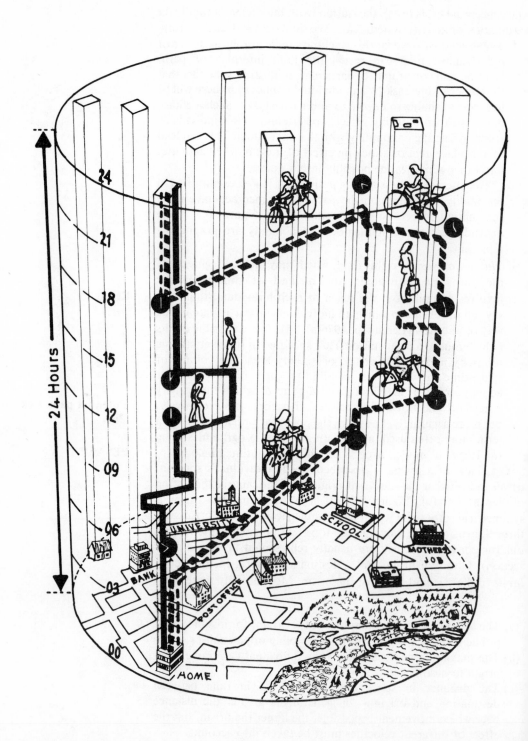

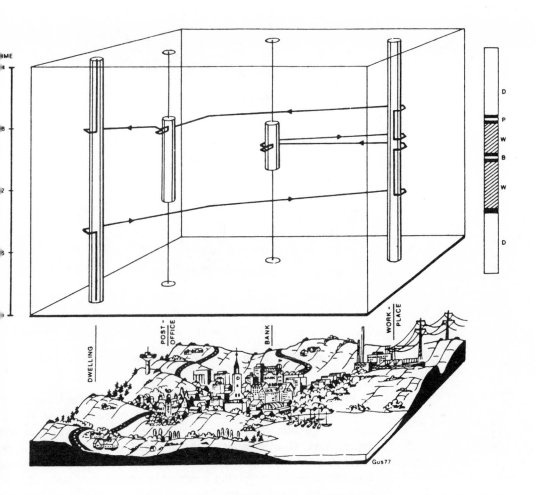

Figure 6.6. (a) Activity programmes of the individuals in a Swedish family over a single day. (After Dagens Nyheter, 1976)

Father:	0900 leaves home	Mother:	0750 leaves home
	0930 bank		0820 drops off child at school
	1000 post office		0830 work
	1015 home		1100 shopping
	1210 lunch at university		1130 returns from shops
	1300 home		1700 leaves work
			1720 picks up child
			1800 home

	Child:	0820 school
		1720 leaves school
		1800 home

(b) An individual's path in a time-space co-ordinate system. (Lenntorp, 1978, 164. Reproduced by permission of Carlstein *et al*. Edward Arnold (Publishers) Ltd)

254

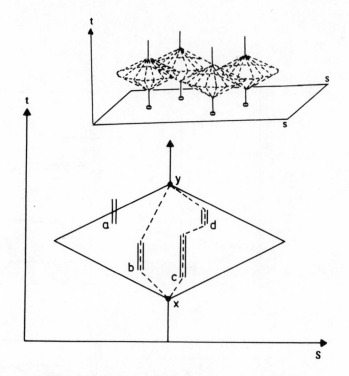

Figure 6.7. Prisms in two and three dimensions and limitations within

In Figure 6.7 the letters a, b, c, d represent tasks which have a pre-determined duration and fixed location in space and time, (i.e. a station). An individual cannot participate in all the tasks which are represented. One or more of the activities is out of reach either because it overlaps the bounds of the prism or because of coupling constraints which mean that a particular sequence of tasks cannot occur within the prism. For instance, (a) must be excluded because of its duration: you will see that it overlaps the limit of the prism. It is not possible to do both (b) and (c) because their time locations overlap. Similarly it is not possible to do (b) before (c) because the velocities required to get from (b) to (d), given the finishing time of (b) and the starting time of (d) would be higher than the maximum speeds which are set by the slope of the prism. However, (c) can precede (d) as a sequence of tasks. The activity programme which is ultimately adopted will depend on the importance attached to each task. In Figure 6.8, the prism in an individual's activity programme, with various stations involved, is illustrated in more detail. (Lenntorp, 1976(a), 34-36). So, during a day, a person can act within several prisms centred on particular stations. These prisms represent the degree of possible choice open to an individual. In Figure 6.8 the prisms and activity areas have been derived with a set of assumptions that are applicable to daily programmes. Here the activity programme, from midnight to midnight, is divided into nine time intervals during which the individual has to undertake particular activities at particular stations — in this case either home or workplace.

iii. *The population or activity system scale:* At this level the time-geographic approach is built around the concepts of time supply and time demand. In order to deal with the complex of demands which face an entire population a distinction must be made between the population system and the activity system (Figure 6.9).

Time supply and time demand

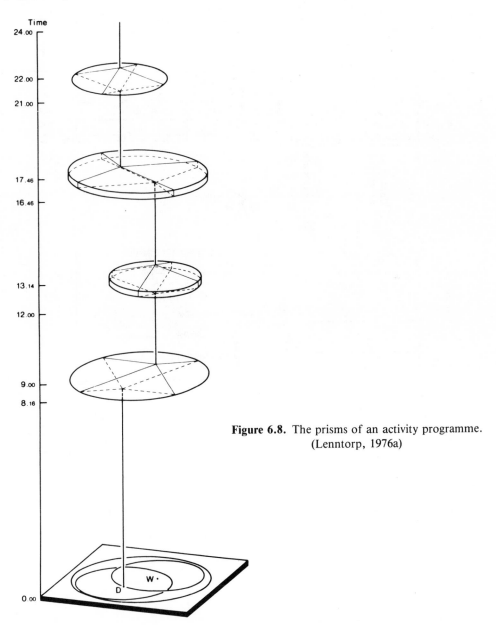

Figure 6.8. The prisms of an activity programme. (Lenntorp, 1976a)

256

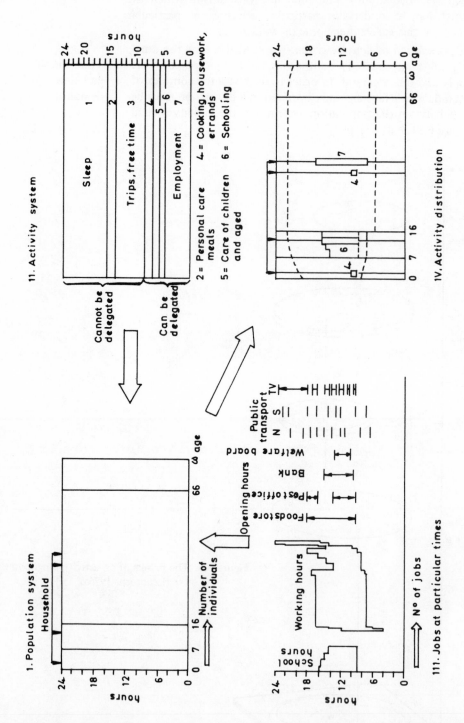

Figure 6.9. Mutual adjustments between population system and activity system. (After Hägerstrand, 1972, 147. Reproduced by permission of T. Hägerstrand)

The population system consists of all individuals in a bounded region of space-time (the aquarium). It depends for its dynamics on that particular population's characteristics, like its distribution of ages and skills (these are equivalent to Chapin's preconditioning factors) as well as other capability constraints. These characteristics are important because 'it is the human population that supplies time' (Carlstein, 1975, 6). The total daily *supply* of time for the population system will be 24 hours multiplied by the number of people. Thus a population of 10,000 has a total daily time supply of 240,000 hours. The activity system consists of all the individual activities and grouped activity bundles which are carried out and which exist within that region. Time *demand* is then the interplay between the population and activity system as it is represented in the 'multitude of human projects conducted within various groups and organizations (as) a part of the institutionalized activity system' (Carlstein, 1975, 6). The elements of the relation between time supply and time demand are shown in Figure 6.10.

Population system and activity system

Institution

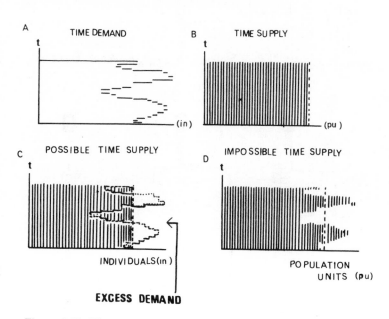

Figure 6.10. Time supply and time demand. (After Carlstein, 1975)

6.4 Examples

So far we have looked at some aspects of time-geography in terms of major principles. However, the approach is not just a theoretical exercise. It involves a large body of empirical work as well, and some examples are now presented.

Category space

(i) As a simple demonstration of the time-geographic approach in relation to the individual scale, consider the dynamic map of Figure 6.11. (Here metric space has been replaced by category or functional space and the paths or trajectories of individuals link functional categories. However, the time axis remains as a universe time axis.) The category space maps immediately reveal different profiles. Contrast, for instance, the typically simple map of the elderly couple with the complex map of the larger, younger family.

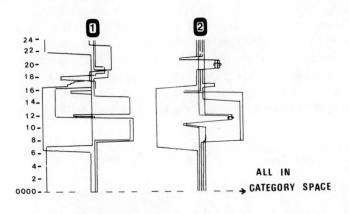

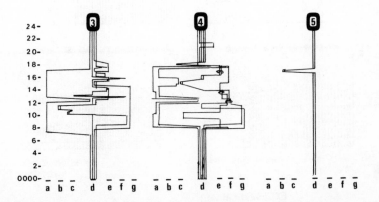

Figure 6.11. Category space as an analytical adjunct. The daily path of each member of five households. Seven categories are recognized: (a) place of work (b) place of services other than commercial (c) commercial services (d) home (e) recreational space (f) other houses (g) schools. Members of each household are considered by age with the oldest one to the left

1. Man 43, car; wife 38; boy 10; boy 8
2. Man 36, car; wife 36; boy 12; girl 10; boy 3
3. Man 44, car; wife 38; car; boy 11; boy 7
4. Man 37, car; wife 34; car; boy 9; girl 7; boy 5
5. Man 81; wife 76.

(After Mårtensson, 1975)

Figure 6.12 shows the results of another type of exercise in the use of dynamic maps. Here the object was to evaluate the impact of summer time (a one hour forward adjustment) on the paths of a farming family in New South Wales, Australia (Thomas, 1977).

ii. At the station level, prism computation apart, the main tool of analysis is simulation. Simulation is seen as a means to a sharper appreciation of the possibilities open to individuals and population aggregates and is generally preferred to inductive, sample survey techniques.

PESASP is a simulation model designed by Lenntorp to evaluate alternative biographies and is an acronym for Program Evaluating the Set of Alternative Sample Paths. It has been under development since about 1967.

Figure 6.13 illustrates a simplified flow diagram of the model. A detailed account is available in Lenntorp (1976). Here we provide only a summary. Unlike similar approaches, for instance Stephens' Monte-Carlo simulation of human activity which we introduced at the end of the previous chapter, or Orcutt *et al.'s* (1962) microsimulation approach, PESASP is a deterministic model. However, it could be easily converted into probabalistic form. It 'was originally designed for use in ... theoretical contexts [and] was principally intended for the analysis of possible combinations and permutations of activities in time-space and as a means of recording differences of another magnitude than many practical investigations require' (Lenntorp, 1976(b), 46). However, it has been used recently in a number of very practical situations relating to use of public transport in Swedish towns and for evaluating the possibilities of access to medical care, child care, etc.

The input to PESASP is in three parts:

(i) Either one or a series of daily activity programmes of individuals.

The space-time environment

(ii) The stations at which activities take place, defined in space by geographical coordinates and in time by length of availability.

(iii) The transport system of the region, characterized as lines or nodes.

In effect the input is a hypothetical individual project, specified in detail as an activity programme. It must include all relevant limits like the earliest possible starting time and the latest permitted arrival time, the duration of opening hours, and the velocity of movement. It must also include the likely permutation of tasks within a project and their degree of elasticity. With this sort of input data specified the simulation of an individual day path can begin.

The output is the completed space-time budget of the individual. It indicates the space-time allocations which would have to be made in order to achieve the project goal. The simulation therefore

PESASP output

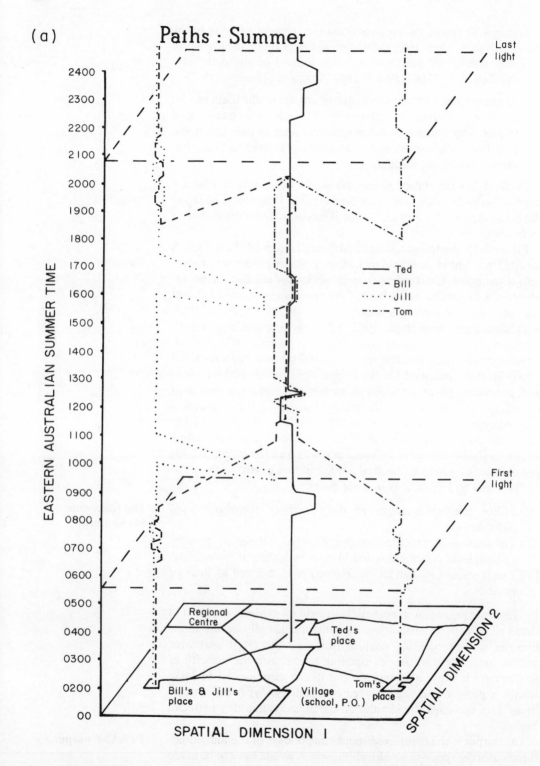

Figure 6.12 (a-d). A comparison of four daily paths, NSW, Australia. (Thomas, 1977)

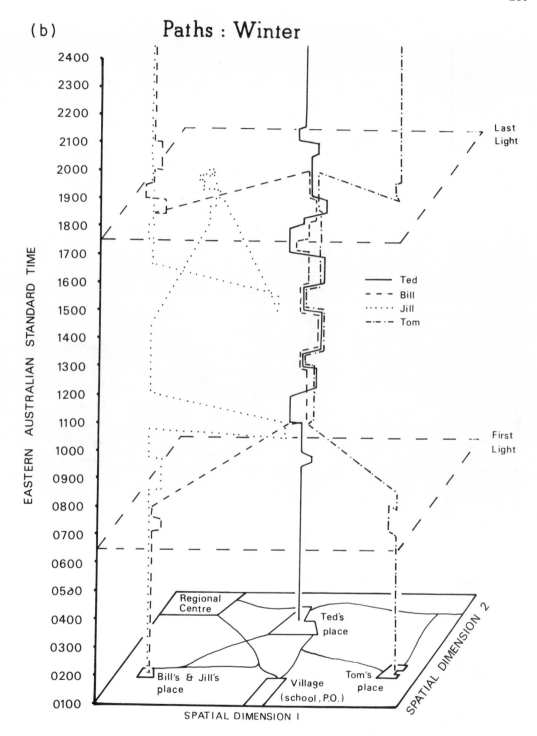

(b) Paths : Winter

(c)

Category Space: Summer

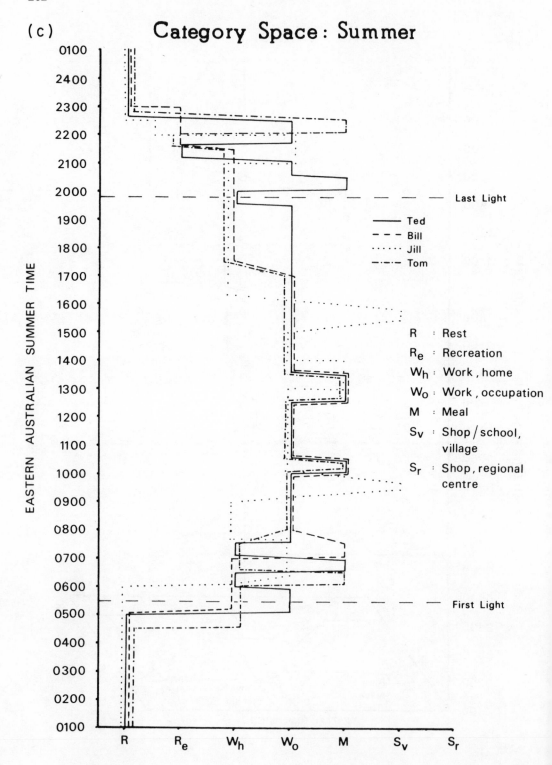

EASTERN AUSTRALIAN SUMMER TIME

Last Light

——— Ted
‐ ‐ ‐ Bill
······ Jill
‐·‐·‐ Tom

First Light

R   : Rest
R_e : Recreation
W_h : Work, home
W_o : Work, occupation
M   : Meal
S_v : Shop / school, village
S_r : Shop, regional centre

R R_e W_h W_o M S_v S_r

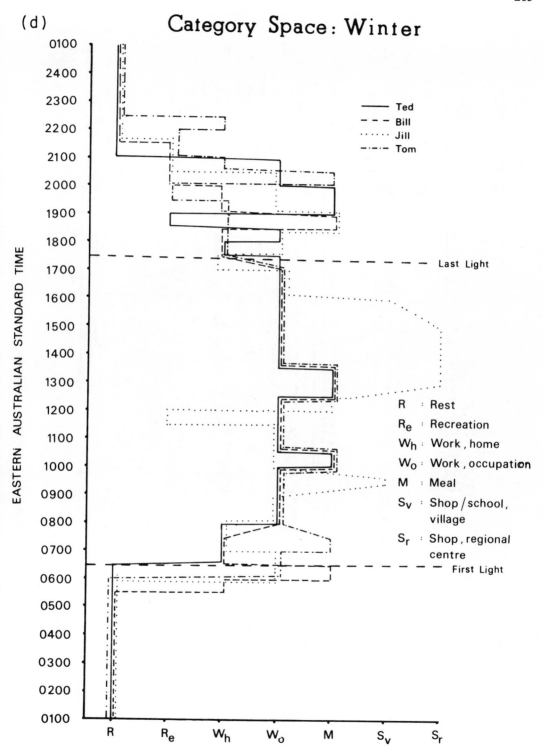

(d)

Category Space: Winter

EASTERN AUSTRALIAN STANDARD TIME

Ted
Bill
Jill
Tom

Last Light

First Light

R : Rest
R_e : Recreation
W_h : Work, home
W_o : Work, occupation
M : Meal
S_v : Shop/school, village
S_r : Shop, regional centre

R   R_e   W_h   W_o   M   S_v   S_r

264

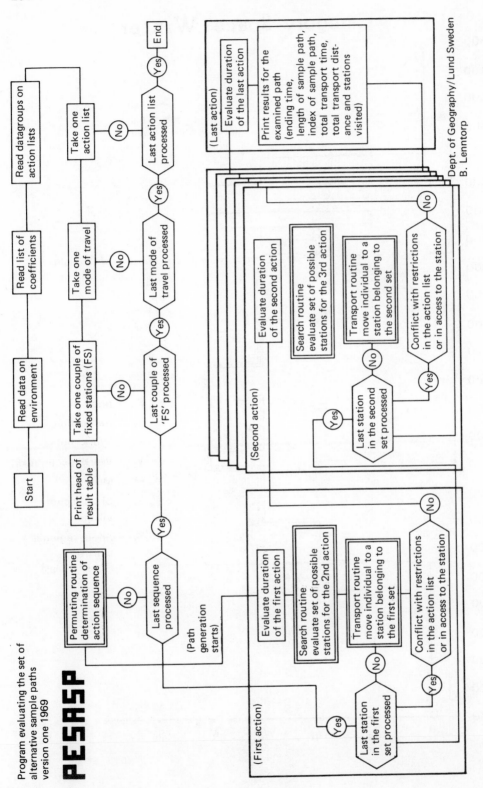

Figure 6.13. PESASP: flow diagram. (Lenntorp, 1969)

attempts to provide answers to the question, 'Which activities would it really be possible to carry out under different conditions and which are impossibilities in a given environment for a give time period?' (Mårtensson, 1975).

The output from a PESASP simulation will, of course, vary according to the objectives involved which, in turn, determine the inputs. The constraints, the velocities and the number of different activities which go to make up a daily activity programme each influence the detail of the output. For example, in one study reported by Lenntorp (1976(a), 111), the data on the environment, in which 230 activity programmes were to be simulated, amounted to 14,000 items on stations, transport networks and other elements and produced 3,500 pages of computer output! Whatever the details the output indicates the number of ways in which a given individual's activity programmes can be carried out, under specified conditions, and indicates the flexibility of the space-time environment. In particular, by identifying impossible tasks, PESASP provides planners with an invaluable means to increase choice in their particular space-time environment by suggesting appropriate strategies of intervention.

Impossible tasks

Consider the following example of the use of PESASP. Lenntorp (1976(b)) studied the travel possibilities of public transport passengers in the Swedish city of Karlstad which in 1973 had a population of 75,000 people and occupied an area of about 35 km². In considering the existing eight route bus service, Lenntorp used PESASP to ask — 'How would alterations of timetables, a re-designed bus route network and the addition of certain social facilities affect the activity programmes of individual inhabitants who have particular recurrent tasks to perform? Would the changes make certain tasks impossible? Which tasks were now impossible?' The underlying rationale Lenntorp took was to simulate these changes for the proposed environment. He then evaluated the similarities and differences. The evaluation involved various theoretical and political assumptions, and therefore no final solutions were looked for.

PESASP and Karlstad

Here is one example of the use of PESASP in Karlstad.

The postulated activity programme: an individual leaves home 40 minutes before work commences. Five minutes of this time involve leaving a child at a day nursery. During the journey home after work the child is collected. The same journey times are involved in both cases. Walking and public transport are the chosen modes of travel. *The space-time environment:* Figure 6.14(a) shows the six major work places in the city and the distribution of existing day nurseries. Two proposed nurseries are shown as (**X**) in the top right of the map. The shaded areas represent rivers and other 'non-urban' space. There are 62 residential test points, marked as dots centred on each 500 metre grid cell they are closest to.

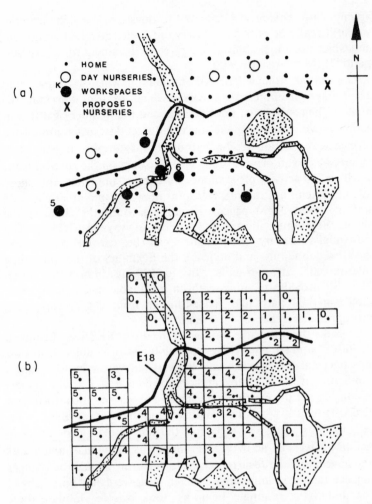

Figure 6.14. (a) Karlstad: location of major workplaces and day nurseries. (b) Karlstad: number of major workplaces accessible from each test point. (After Lenntorp, 1976(b))

The possibilities of performing the intended activity programme were assessed for each nursery. In Figure 6.14(b) the number of workplaces accessible from each test point given the constraint that a nursery must be visited, is shown. (The maximum number of possible work places accessible from any one test point is of course six).

Now if the existing environment is altered, how do the possibilities change? A change in the frequency of bus services is made. Four departures are now made hourly, instead of three. The graphs (in Figure 6.15) show the number of test points from which the activity programme can be completed, in relation to each workplace. The visit to the nursery is still a constraint. The upper graph (a) shows the situation before the frequency of bus services is increased, the lower graph (b) shows the situation following the

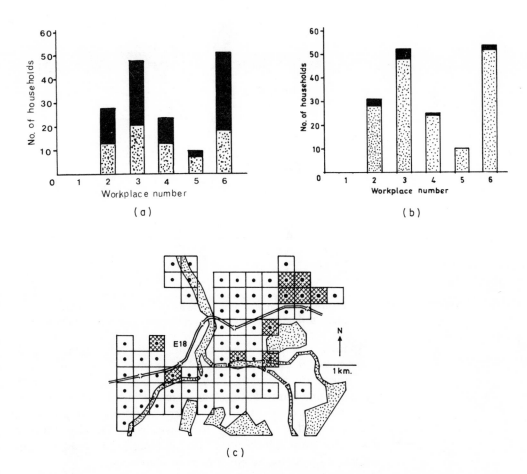

Figure 6.15. A PESASP example, Karlstad. (a) Test points and workplaces before the increase in service frequency. (b) Test points and workplaces after the increase in service frequency. (c) Test points accessible to one additional workplace after the increase in service frequency. (After Lenntorp, 1976(b))

increased bus service frequency. In Figure 6.15(c) residential test points now accessible to one additional workplace as the result of the new environment, are shown as hatched cells.

What happens to the possible individual activity programmes if a new bus route is now added to the existing network? A new transverse bus route linking the northeastern areas of Karlstad with the western parts has been added going from the residential area in the northeast of the city through the northern areas, crossing the river on the bridge carrying the European Highway (E 18) and then following the motorway westwards. The new route is shown as a pecked path in space. The amount of improvement the new route offers is not great but when the frequency increase figures are considered three additional cells now have access to one more workplace (Figure 6.16 shows this situation).

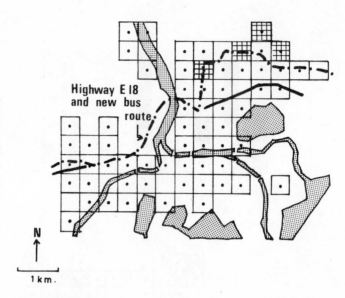

Figure 6.16. Change in possible activity programme following the introduction of a new bus route. (After Lenntorp, 1976(b))

Another simulation experiment, with the same activity programme, involved increasing the number of day nurseries by two (these are shown in Figure 6.14(a) by two crosses in the northeast quadrant of Karlstad). Located in the suburban areas, these new nurseries can attract custom from nine test points, but this is only the case if the workplace is in the city centre at workpoint A6 (Figure 6.14(a)). For all other residential test points and work locations the constraints are too strong.

Many other alternative strategies could have been tested in Karlstad using PESASP. These would include adjusting the existing possible travel times, increasing velocity, changing mode-split relations, altering work or nursery hours and so on. The range of possibilities is as immense as the range of policies.

Potential activity programmes

PESASP has been 'constructed primarily to analyse potential combinations and permutations' (Lenntorp, 1976a, 138) among potential (possible) activity programmes given certain constraints. As a 'method for handling individual-paths operationally' it can be used to supplement or even replace the conventional transport studies based on surveys and macro-scale models. It is also clearly of value in the field of physical planning in general. The approach adds a time dimension to a planning system too often obsessed with spatial variability. 'Much of the complex analysis of behaviour can be set aside if we can show that the planned environment is flexible (' ... it would be much more fruitful for the analyst to give up his efforts to predict behaviour directly ... and instead focus his attention on how to find out in what ways limits to freedom of action come about' (Hägerstrand,1975, 3) (Lenntorp,1976a, 137). This is the point where conventional inductive survey methods fall short; they do not inform us about non-behaviour, or about impossible action.

The social inequalities between men and women have provided Palm and Pred (1974) with an opportunity to apply some of the Lund notation to contemporary issues. Their study reveals how much information of economic, political, and ideological significance can be kneaded out of contemporary social conditions by adopting the time geographic approach. Women, especially unmarried working mothers with children, are the subject of numerous constraints in their choice of activity programmes. If they are also poor then the constraints on movement tighten. Consider the case of 'Jane', an unmarried mother of a two-year-old child, living at A. Jane's activity prism and dynamic map is traced out in Figure 6.17.

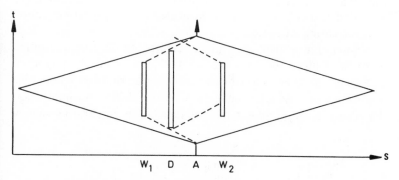

Figure 6.17. Prism for an unmarried working mother. (Palm and Pred, 1974)

As can be seen Jane is unable to leave home for work before a certain time because of her child's dependence on her for feeding, dressing and other personal needs, and also because the only accessible nursery is not yet open. *Geographically* it does not as yet exist. Jane is therefore faced with severe coupling and authority constraints, which determine the stations she can reach and the paths she can take. Since Jane is also trapped within the authority domain of the social role of a mother of small children, her choice **Domains** of workplace is limited because she has to collect the child from nursery school in the afternoon before the nursery closes. What happens when Jane is offered the choice of two jobs (at W1 and W2 in Figure 6.17)? Both involve the same number of working hours and are within the boundaries of her prism. The job at W2 is preferred because it takes into account her qualifications, pays a higher wage and seems to offer more chance of promotion. But because of the workplace's location in space relative to the nursery D (which would mean arriving late to work after dropping off the child at the nursery, and arriving at the nursery in the afternoon after it had closed) this option is denied her. She is therefore forced to take the job at W1 in spite of her qualifications and the fact that relative to her home W2 is suitably accessible. And the job at W1 has its own inconveniences; for instance, shopping cannot be fitted in unless the lunch break is taken up.

Intended projects

We can now see that the space-time environment discriminates against this individual because it does not allow her to realize intended projects, from the day-to-day problem of when to shop to the lifetime problem of building a career. Many surveys have shown that women seem to take positions of lower status and responsibility than their abilities would suggest are open to them, above all because of their role as mothers (Tivers,1977). The time-geographic approach pinpoints the space-time environment as one of the major culprits in generating the problem. In turn many other problems stem from the frustration experienced in trying to reconcile the role of mother with the role of wage-earner. Child care may be threatened, work efficiency may be reduced and there is a hidden social cost which is measured out in opportunity costs and increasing alienation.

Similar time allocation conflicts created by the role of women have been identified by Carlstein (1975) in a different type of economy, this time following the introduction of a new school into the village of Tepoztlán in Mexico, in the 1950's. An innovation like this is a major source of time demand in an agrarian society

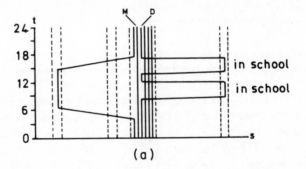

(a)

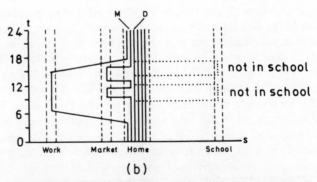

(b)

Figure 6.18. The day paths of a Tepoztlán family and the impact of a new school. (After Carlstein, 1975)

(Carlstein,1978). In Figure 6.18(a) the day paths of one Tepoztlán family are displayed as a simple two-dimensional dynamic map. There are six members of the family, each one represented by a single and continuous line (the dotted vertical lines define the boundaries of the four relevant stations). From left to right the individual members of the family are the father who works land some two-and-a-half hours away from home, the mother, a child of school age and three children yet to reach school age.

Figure 6.17(b) shows the dynamic map of typical day paths before the introduction of the new school. In this situation it was possible for the mother to leave home and go to the market to buy and sell, leaving the school age child in charge of the children. The pecked line shows that this is precisely the time when the child would now be at school. There is, therefore, an obvious conflict between traditional roles and the 'modernizing' roles induced by the time-demanding innovation. If the family had had any sons then the conflict would have been even more extreme as they would have been working in the fields at the time now to be allocated to school.

The traditional education system in Tepoztlán was well integrated into the 'ordinary life of individuals and households' (Carlstein,1975, M3), it was spatially and temporally flexible being usually conducted 'at home' and not precisely fixed by invariant clock and calendar time. The result was that the 'knowledge and skills transferred were locally accessible and locally applied' (Carlstein, 1975, M5). But with the introduction of the new school time allocation conflicts were introduced, seriously affecting the daily round of activity in Tepoztlán. In this situation, using a time-geographic analysis, it becomes obvious that the western concept of a school as a spatially and temporally fixed place was inappropriate to local conditions. Inevitably high rates of truancy resulted. More thought should have been given to a 'school' modelled on the traditional education system.

As a final example of the station-level of analysis consider the use of the time-geographic approach in Newcastle, Australia. Here the approach was used to study the day paths of ten employees of a newly established Community Health Care Centre (Mitchell 1975). The centre is located about 20 kilometres from Newcastle's CBD in a suburban location in an area of generally low socio-economic status and high family status (Parkes, 1971, 1973, and 1975). At the time of the study the Centre had only been in operation for four weeks and the full complement of staff had not been appointed. The Centre faced a range of problems from the outset. One problem that became apparent was that the number of employed persons, male and female, who lived within six kilometres of the Health Centre and who were also on shift work was considerably higher than the city average. Therefore, it soon became obvious to the community nurses that many people in the catchment area of the Centre were in need of aid before 0700 and after 1700. For instance, for family group therapy sessions the weekend became the only time that appointments could be made for the whole family. But as the population at large also expected normal hours to be available to them it soon became clear that staffing requirements needed to be increased in order to accommodate abnormal activity programmes.

Coupling constraints

Packing

It also became apparent as a result of the time-geographic study that it was with coupling constraints that the Health Centre staff were to have their greatest problems. These coupling problems were exacerbated by a packing problem which was evident as an escalation in population time demand towards the end of the week, particularly on Fridays, when 'things just seem to flare up'. Many of these difficulties were associated with the needs of children and in particular with delinquency, truancy and general misbehaviour which had become 'too much ...' — the Youth and Community Service Officer 'can sort them out.' This sort of packing problem not only put pressure onto other services but was also a part of the pressures which seem to be put on working mothers. By the end of the week they are 'at the end of their tethers.' (See also Palm and Pred above and Mårtensson, 1975, 1978).

The use of the time-geographic perspective in considering this range of problems involved ten individuals. Their day paths were plotted from diary records kept over a four day period; two consecutive days in late July and two consecutive days in August, three weeks later. The use of transparent overlays allowed extremely accurate comparison of paths from individual to individual, and from day to day, as well as allowing the evaluation of sets of individual paths when the combination of individuals was changed. Such comparisons allowed a rapid assessment of coupling and other constraints on the daily activity programmes of employees and proved especially informative in plotting day paths in advance from a study of appointments. The anticipated prisms of the nurses and other employees could also be plotted. In this way dynamic maps of the individual programmes in the future were able to be adjusted to allow better synchronization and synchorization.

Synchronization and synchorization

After a period of time when *actual* day paths were plotted and compared with the *anticipated* day paths, it was possible to come to a heightened awareness of the operation of constraints and limitations to the realisation of the Centre's expectations. Adjustments were then possible, for instance by altering work schedules and improving accessibilities by more rational allocation of vehicles, telephones, and operating hours.

(iii) At the activity and population systems level the time supply and time demand of a population has been considered in a time-geographic framework by Ellegård, Hägerstrand and Lenntorp (1975, 1978) as a way of assessing alternative futures for Sweden.

How, for instance, does the contemporary context of time supply and time demand differ from a simulation of the context of time supply and time demand in the year 2000; one in which there has been a marked change in favour of capital-intensive industry, shorter working hours, and an expansion of the further education system?

The model adopted to investigate this 'future' population context had six stages (Figure 6.19). In the first stage (S1), an appropriate region and time interval was chosen, in this case a day. In stage two (S2) a forecast of population time supply (the number of people forecast to live in the area multiplied by 24 hours) and time demand was calculated. In stage three (S3) the total time available was divided between activities classified into three categories:

(i) Time devoted to production of goods
(ii) Time devoted to consuming services
(iii) Time devoted to vital human activities (e.g. sleep)

Coincidentally the population was divided into three categories which corresponded with the division of activities outlined above. Each population category was allotted a quantity of time to devote to vital needs. In line with projections of manpower one group within the population was assigned the role of goods production, another service consumption and the third a mix of production and consumption activities. In stage four (S4) the activities were distributed over time and within the population, so that the actual time demand and time supply of the population could be evaluated. In stage five (S5) this distribution was related to the temporal and spatial organization of collective activities and the age structure, household composition, and transport opportunities of the population. Finally, in stage six (S6), the mix of activities in space-time given by (S5) made it possible to estimate the journeys generated by the need to move between these activities. In addition a number of additional sub-models were used at each stage; (for a full account reference must be made to Ellegård, Hägerstrand, and Lenntorp, 1975). The model was run using a number of 'alternative formulations in an attempt to capture general tendencies of development within a number of sectors of society ... and so ... provide the background to more detailed future pictures of the year 2000' (Ellegård, Hägerstrand, and Lenntorp, 1975, 12). Changes in values in society as well as trends in population development in decentralization or centralization of population and the development of a self-service economy, all supported by new, sophisticated techniques of production, must lead by the year 2000 to new activity organizations. These new organizations can be determined in a useful way by the Ellegård-Hägerstrand-Lenntorp model.

274

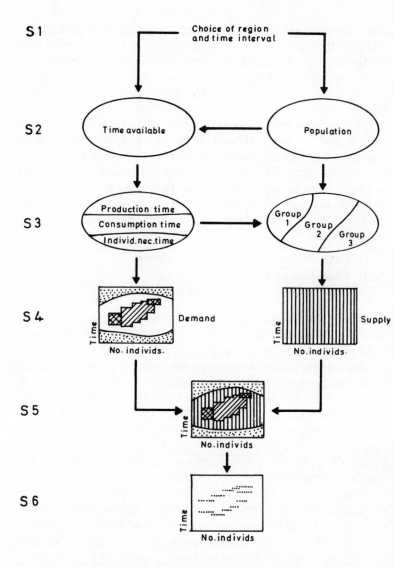

Figure 6.19. Population time supply and population demand. (Ellegård, Hägerstrand, and Lenntorp, 1975, 1978)

6.5 Concluding comments on the method of time-geography

The time-geographic approach has now been introduced. To date most of the references to this scheme have been concerned with either introducing the approach or with reporting applications of its general principles. Few authors seem to have offered a critical appreciation of time-geography. The exceptions include Van Paasen (1976), Pred (1977, 1978), Thrift (1977a), Gregory (1978), and Rose (1977). In this last section we shall take one of these critiques (Rose, 1977) and use it in two ways. One is as a basis for answering some of the more general points often made in connection with time-geography. The second is to draw attention to some of the misconceptions that are constantly in evidence concerning time-geographic method.

Reflecting 'on the notion of time incorporated in Hägerstrand's time-geographic model of society', Rose (1977) recognizes that the approach 'represents a break from the decidedly spatial traditions with which geography has been previously concerned'. He wonders whether geographers will ever feel sufficiently comfortable with the notion to adopt time as 'one of their suitable objects of study'. Rose makes a number of interesting observations but we have isolated six points, in particular, that need to be discussed.

Social time and time geography

The first point Rose makes is that Hägerstrand and his colleagues are not alone 'out there in space-time'. This may indeed be beyond dispute, but there then follows a widely held and rather fundamental misinterpretation of why this is so. This is that the time geographic approach has as its space-time companion the work of the multinational time budget study group and time budget study in general. In terms of method this is just not correct. In time-geography the time budget is an *output* of the method: for time budgets and space-time budgets it is an *input*. Similarly time-geography concentrates on *allocation* of time and space whilst time budgets and space-time budgets have, until recently, concentrated on *use* of time and space. These are most important differences to keep in mind.

Secondly, whilst accepting the demonstrated efficacy of the notion of physical (clock) time used in time-geography, especially in relation to movement, Rose argues that there remains a need to assess the way in which 'lived time' and 'experience of time' relate to the 'physical' time adopted in the time-geographic scheme: the universe time of our earlier discussions. Rose puts it like this — 'in brief, the question is, will the time-geographic model provide us with a credible isomorph of human experience of time?' (Rose, 1977, 43). The short answer to his question is 'No, not yet'. This is because time-geography was never explicitly designed to be isomorphic with 'lived time' in the sense of experiential psychological time. Which is not, of course, to say that it cannot be. (Buttimer and Hägerstrand, forthcoming).

A third and associated proposition made by Rose (and others) is that determinants of behaviour like attitudes, motives, and choices create problems for the 'physicalist' approach of time-geography. This is not necessarily true. As yet there is little evidence that attitudes and motives can be explicitly linked to behaviour. Quite often people think and say one thing but do something else because they have no choice. Attitudes, motives and so on are only important when choice exists. At present in, for instance, transportation modelling, a major research effort is aimed at mapping out areas of human life where choice exists and where, therefore, standard attitude-behaviour-choice models are relevant (Burnett and Thrift, 1979). But then again, time-geography does not deny a role for intention and experience (see back to Figure 6.2), rather it sees them as sufficiently elusive to be rather difficult to handle at the moment (Hägerstrand, 1974).

The fourth issue relates to Rose's comments about the familiar space-time and substance language problem in science (Harvey, 1969). The implication seems to be that the time-geographic scheme is naive in its conception of these language difficulties. For instance, it is stated that the antecedent of the language used by Hägerstrand is 'Minkowskian space-time language'. But this is true only to a point — the space-time of the Lund scheme is much more closely associated with Newtonian space and time, treated together so that all things occur, quite explicitly and always, in space and time. The elimination of the height dimension of space, in the time-geographic scheme also gives Rose some concern. In fact it is not eliminated in the model at all, only in the diagrams of dynamic maps, and there only for obvious practical, cartographic purposes. And time is most certainly not substituted for the height dimension. This would be absurd (Hägerstrand,1974, 1975, 1976).

A fifth concern about the Lund approach, and one which is often heard in human geography, is that there is an 'overemphasis of the time dimension, sometimes at the expense of the spatial dimensions'. This is a fundamental misconception because the whole rationale of the time-geographic approach depends on the inseperability of space and time. They are one. There is no reification of *time* in time-geography. Connected with this argument (and in the same vein) Rose also takes up the issue of the packing problem, 'although there is a mention of certain "packing" problems in the model ... emphasis is clearly laid on the temporal congestions.' In fact, the time-geographic literature is replete with the notion of *space*-time packing for 'Even the simplest operation, ... consumes some space and some time' (Hägerstrand,1974, 6).

Finally, and in a more positive vein, Rose's suggestions about 'new types of time for the Hägerstrand model' are provocative. Gioscia's (1972) notions (refer back to Chapter 2), are interpreted in terms of synchronization, vectoriality, acceleration, time compression and finitude and are proposed as a basis for an experiential model of time. Each term is the foundation for a co-ordinate system of lived time. The sense of change which is proposed as the basis for our sense of time arises not through motion but from the motion of a 'thing' with respect to its own framework. Acceleration and deceleration and the forces which bring 'it' about are seen to be at the root of time sense. However, Rose is unable to suggest 'how these properties of socially experienced time should be operationalized within the Hägerstrand model' and this is just where the time-geographic scheme has its most distinct advantage because it is to some degree predicated on this observation, as we have shown above! It is *intentionally simple and physicalist,* (it has also been operationalized). And, as exemplified in Rose's paper, much of the criticism of time-geography can therefore been seen to be misplaced. It either arises from misconceptions about what time-geography is or from expectations that time-geography is a nicely worked out and completed theoretical schema, which explains *all* aspects of the social use and allocation of time-space. About the first type of criticism time-geography can do little; about the second it continues to do much. Time-geography is a developing subject, its aim is a complete and unified 'social ecology' (Hägerstrand,1974).

In this introductory chapter we have only considered the bare bones of time-geography. And we have restricted the discussion to topics relating rather directly to human activity. We have not touched on a multitude of subjects to which the time-geographic approach has been applied, yet these are as diverse as the study of social science and include industrial location theory (Erlandsson, 1975), alienation (Pred, 1977, 1978), manpower studies (Pred, 1978), technological impacts (Pred, 1978), historical demography (Hägerstrand, 1978) and social structure in general (Wallin, 1974). Nor have we considered some of the new and useful theoretical advances coming from Lund and elsewhere, for instance Hägerstrand's ideas on niches (Hägerstrand,1974), Buttimer's ideas on humanistic orientations like home and reach (Buttimer, 1978), Carlstein's (1978) ideas on packing or Wallin's (1977) ideas on the use of future time through 'formation capital'. These await a compendious volume of their own.

'The time-geographic framework ... holds the promise of identifying new questions of social and scholarly significance. It may also satisfy the longing shared by many for a more humanistically oriented geographic approach to some of the more complex and frustrating problems of modern society. In addition, in so far as time-geography has something to say about the deep structure of all interactions and transactions between man and man, man and elements of the natural environment and man and man-made objects, it has the potential for enabling geographers to attain a new level of intellectual maturity.'

(Pred, 1977, 207).

Notes

1. The determination of these paths and prisms (see below in text) is outlined in greater detail in Appendix A.
2. The determination of prisms will be considered in Appendix A.

7
Space-time signatures on the land: Diffusion, convergence, and periodicity

'The task of coping with spatial pattern (in human geography) has in most cases precluded true concern with process, or *vice versa*. We have long remained constrained by the old high school schema which saw history and geography as two orthogonal planes intersecting only in a Huxleyesque timeless present. Certainly the notion that space itself changes dramatically through time has only recently received sufficient attention and the notion of producing dynamic models in which space and time are both central and interlinked is in its infancy.'

(Forer, 1978b, 232)

7.1 Preliminaries

With movement, space and time become coincident, as space-time. The study of movement is distinct from the study of spatial distribution. It takes place in the domain of space (Chapter 1) as a relative linkage of spatially and temporally distributed items. In this chapter we study three categories of movement from human geography. The first is diffusion, the study of which was initiated by Hägerstrand. It is no accident that his interest in the theory of diffusion has more recently developed into the time-geographic approach which we discussed in the previous chapter. We shall consider diffusion through examples without entering into a detailed presentation of the 'theory' of diffusion; which is well covered in recent texts such as those by Abler, Adams, and Gould (1971) and Haggett, Cliff, and Frey (1977).

Convergence and divergence are in the second category of movement which we shall study, in terms of space-time metrics.

Movement and space-time

Diffusion

Convergence-divergence

Once again, by the use of relevant examples, the idea of convergence-divergence should become clear. The third category of movement we shall study is associated with periodicity, in particular as it is manifested in periodic markets.

Periodicity

With diffusion, convergence-divergence and periodicity the place of time in human geographic studies is also ensured at larger scales of space and/or time. Once again time is shown to be an object of human geographic study co-equal with space.

7.2 Diffusion

Items

To diffuse means to spread out. It refers to a process whereby items become relatively less concentrated over space but relatively more concentrated over time. Interestingly, Abler, Adams, and Gould (1971) index diffusion as *spatial* diffusion. This is an incorrect description of the process of diffusion. Taking up the argument introduced in Chapter 1, where geography was distinguished from geometry because in geography time is unavoidably coupled with space, geographic diffusion becomes a space-time process.

Diffusion processes are usually considered within a distance/universe time framework but just as there are alternative notions of distance to the normal Euclidean notion (for instance, social distance) so too the time component in space-time or distance-time could be transformed to incorporate one or other of the paratimes discussed earlier. Although we are unable to make much progress with such a notion of time because of the existing state of the art of time metrication, there is no fundamental conceptual difficulty. The spread of items (especially of ideas, conventions and values) in para-space-time presents a challenge to human geographic research.

In the so-called 'Panic of 1837' banks in the United States of America stopped converting bank notes into specie (i.e. into gold and silver) but nevertheless remained in operation. The spatial characteristics of the spread or diffusion of this event across the USA have been studied geographically by Pred (1973, 1977). Newspapers of the time reported public resolutions in favour of the suspension of conversion to specie and this provided the data base for the mapping of the diffusion of this new banking process. Difficulty is often associated with identifying the source of an innovation, in both spatial and temporal terms. Further difficulty often lies in identifying the end of the process; the state in which there is no further spread. The Panic of 1837 had the advantage of a clearly identifiable space-time source and a relatively short diffusion duration; in effect about one month, May 1837.

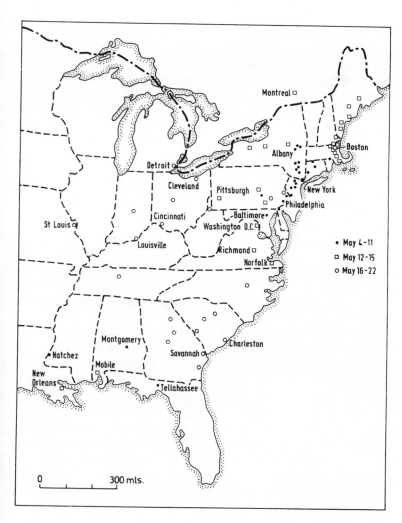

Figure 7.1. The pattern of diffusion for the Panic of 1837. 1. Suspension dates May 4-May 12. 2. Suspension dates May 12-May 15. 3. Suspension dates May 16-May 22. (After Pred, 1977. Reproduced by permission of Hutchinson Publishing Group Limited)

At the time of the Panic no media existed to enable near instantaneous transmission of information, as now with telephone, telex, and radio, and so the primary channel of information transmission was word of mouth and, subsequently, newspaper. Figure 7.1 captures some of the movements that took place (Pred, 1977). The practice of suspending specie payment seems to have first occurred in Natchez (Mississippi) and two days later in Tallahassee (Florida). However, the practice occurred on the same day in Montgomery (Alabama) as in New York. The diffusion was complex in its geometry. But an additionally interesting feature is the compressed time interval over which the new practice spread across a sub-continent, at a time when the movement of information was slow and hazardous.

Another example of the diffusion process has been studied by Robson (1973) in England. In this case the time period involved was twenty years rather than the thirty days or so required for the spread of specie suspension in Pred's study. Until the introduction of gas-light, the night hours were a severe limitation to human activity. With the coming of coal gas lighting, as represented in the spread of gasworks, a colonization occurred which was almost as dramatic as the better known colonization of the New Worlds — the colonization of night time (this is a theme we take up in more detail in Chapter 8). It appears that the first gas lights appeared in Pall Mall, London, in 1820 and over the next twenty years or so they spread to all the major cities of Britain. Gas lighting was to prove an important factor in keeping the new industrial plant operational, not only at *night* but also during the short 'winter days'. Identification of the diffusion of these gasworks is an important aspect of the geography of the early industrial revolution and of the move to *urban* lifestyles as the constraints imposed by natural light-dark and climatic cycles were overcome.

A third and final example is provided by Huff and Lutz (1974). They have documented the spread of political unrest, between 1960 and 1972, in Northern Africa. In Figure 7.3 the changing spatial pattern of coups d'etat or attempted coups d'etat is shown by means of graphs, with connections drawn only if a coup is spatially adjacent to the countries and to a coup in the previous time period. (You might also wish to refer to the original maps drawn in conventional choropleth by Huff and Lutz. The graph representation is our simple alternative; it seems to carry more space-time information.) The spread or diffusion of coups d'etat in this example can be related to three factors:

(i) Spatial proximity to countries which have experienced coups.
(ii) The centrality of a nation within the defined region, with the implication that the greater the centrality, the greater the probability of a coup.
(iii) The level of urbanness of the nation.

In Figure 7.3 the initializing letter on some of the vertices is simply the first letter of the name of the nation or nations which are added to the graph because they are recorded as having experienced their first coup or have experienced a coup after a period of stability. The directed lines simply indicate space-time succession and do not mean there is a causal connection. The thick pecked lines indicate a possible barrier to the spread of coups. By reference to the key map in the centre of the 'clock' you will see that the lines run through nations which did not experience a coup during the period 1960-1972; or at least for which no coup is recorded.

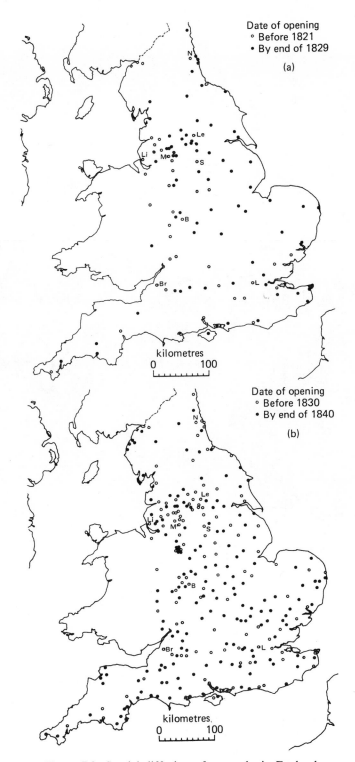

Date of opening
○ Before 1821
● By end of 1829

(a)

kilometres
0   100

Date of opening
○ Before 1830
● By end of 1840

(b)

kilometres
0   100

Figure 7.2. Spatial diffusion of gasworks in England.
(a) 1820-1829. (b) 1829-1840. (After Robson, 1973.
Reproduced by permission of Methuen & Co. Ltd)

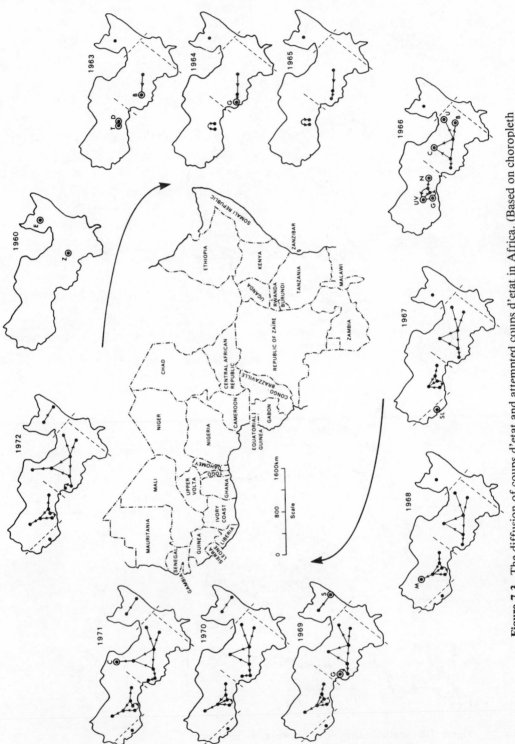

Figure 7.3. The diffusion of coups d'etat and attempted coups d'etat in Africa. (Based on choropleth maps by Huff and Lutz, 1974)

Many other examples of diffusion could be given, ranging from the spread of agricultural innovations through to the spread of cable television; from the diffusion of highway improvements to the diffusion of Rotary Clubs; from the spread of the Russian agrarian riots through to the spread of the 'Captain Swing' riots in England at the turn of the nineteenth century, from the spread of cholera or measles to that of divorce reform, and so on. Can such a diverse range of examples have anything in common? The answer is 'yes'. At least three kinds of diffusion can be identified.

Consider again the diffusion of the Panic of 1837 and the diffusion of gasworks, both briefly illustrated above. The common feature here, in spite of the dramatically different spatial and temporal scales, is the demonstration of hierarchical diffusion. **Hierarchical diffusion** Large places, important people or powerful institutions tend to get word of new information or innovations first and then transmit this information, usually selectively, down the hierarchy of places, people, or institutions. Such a process occurs because many innovations (as information at least), diffuse initially in relative space (see Chapter 1) rather than absolute geometric space.

Once an innovation has reached a particular location (in territorial or socio-economic space) the likelihood of a localized neighbourhood effect develops, although barriers can interrupt this movement. The neighbourhood effect is more usually associated with the spread of information at the individual scale as in the classic studies of adoption of agricultural schemes by Hägerstrand. 'Localized' diffusion through the neighbourhood effect is usually **Neighbourhood effects**

Barriers

Expansion diffusion

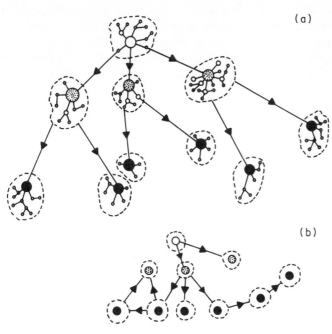

(a)

(b)

Figure 7.4. Hierarchical and neighbourhood diffusion effects

Contagious diffusion

known as expansion diffusion, except when the spread of a disease is being studied, as in epidemiology, in which case this localised diffusion is known as contagious diffusion (e.g. Cliff *et al.,* 1975). Figure 7.4 illustrates these diffusion types schematically.

Sequence of linkages

Adoption

However, in reality, diffusion processes usually display both hierarchical and neighbourhood effects, and common barriers also exist. As Pred (1977) has pointed out, until recently there has been little work done on the direction and sequence of linkages through which adoption-influencing information passes, as for instance from one city to another. The usual practice is to compute rather imprecise correlations between data and city size. But when the sequence of linkages, an inherently temporal component, is investigated it appears that information does not pass neatly and hierarchically from a larger city to a smaller one. More sensitive and theoretically sound indicators of city prominence than size alone are therefore required, and this means that socio-economic space indicators and population size are required. In fact a number of dimensions of para-space might have to be used, and if diffusion is a 'space-time meshing' (Abler, Adams, and Gould, 1971), then it might also be necessary to develop para-time indicators.

Para-spaces and para-times

Apart from classification of diffusion processes according to the dominant effect which appears to have been operating, there are certain characteristics of the diffusion processes which can be summarized by a rather simple diagram such as that illustrated in Figure 7.5. Any region through which an innovation passes or, more precisely, through which information about an innovation passes will contain some people who will adopt it readily and others who will be late to adopt or who will possibly even reject the innovation. A sequential process is operating, marked by the changing rate of the proportion of new adopters — a diffusion time. This will vary from space to space according to political, socio-economic and demographic profiles which act as stimulators and barriers.

Diffusion time

Innovators
Early majority
Late majority
Laggards

The first adopters are usually called innovators. Their example is quickly followed by an early majority of the region's population and then by a so-called late majority: leaving only the late adopters or laggards. The proportion (per cent) of people adopting an innovation is usually plotted along the vertical axis of a graph and the time or date of adoption is plotted on the horizontal axis. A logistic 'S' shaped cumulative proportion curve acts as a useful model of the process. The territorial geometry of the various categories of adopted is not so easily modelled, however.

The lower curve can be pictured as a single wave with the various adopters distributed over some bounded territory or region; it then moves on, in a number of directions, where new innovators, early majorities, etc. take it up again. However, the picture is complicated because the innovators in any one region may well adopt an innovation before the late majority or laggards in another

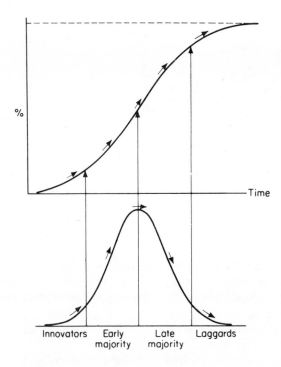

%

Time

Innovators   Early   Late   Laggards
             majority   majority

Figure 7.5. The distribution and accumulation of
innovation adopters

neighbouring region. The pattern of waves therefore appears more
like a storm-tossed sea than the ripple or wave pattern induced by a
single pebble falling into a still pond.

This idea of propagating waves of innovation was first
formalized as a four-stage inductively-derived model by
Hägerstrand in 1952 (cf. Haggett, Cliff, and Frey, 1977, for a
recent summary). Consider Figure 7.6. Part A shows the
hypothetical situation with distance on the horizontal axis and the
innovation ratio on the vertical axis. Part B shows an actual
example based on the diffusion of radio receivers from Malmo to
Hässleholm in southern Sweden, a distance of about 90 kilometres,
and a period of 22 years between 1925 - 1947. When the proportion
of people in an area, who have adopted an innovation (the
innovation ratio) is plotted logarithmically against the distance
from the centre of the innovation, four quite distinct stages appear
to be identifiable. The primary stage (I) is marked by a strong
contrast between the innovation centre itself and peripheral areas.
The diffusion stage (II) is characterized by rather more rapid
adoption rates in the peripheral areas so that, as an example, the
Malmo-Hässleholm study areas, more than 80 kilometres from the
innovation centre, achieved innovation ratios which were as high as
existed in Malmo only ten years previously. The third stage (III) in

Propagating wave

Innovation ratio

Primary stage

Diffusion stage

288

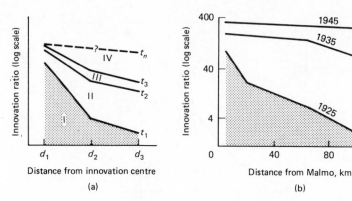

Figure 7.6. The four stages in the passage of an innovation wave. (After Hägerstrand, 1952, 13, 17. Reproduced by permission of T. Hägerstrand)

Condensing stage

Saturation stage

Mean information field

Paracme

this diffusion model is called the condensing stage, and exhibits a slowing down in the diffusion rate, with each area having an innovation adoption ratio of rather similar value. The final stage (IV) is the saturation stage. There is now little evidence of a distance decay in adoption rates as distance from the original centre of innovation increases.

In Hägerstrand's earliest diffusion modelling it was the neighbourhood effect that was considered to be the primary means of diffusion. This was modelled by Monte-Carlo simulations based on spatial, neighbourhood constraints associated with the notion of a mean information field (MIF). For a discussion of these and other aspects of diffusion modelling, refer to Haggett, Cliff, and Frey (1977), and Blaikie (1978).

Barker (1977) has suggested that many of the existing models of innovation diffusion should be supplemented by study of what he called the paracme. A paracmastic process involves individuals or institutions in the shedding of items which were once adopted in an innovation diffusion, as they become old-fashioned or are technically superseded (gas lights for instance).

Jones (1978) has given a good example of innovation paracme in his study of the life-span of the urban tram. In essence, the paracme is a diffusion of rejection but it is definitely not just a simple reversal of the original innovation diffusion, because this would lead to a concentration of the now out-moded item in just those areas which first adopted it!

With the adoption of new modes of communication like television, telephone and telex, information about 'new' innovations can be almost instantaneously transmitted through space. (The space-time paths of such movements when represented as in Figure 1.1 are more or less horizontal). But there are still many circumstances in which this does not occur. Industrial research and development is an obvious example. Here the intention is to restrict information movement so that space-time paths become vertical within specified authority domains and defined areas.

Even so the Panic of 1837 could not have lasted for as long as a month in 1979.[1] Our world has 'shrunk'. In the next section we consider this notion of a 'shrinking earth', as represented by convergence and divergence: a 'plastic' space-time as Forer (1978) has called it. Not only has our world shrunk in knowable terms, it has also been transformed from a more or less rigid, inflexible space-time, based on a very limited number of movement modes, into this plastic space-time by the bewildering variety of modes of communication now available.

Plastic space-time

7.3 Convergence, divergence, and space-time[2]

In 1952 Ellsworth Huntington noted that 'the size of the earth as measured by a day's journey has shrunk so much that the approximate time to travel from Portland, Maine, to San Diego, California, may roughly be reckoned as follows: on foot ... at least two years; on horse ... at least eight months; by stage coach and wagon ... at least four months; by rail in 1910 ... about four days; by air today (1952) ... about ten hours' (Huntington, 1952, 529). Since 1952 this journey by air has decreased to just over three hours. Whilst such dramatic decreases in time have not been uniform they illustrate a general tendency which has become a major area of research in human geography.

The process by which the travel time required to reach one place from another place decreases over time is called time-space convergence (Janelle, 1968). The most significant factor in achieving time-space convergence has been technological improvement in modes of transport. As an example Figure 7.7 illustrates the change in the range of speeds of civil aircraft since the end of the First World War in 1918. Note the increasing range of speeds, because the overall impact of a transport mode on time-space convergence depends on the range of possible velocities as well as the frequencies of departure and not simply on the absolutely fastest mode. However, it is instructive to speculate on the passenger air transport situation, in terms of the absolutely fastest vehicles, sixty years from 1978. If we could assume another eight-fold increase in airspeed, from Concorde's present 1350 mph to eight times this speed, a flight from London to Sydney, assuming no fuelling stops, would take about one hour. The implications for flight dysrhythmia (Chapter 2) and social time phase shifts are dramatic indeed!

Time-space convergence

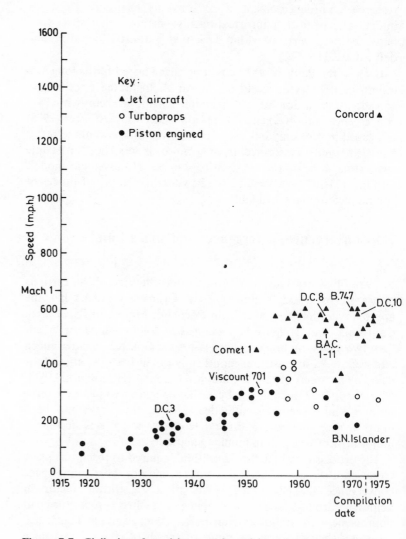

Figure 7.7. Civil aircraft cruising speeds and introduction rates. (Forer, 1974. Reproduced by permission of Pion Ltd.)

In a general way the impact of such rapidly changing speeds can be illustrated as a time-space convergence graph between any pair of places. On a semi-logarithmic graph, time in minutes is represented on the vertical axis in order for the slope changes in the curves to be directly comparable with the large time ranges on the horizontal axis. Figure 7.8 shows some examples of these graphs, and the horizontal axis represents the passage of time.

In addition to this graphical method it is also useful to calculate a simple measure of the average rate of time-space convergence between spaces.

$$AR_{tsc} = \frac{\text{Travel time (in } Y) - \text{Travel time (in } Y+k)}{\text{Interval in years between } Y \text{ and } Y+k}$$

where AR_{tsc} is average rate of time-space convergence

Y is the first year being considered

$Y+k$ is the last year being considered

Substituting real values into the formula above, the average rate of time-space convergence between Edinburgh and London from 1776 to 1966 (Janelle, 1968) is given by:

$$\frac{5760 \text{ min} - 280 \text{ min}}{190} = 28.84 \text{ min per year}$$

When rail times alone are considered for the period 1850-1966 AR_{tsc} becomes 3.4 minutes per year.

Time-space convergence diagrams can also be used to illustrate the closure of time distance between places based on telephone and postal communication between them, as in Figure 7.9, in the same way as transportation modes were depicted in Figure 7.8. A point to note here is that divergence in the travel or communication time necessary to reach one place from another place is also possible: the best example of a situation where this divergence occurs is traffic congestion.

Janelle (1968, 1969) has suggested that the phenomena of time-space convergence and divergence can be linked in a model of spatial re-organization. In this case spatial re-organization does not have the usual connotation of moving objects around on a Euclidean plane, rather it implies re-organization of space itself. Janelle's expanded scheme is shown in Figure 7.11. In essence this expanded version of Janelle's original model is based on two outcomes of increased interaction, a convergence outcome represented by the right hand loop in the diagram and a divergence outcome represented by the left hand loop of the diagram.

Convergence outcome

Divergence outcome

The relevance of diffusion concepts to time-space convergence and divergence studies has been well illustrated by Clark (1974) in his study of the diffusion of STD (subscriber trunk dialling) telephones in north-east England. Here the pattern of diffusion (Figure 7.12) also indicates an increasing time-space convergence between places as the STD service makes communication cheaper and faster.

At the same time Clark's study provides an example of how time-space convergence can be linked, however summarily, to variations in time sense. Figure 7.13 shows how access to a very accurate clock time (the STD speaking clock service) varies regionally. Part of the explanation for this variation may lie in how differing patterns of diffusion and therefore convergence of STD telephones have evolved.

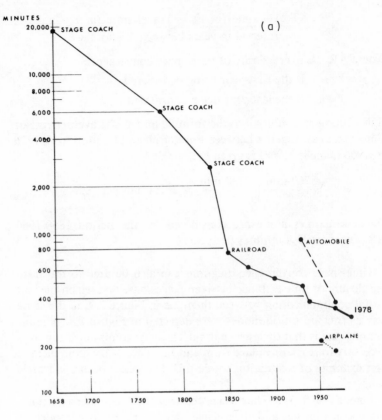

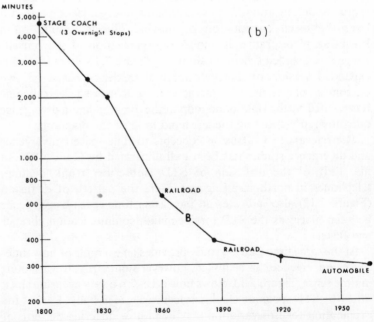

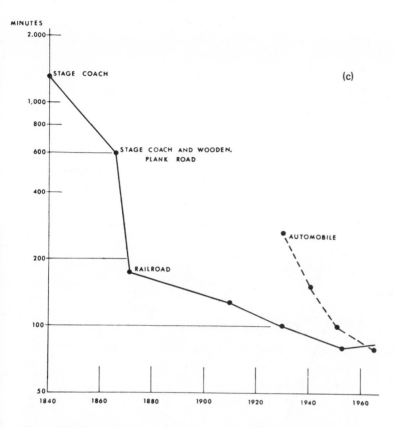

Figure 7.8. Examples of time-space convergence. (Reproduced by permission from *The Professional Geographer* (the *Annals*) of the Association of American Geographers, Volume 20, 1968, D.G. Janelle)

Generally time-space convergence has been studied at the larger spatial and temporal scales — in the terms of earlier chapters, in Big Space and Big Time. But recently there have been attempts to understand convergence in terms of smaller spaces and times within relatively small spatial regions such as a single city, or over relatively short time periods, for instance a day or a week.

One set of plastic spaces that comes into the latter category and which are increasingly studied are those spaces that are known as 'palpitating' (Forer, 1978). These include situations where convergence or divergence takes place between two cities during a period of twenty-four hours, as transport timetables permit. Bristol, for instance, is about 5 hours by train from Leeds, but services are available only four times in a day. If the last of the four trains is missed, the next train is fourteen hours away — the destination is now nineteen hours away by train and not five. Jannelle (1976) has illustrated this phenomenon in terms of the weekly timetables of stagecoaches in the United States in the nineteenth century. Then again time-space convergences and

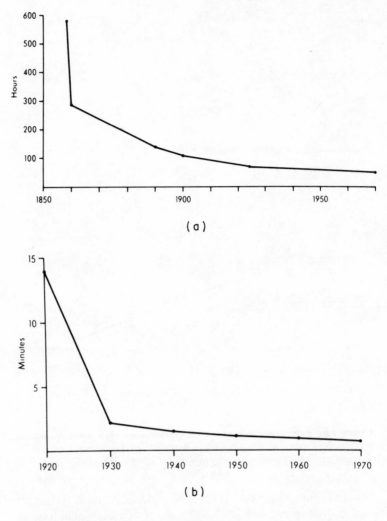

Figure 7.9. Time-space convergence between New York and San Francisco. (a) Postal 1850-1970. (b) Telephone 1920-1970. From Human Geography in a Shrinking World by Abler *et al.* © 1975 by Wadsworth Pub. Co., Inc., Belmont, Calif. 94002. Reprinted by permission of the publisher, Duxbury Press

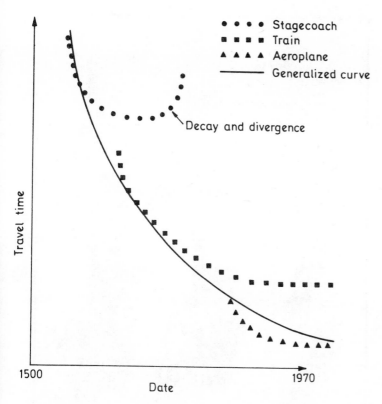

Figure 7.10. Generalized time-space convergence curves. (Forer, 1974. Reproduced by permission of Pion, Ltd.)

An Expanded Model of the Process of Spatial Reorganization

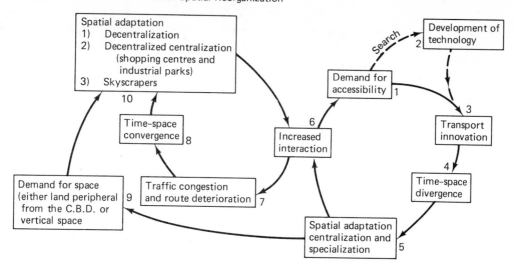

Figure 7.11. An expanded model of the process of spatial re-organization. (Reproduced from *The Professional Geographer (the Annals)* of the Association of American Geographers, Volume 58, 1969, D.G. Janelle)

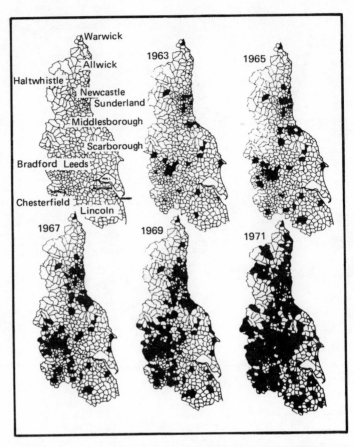

Figure 7.12. The spread of STD telephone: exchange areas in north east England. (Clark 1974)

divergences are not always symmetric because time distances between two points may be different depending on direction. Janelle (1973) cites the example of time-distance by barge on the Volga River in Russia. The journey is 102 hours downstream from Gorki to Astrakhan but 220 hours from Astrakhan to Gorki in the upstream direction (Janelle, 1973). Technological improvements in barges will still reflect this friction of convergence.

Friction of convergence

Convergence and divergence can also be considered within a small spatial region. For instance, Forer (1978a) has studied the process at the city scale by evaluating the effects of nine modes of transport and communication which have operated in Christchurch, New Zealand, between 1880 and 1970. Some of these were innovations within the time period, (e.g. the motor car), whilst others had been shed as part of the paracmastic process (e.g. tram cars). Figure 7.14 illustrates this gradual change in mode availability.

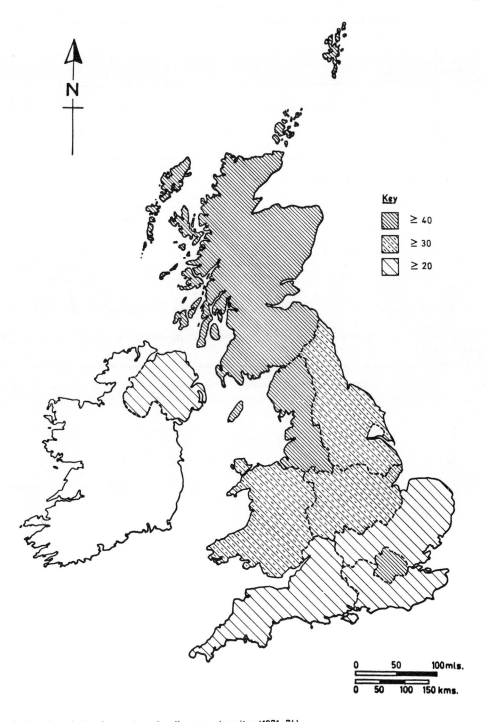

Key

▧ ≥ 40

▨ ≥ 30

▱ ≥ 20

Regional variation in number of calls per subscriber (1971-74)

Figure 7.13. Regional variations in calls to telephone-signalled time in Britain

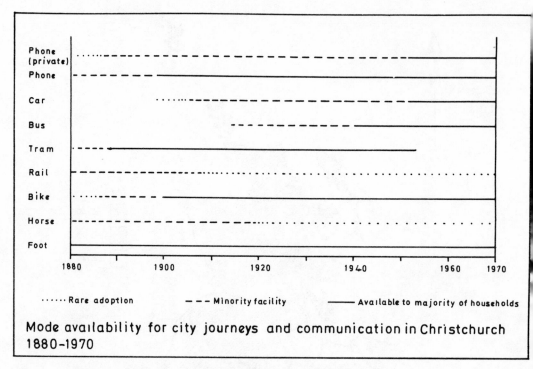

Mode availability for city journeys and communication in Christchurch 1880-1970

Figure 7.14. Mode availability in Christchurch, New Zealand, 1880-1970. (Forer, 1978, 108. Reproduced by permission of Carlstein *et al.* and Edward Arnold (Publishers) Ltd)

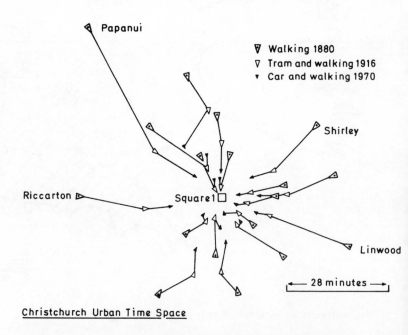

Figure 7.15. Christchurch New Zealand, a portion of its urban time-space. (Forer, 1978a, 117. Reproduced by permission of Carlstein *et al.* and Edward Arnold (Publishers) Ltd)

Forer (1978) sampled a number of points in the city area of Christchurch to assess the resultant adjustments in size and shape of the city's plastic space induced by these mode changes. 'What is the extent of warping which such changes induce? How plastic is the space of Christchurch?' Using multidimensional scaling methods, Forer was able to answer these questions. Figure 7.15, for instance, shows the movement of the sampled points in time distance according to mode of travel.

From the foregoing discussion and examples it is now possible to draw some general conclusions about the nature of the plastic spaces of convergence and divergence. One is that, whereas the world may have 'shrunk' in terms of the fastest possible speed at which people are now able to move and communicate, up to the boundaries represented by Concorde and by the telephone, the notion of a 'global village' only has a limited validity. Even at the intra-urban scale it is clear that movement and communication depend on *access* to particular modes. This in turn depends on a number of factors, of which ability to pay is the most obvious. The results of this differential access mean that for many the world is still as large as it has always been (Figure 7.16), dependent upon socio-economic class. To gain access to the smaller world of the

Access

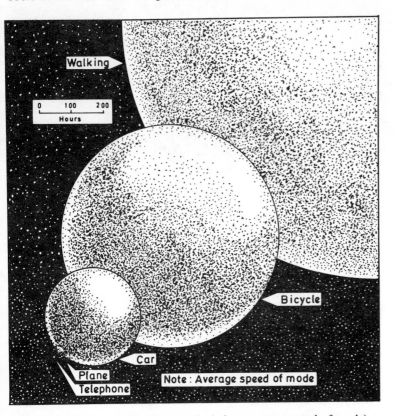

Figure 7.16. The shrinking world. (Scale is to average speed of mode)

aeroplane or the instant world of the telephone requires hard cash.

Secondly, and as a result of the above, each mode of movement and communication has its own domain and although we can surely conceive of shrinking cities, such a shrinkage is not even. As Forer's (1978a) study shows we must think in terms of an aggregate convergence area. In addition it must be realized that the process of uneven shrinkage operates over the entire city area at all scales. For instance, when the lifts break down in high rise buildings or when there is congestion and queuing there is an expansion of space-time; by contrast installation of a new lift or a pedestrian crossing over a busy street will bring about contraction. Figure 7.17 illustrates this notion (Forer, 1978). A 6-level building, which has a fast lift installed, is smaller in space-time than a 4-level building without one. A similar transformation of space scales by time distance can be applied to road widths: a narrow road of 10 metres may be wider in time-space than a road which is 30 metres wide if the latter has a pedestrian crossing whereas the former carries high traffic densities and has no crossing point.

Shrinking cities

Aggregate convergence area

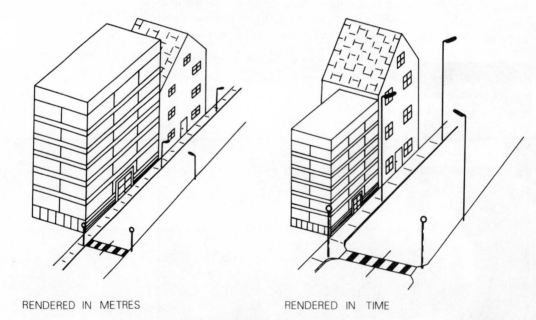

RENDERED IN METRES   RENDERED IN TIME

Figure 7.17. Urban physical structures in space and in time metrics. (Adapted from Forer, 1978, 102. Reproduced by permission of Carlstein *et al.* and Edward Arnold (Publishers) Ltd)

This brings us to our third point. A map transformation of all relevant widths and dimensions from a purely Euclidean geometric rendering into a time-space metric of some sort undoubtedly provides a more realistic model of the relations of settlement structure to movement and communication possibilities. Such a metric could also be weighted according to time of day as the perceived value of 'minutes saved' changes.

Finally we can also conceive of areas as having a time density (Forer, 1978), due to the fact that 'many of our everyday affairs require the maximisation of interaction possibilities (and) if we accept density as a measure of likely interaction and time as a suitable metric then the time-areal pattern ... can be translated into one of densities related to process. High densities would indicate either high opportunity or high stress and probably both' (Forer, 1978, 104). The time density of an area, in Forer's terms, is akin to the space-time packing concept in the Lund time-geographic approach. Offices, equipped with telephones and other means of communication, have a high time-density; an urban park, on the other hand, has a low time density. Residential areas will also vary in time density. For instance, although it is generally the case that the correlation between the social status of an area and the net or gross residential densities is negative, when the densities have been calculated in the conventional manner of people or dwellings per unit area, the correlation is likely to be positive when time densities are considered. High socio-economic status will be associated with high time-density because high status area residents engage in a greater variety of activities per unit time (Parkes, 1974), i.e. their space-time density is higher (Chapter 8 following). It is also possible to identify 'stagnant' or slow moving time-spaces; for instance those where walking is the only possible mode. Other time-spaces are 'palpitative', being sometimes fast and sometimes slow, as for instance in the example of the difference in pace of the CBD of any large city when comparing 1400 with 0200. At 1400 activities per unit small time are much higher than at 0200. Many differences of pace between urban subareas will, of course, be induced by the time-tabling of public services (Forer, 1978). Rather similar notions of city-fast and city-slow areas have been outlined elsewhere by Parkes (1973, 1974).

We can now see that the warping of time-space means that it is not always sensible to persist with the use of insensitive Euclidean measures of distance. The same conclusion has been reached in everyday life. In Milan, for example, the ticketing system for buses allows the passenger to buy 'time' rather than distance. A seventy minute ticket is illustrated below.

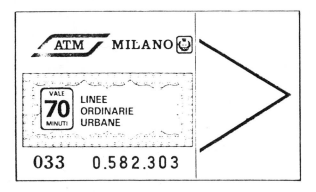

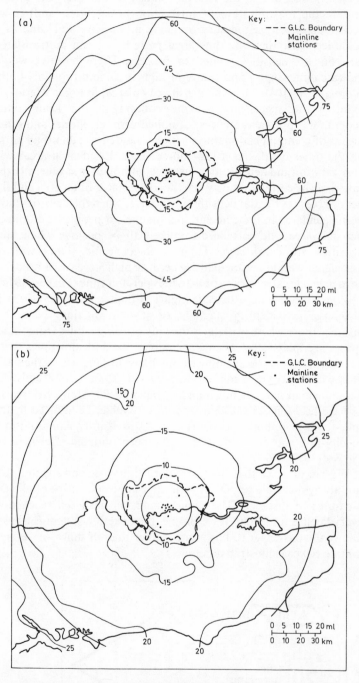

Figure 7.18. Accessibility to central London 1972. (a) Net distance (in miles). (b) Cost distance (in £). (c) Time distance (in minutes). (d) Generalized time distance

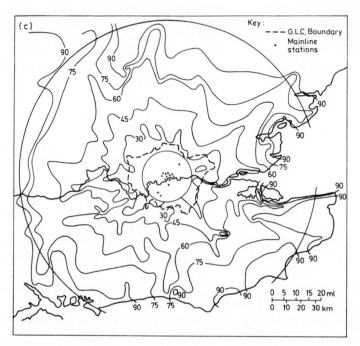

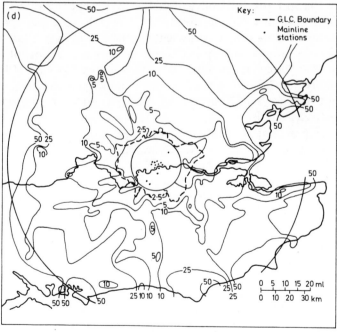

In the same way people often talk quite naturally about a place being 20 minutes or 20 hours away. There are many metrics which can form an alternative to the simple one of spatial distance. In Figure 7.18 accessibility to the centre of London by rail in 1972 is shown in the normal, inelastic Euclidean distance terms. It is then shown contoured according to a variety of other metrics. The conclusion which may be drawn is that a Euclidean measure of distance does not always carry as much information as these alternative measures. However, this kind of simple contouring is only the first step that can be taken in representing the warping of space. Apart from the use of map projections it is also possible to use 'empirical' map projections to depict plastic space. These rely on the use of a technique called multidimensional scaling (Golledge and Rushton, 1973), which takes a set of distances between points and from them produces a 'map' of the points, dependent on the number of dimensions required, but still retaining the order of distances. Using this technique it is possible to show how spaces are warped by use of measures other than Euclidean distance. Consider an equal area planimetric map of the city of Toronto, as in Figure 7.19(a). Multidimensional scaling allows the city to be represented in travel time-space with units of distance based on minutes, as in Figure 7.19(b) (Ewing and Wolfe, 1977). Note how the 'physical' Euclidean space map of regular grid squares and the superimposed street network of the conventional map (a) are distorted (or warped) when they are transformed by a time distance metric (b). In Toronto the areas which experience the most warping are located in the most densely built-up and congested 'city fast' sectors of the inner city. Similar maps have been produced by Golledge (1978) and Forer (1978a, b) in connection with the representation of subjective space and distortions by different modes of transport, respectively.

This representation of 'physical' Euclidean space in time distance terms using scaling becomes even clearer in the time-space 'map' of New Zealand, illustrated in Figure 7.20(a), when it is compared with, say, an orthomorphic or conformal projection like that found in any atlas. (Strictly speaking the coastline cannot be drawn in).

Field

But the representation of the plastic space of convergence and divergence has by no means been solved by the techniques of multidimensional scaling and other methods have been suggested. These revolve around the idea of a field — a continuous distribution of values which occur on a Euclidean plane. One approach — the analytic — has been proposed by Angel and Hyman (1970, 1976). A conventional map of the Greater Manchester area of England shown in Figure 7.21(a) suggests that the time-space of this conurbation is symmetric in all directions around the CBD. The map provides no indication at all that this is not the case. Angel and Hyman proposed a procedure for obtaining good estimates of travel times between different locations in urban

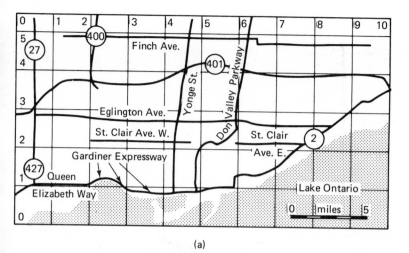

(a)

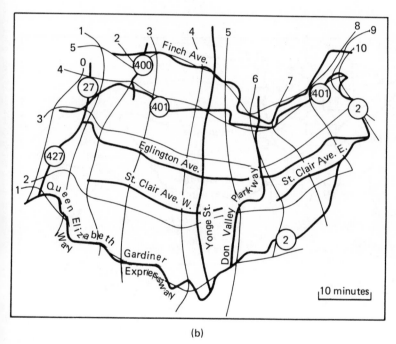

(b)

Figure 7.19. (a) Toronto in 'physical' space. (b) Toronto in time-space based on displacement vectors. (Ewing and Wolfe, 1977. Reproduced by permission of Pion Ltd.)

areas and applied their proposals to the Greater Manchester area, based on 1965 street systems. Their approach is based on the idea of velocity fields. The urban traveller is assumed to move over a continuous isotropic terrain and to want to minimize travel time between origin and destination by following the shortest time path (or isochrone).[3] The velocities of travel vary only as a function of distance from the city centre, that is they will be lowest near the city centre (where speeds are lowest) and highest at the perimeter (where

Velocity fields

306

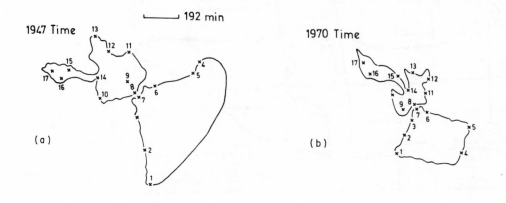

1947 Time

192 min

1970 Time

(a)

(b)

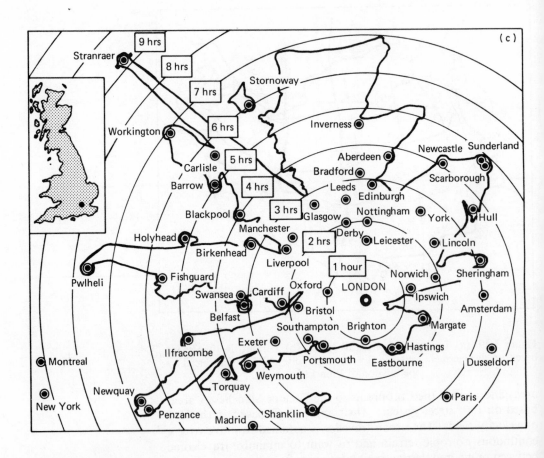

Figure 7.20. (a) New Zealand in time-space. (After Forer, 1975). (b) England, Wales, and Scotland (1978). (*Sunday Times,* 10 September 1978 from *New Society,* Nathiel. Reproduced by permission of *New Society*)

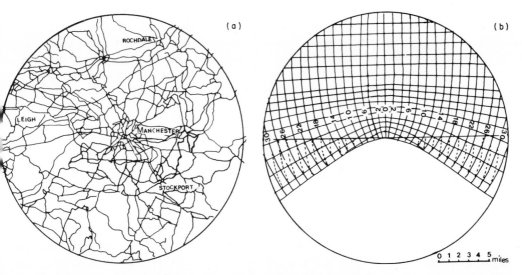

Figure 7.21. Greater Manchester and its Chronograph (1965). (a) The conventional map (b) The chronograph with time contours drawn at two minute intervals from the original radial, 0.

The chronograph may be rotated about the centre of the city to provide measures of travel time between any two points. The lines on the chronograph are calculated from the velocity field value at a number of points throughout the area in relation to an original radial which acts as the datum for the calculation of all the time contours. 'To obtain readings of travel times between two points, we rotate the chronograph about the centre of the city until the two points (origin and destination) lie on one minimum path, or in a band between two minimum paths. We then simply add or subtract the times from the original radial to each of the points to obtain the correct travel time.' (Angel and Hyman, 1970) (for 7.21b only)

speeds are highest). Variations in travel speeds in different directions are ignored. 'Thus suburban areas where densities are low allow for considerably higher velocities than congested urban centres' (Angel and Hyman, 1970, 7). The model, therefore, characterizes variations in urban traffic speeds as a velocity field. Angel and Hyman estimated a radially symmetric velocity field by calculating the average link speeds at different distances from the centre of Greater Manchester. A best fit curve was estimated for this empirical data. Their model results were then compared with the results obtained by a major transportation study of south east Lancashire and north east Cheshire (SELNEC), and appeared to show a good correlation. Having estimated an apparently realistic velocity field three interrelated methods for computing travel time on velocity fields were considered. One of these methods is a graphical device called a chronograph. It is illustrated in Figure 7.21(b).

Chronograph

The chronograph reinforces the significance of the time-space dimensions or urban scale for everyday behaviour in cities. In addition in some cases transformations can be found which enable velocity fields to be transformed into a time surface which describes the city as an isochrone map of all points in the city which are an equal time away from each other (Angel and Hyman, 1976).

Time surface

Elsewhere we have discussed some further notions about convergence and divergence (Parkes and Thrift, 1978) as they might be applied to time-time and space-space convergences and divergences, but they are beyond the scope of the present discussions which now move to the introduction of space-time substitution as a process which produces certain forms of land utilization, perhaps the best known examples of which are periodic markets.

7.4 Space and time as substitutes in land utilization

**Substitution of
space and time**

We have seen that additional information about places, and especially their geographical (space-time) relations, can be obtained by using non-spatial metrics. In one of the last major papers he wrote before his death, Edward Ullman presented a closely argued essay on the substitutability of space with time. 'To use the earth requires the organization of space and time. They (space and time) provide the setting and the coordinates and they are interrelated; one can often be substituted for or measured in terms of the other' (Ullman, 1974, 125, our emphasis).

Geographers have certainly been aware of this interrelation, a fact apparent from published work in a variety of fields. We shall consider only three examples of substitutes, these being field systems, land use, and periodic markets.

Many systems of land use are founded upon space-with-time substitutions of one sort or another involving periodic use of land. Bradley, Raynault, and Torrealba (1977) suggest that there are at least two perspectives which can be adopted. One is the 'articulation of agricultural operations' and the other is 'the staggering of harvests and therefore the staggering [or timing] of the supply of food'. Figure 7.22 illustrates the major land systems of the Guidimaka region of Mauritania in North Africa where their research was centred. 'The time factor is particularly important from the point of view of the overall coherence of the (Guidimaka) agrarian system' (Bradley, Raynault, and Torrealba, 1977, 73) and Figure 7.23 shows the calendar of agricultural activities in different areas of the region.

To cultivate this region creates many problems for the local people, intricately associated with the meshing of space and time. Most obviously these problems occur in relation to climatic factors, but recently migration to the industrial regions of France, where Guidimakan migrants stay for perhaps three years, returning intermittently, has produced new difficulties. The migrant workers' cash inputs disrupt the balance of the local economy, creating phase shifts in periods of plenty and periods of shortage. This intermittent migration also reduces the time supply of the

A   B   C   D   E   F   G

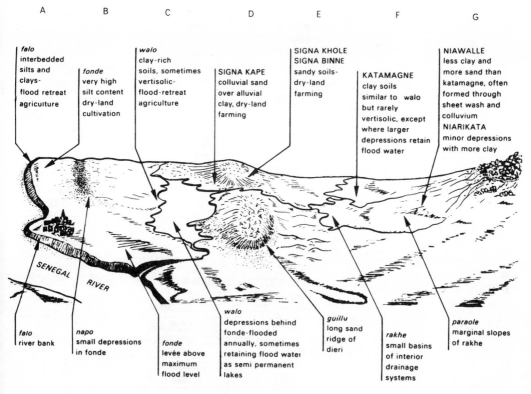

falo
interbedded
silts and
clays-
flood retreat
agriculture

fonde
very high
silt content
dry-land
cultivation

walo
clay-rich
soils, sometimes
vertisolic-
flood-retreat
agriculture

SIGNA KAPE
colluvial sand
over alluvial
clay, dry-land
farming

SIGNA KHOLE
SIGNA BINNE
sandy soils-
dry-land
farming

KATAMAGNE
clay soils
similar to walo
but rarely
vertisolic, except
where larger
depressions retain
flood water

NIAWALLE
less clay and
more sand than
katamagne, often
formed through
sheet wash and
colluvium
NIARIKATA
minor depressions
with more clay

falo
river bank

napo
small depressions
in fonde

fonde
levée above
maximum
flood level

walo
depressions behind
fonde-flooded
annually, sometimes
retaining flood water
as semi permanent
lakes

guillu
long sand
ridge of
dieri

rakhe
small basins
of interior
drainage
systems

paraole
marginal slopes
of rakhe

SENEGAL RIVER

Figure 7.22. Major soil and land systems of Guidimaka. (After Bradley *et al.*, 1977. Reproduced by permission of P.N. Bradley)

population (Chapter 6), an important factor in the size of the food supply.

Apart from the timing problems created by the shortage of labour (which is not balanced by a reciprocal lowering of demand for food), time-space competition exists as a result of the variation in soil and climatic conditions. There is, for instance, competition between cultivation of the dieri (the non-alluvial areas) and fonde (the levee) (Figure 7.22), particularly at the start of the rainy season when the sowing of sorghum and maize on the fonde coincides with the beginning of weeding of the maize, millet, and sorghum in the dieri. Another 'bottle neck' arises in a year of normal flooding, between the harvesting of dry-farmed crops ((b), (d), and (e) in Figure 7.22) and the sowing of flood retreat fields (a) and (c), because delay in either case could be catastrophic. A third conflict occurs between farming on the walo (sedimentary basins below the levee) and the falo (upper segments of the river bank). In the interior there are also conflicts between harvesting and weeding. The land-use system which results is therefore a reconciliation of the differing space and time supplies and demands. The Guidimaka farmers have produced a delicate agrarian balance by 'adjusting the

310

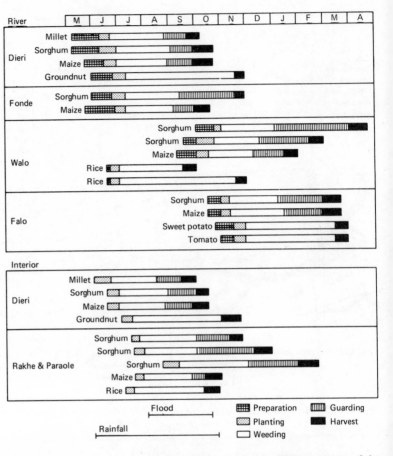

Figure 7.23. Calendar of agricultural activities in the different zones of the riverine and land systems, Guidimaka. (After Bradley *et al.*, 1977. Reproduced by permission of P.N. Bradley)

relative size (space) of the areas cultivated in the different zones of the land system ... by adapting their agricultural calendar (time) according to growth cycles' (Bradley, Raynault, and Torrealba, 1977, 76).

Seasonal round

Rather similar adjustments in space and time can be identified in more 'advanced' farming economies. Mead (1958) has discussed the seasonal round in Finland, based on data collected using diary surveys as a basis for compiling time budgets for individual farmers. Figure 7.24(a) shows the seasonal rhythm of activity in the Lapinlahti area of Finland and Figure 7.24(b) illustrates the time-space scheme in rather more detail for January and for July. Field rotations are a familiar method of increasing effective land use, but are they usually interpreted in essentially spatial terms? But in fact, such periodic adjustment is a time spacing factor (Chapter 3) and aims to optimize the use of time as well as the use of land. One example (which has been used elsewhere to exemplify the idea of time spacing and space timing (Parkes and Thrift, 1975)) is provided by the work of the Danish geographer, Hastrup (1970). The 'vangelag' field system of the Antvorskov district of Western Zealand in Denmark is a response to a space-time problem which was first confronted by a rural community in this area some five hundred years ago. The problem to be solved involved allowing as much land as possible to lie fallow without incurring the financial and temporal burden of boundary fencing to keep cattle and sheep on the grazing (fallow) land but away from the crops. Figure 7.25 is a diagrammatic summary of the principles of the system (Parkes and Thrift, 1975).[4]

In Figure 7.25(a) the initial rather haphazard spacing and timing system is illustrated. Three field systems are independently rotated — some clockwise and some anti-clockwise. Over time the relative inefficiency of this system must have been realized and it was gradually replaced by a system of space-time meshed rotations with

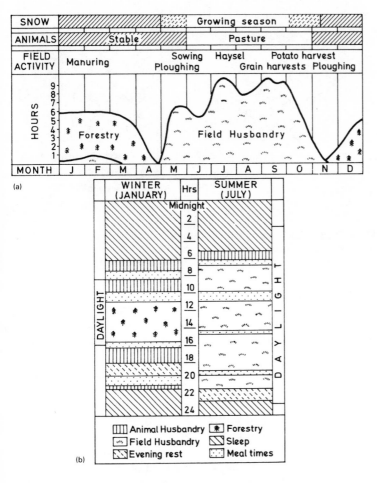

Figure 7.24. The seasonal rhythm of activity on mature farmsteads in the Lapinlahti area of Savo. (Mead, 1958. Reproduced by permission of K.N.A.G. (Royal Dutch Geog. Society))

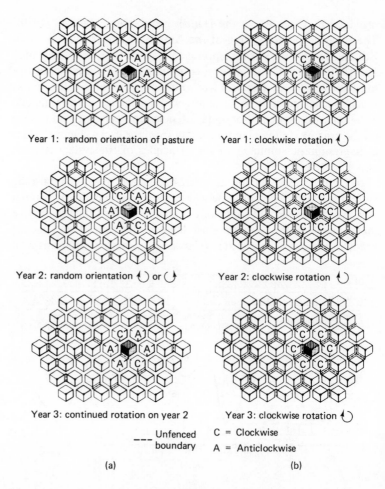

Year 1: random orientation of pasture

Year 1: clockwise rotation

Year 2: random orientation

Year 2: clockwise rotation

Year 3: continued rotation on year 2

Year 3: clockwise rotation

- - - Unfenced boundary

C = Clockwise
A = Anticlockwise

(a)   (b)

Figure 7.25. Time spacing and space timing in an agricultural context. (a) Independent rotations. (b) Meshed rotations. (After Parkes and Thrift, 1975. Reproduced by permission of Pion Ltd.)

a consistent orientation as shown in 7.25(b). In this manner space was timed in its use in such a way that the objective of maximizing fallow land and minimizing boundary fencing could be achieved. Time was spaced so that it could be 'saved' and used or allocated to other needs.

Such periodic systems are not confined to sedentary agrarian societies. Migratory (nomadic) cultures provide obvious examples of a space-time unity in the organization of human society. Barth (1959) studied the land utilization pattern of tribes in south Iran (then Persia), shown in Figure 7.26.

Each tribe lays claim to an 'il-rah' or tribal road which it travels every year. Sections of the road are used at specified and regular (clock-like) intervals. Consider the Basseri tribe as an example. Their il-rah is shown by the thick line (Figure 7.26). It extends from the coastal hills at Lar in the south east, up to Kuh-i-Bul in the

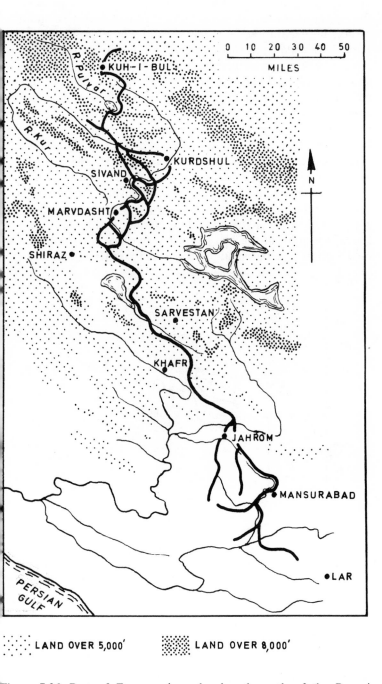

Figure 7.26. Part of Fars province showing the path of the Basseri migration. (After Barth, 1959. Reproduced by permission of Universitetsforlaget, Oslo)

north, a distance of nearly 400 kilometres. In the extreme south is the region of winter dispersal. Here in March each year the tribe gathers in preparation for the northward migration. However, the Basseri must share this route, or part of it, with other tribes, so they can own or have absolute right to this route only at specific times, and even then only in relation to specified sections of it. They must move to the right space at the right time. This means moving at an appropriate velocity, through the space-time corridor which is provided for them; in itself this is an intricate manoeuvre. To the Basseri the month of 'March' is both a space and a time, in a complete sense, a place. Barth used the analogy that each il-rah was, in effect, like a railway schedule — each train (tribe) arriving at certain stations at specified times, occupying the station (as in Chapter 6) for a specified time and then moving on to another one at a prescribed velocity. But in addition, there is also a seasonal and spatial variation in the carrying capacity or packing potential of each station and this is allowed for in the different tribal time-

Carrying capacity and packing potential

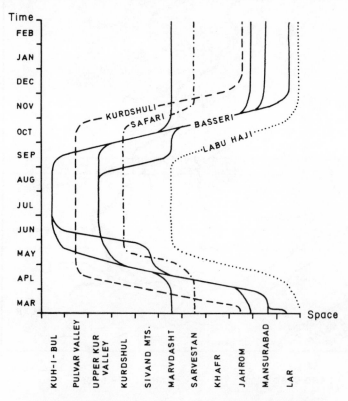

Figure 7.27. Space-time paths of tribal groups in the strip of land occupied by the Basseri during a yearly cycle. The dashed lines show the paths of other tribes and the arrow heads the 'place and time where other tribes enter or leave the Basseri strip' (Barth under Figure 2, p.6, 1959). (Adapted from Barth, 1959. Reproduced by permission of Universitetsforlaget, Oslo)

tables, Figure 7.27. We have rotated Barth's original diagram in order to make it easier to recognize the (implicit) time-geographic principles involved. The paths or trajectories are akin to the paths of the Lund scheme (Chapter 6).

Periodic markets are another well-studied area of space-time meshing in geography. Their genesis is probably to be found in the seminal work of Skinner in China in the 1930's, and in the work of Dresch (1939) in Morocco. 'In Morocco, the markets are called soukhs and characteristically operate at weekly intervals. In fact, the soukh was named after the day of the week, so that there is a repetition of place-names all over maps of Morocco — Soukh el "Thursday", etc. They are regarded, according to Mikesell (1958) as "events" but not settlements — points in time rather than points in space' (Ullman, 1974, 128).

Periodic markets

Periodic markets tend to have been established when transportation, and mobility in general, was poor. Usually, as a concomitant to this, population was also small and low in spatial and temporal density. A seller may draw custom from a five or ten kilometre radius when there is a single fixed settlement location of sufficient size, but may require all of the trade within, say, 25 kilometres in order to make a normal profit. Therefore one choice which is open to him is to move from point to point within the area of the 25 kilometre radius. With a similar alternative apparent to other merchants a circuit of markets will be established but it will function periodically at different locations. Such a periodic circuit of markets is called a periodic market system. The periodicity or frequency of such markets varies widely and is a function of many factors including population density, culture traits relating to traditional methods of ordering the days, transport availability, and the nature of the goods and services being supplied.

Circuit of markets

Ullman (1974) has attempted to incorporate aspects of the classical central place marketing principle into a model of the timing of markets and their trade areas. The time-market area principle which he developed is shown in Figure 7.28. An increase in the frequency of a market (a shorter period between 'market days') requires an increase in trade area. The diagrams, therefore, illustrate the relationship between the frequency of periodic market and the spacing and size of trade area. For instance, if a 63 km^2 trade area is required to support a weekly market, seven times as much area is required to support a daily one. But such an increase in area means a greater maximum travelling distance; resulting in an increase in the area required to support a market from 4.3 kilometres for the weekly market to 11.3 kilometres for the daily market. There would also need to be an increase in the separation of markets from 8.6 to 22.6 kilometres. It is possible to calculate the relationship between the periodicity of the market and the size of the trade area required (Figure 7.28(b)) as well as the periodicity and the distance between markets.

TIME OF MARKET AND SIZE OF TRADE AREA

Increase in Market Frequency = Increase in Trade Area Required
(Homogeneous Plain, Hexagonal Trade Areas, Market Size Constant)

A. TRADE AREA REQUIRED FOR MARKET MEETING

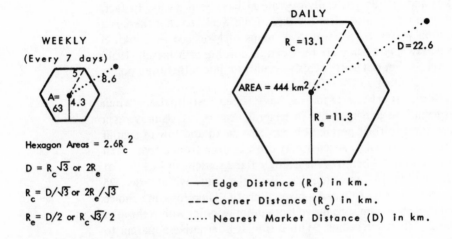

Hexagon Areas = $2.6R_c^2$

$D = R_c\sqrt{3}$ or $2R_e$

$R_c = D/\sqrt{3}$ or $2R_e/\sqrt{3}$

$R_e = D/2$ or $R_c\sqrt{3}/2$

—— Edge Distance (R_e) in km.

--- Corner Distance (R_c) in km.

····· Nearest Market Distance (D) in km.

B. TRADE AREA AND SPACING NEEDED TO INCREASE FREQUENCY OF WEEKLY MARKET

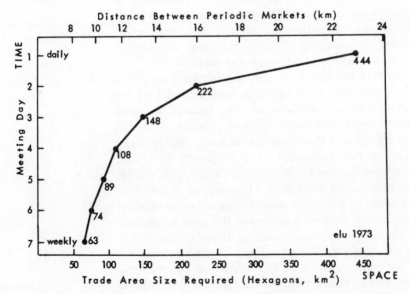

Figure 7.28. Time of market and size of trade area. (Ullman, 1974)

Of course, periodic 'marketing' is not restricted to the rural sectors of small scale societies. In Britain, symphony orchestras and theatre companies market their services through a 'provincial circuit', as do 'pop' groups and the provincial judiciary! Regular Saturday afternoon 'home' and 'away' football matches are also competing for money in a periodic market system. Many county towns in Britain still have market days whilst many towns have Sunday markets (Strachan, 1978). Rural areas are visited by mobile libraries. In Singapore (Yeung, 1974) a system of travelling night markets has evolved over the past two decades, shown in Figure 7.29. On every night about eight markets operate at the peak trading hours from 1900 until 2200. Successive nights reveal the markets located at (successive) different sites. The market cycle is one week. Clearly the temporal spacing of competitors and customers is as important as their spatial distribution is for the determination of the location of a conventional market. Table 7.1 summarizes the mean temporal and spatial separation of the travelling night markets of Singapore. It illustrates the principle (Smith, 1971) that markets held on the same day are farther apart in space than markets on pre- or post- adjacent 'days'.

In general periodic markets seem to conform to many of the principles of classical central place theory. (For an introduction see Beavon, 1977). They confirm the importance of the strategy of space-time substitution.

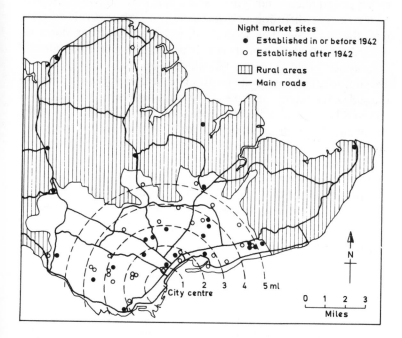

Figure 7.29. Travelling night markets in Singapore. (Reproduced by permission from the *Professional Geographer of the Association of American Geographers,* Volume 26, 1974, Y.M. Yeung)

Table 7.1. Temporal and mean locational spacing (in miles) of Singapore's travelling night markets (Yeung, 1974. Reproduced by permission from the Professional Geographer of the Association of American Geographers, Volume 26, 1974, Y.M. Yeung)

	Temporal separation (nights)			
	Same	Adjacent	One night	Two nights
By market group				
All markets	2·09(1·28)	1·23(1·36)	1·35(1·62)	1·37(1·93)
Rural markets	4·59(1·95)	4·57(1·23)	5·01(2·07)	5·88(2·72)
Urban markets	1·77(0·78)	0·84 (0·65)	0·89(0·73)	0·80(0·59)
By night of week				
Monday	2·22(0·57)	0·95(1·19)	1·27(1·22)	0·79(0·69)
Tuesday	1·41(0·70)	0·70(0·40)	0·70(0·53)	0·70(0·26)
Wednesday	2·58(2·52)	1·59(2·05)	1·54(2·22)	1·51(2·67)
Thursday	1·77(0·50)	0·72(0·38)	0·78(0·56)	0·87(0·61)
Friday	2·53(1·09)	1·48(1·47)	1·72(1·34)	1·61(1·31)
Saturday	1·82(1·30)	1·88(1·51)	2·05(2·45)	2·64(3·47)
Sunday	2·41(0·47)	1·09(1·09)	0·93(0·59)	1·08(1·10)

Source: Night markets survey, 1970.
Standard deviations are shown in parenthesis.

Movement writes space-time signatures on the land surface which geographers study. Both space and time are always jointly involved. The diffusion and paracme processes provide insights into the way the 'landscapes' of today differ from those of yesterday. Through simulation it is possible to gain some idea of the way our world, or parts of it, will look and behave in the future. Technological advances in transport and communication media induce adjustments in the relations between spatially and temporally segregated items. While convergence appears to be the dominant direction of change in relative time-space, divergence also occurs. The measurement of space in metrics which are not purely Euclidean seems to provide more realistic, perhaps even more precise, coordination for the study of human geographies and the places they occur in. The organization of space for such familiar activities as marketing is enhanced by timing the use of space, as in periodic markets.

Notes

1. Refer forward to Figure 7.16, The Shrinking World.
2. In this section we will use in general the term time-space rather than space-time because this has been the convention in the studies which we cite.
3. This was a specific model constraint and should not be considered of

general significance. The details of the analytical methods cannot be illustrated here, cf. Angel and Hyman 1976.

4. We are grateful to Peter Haggett for drawing our attention to Hastrup's work and for his rough sketch which was the basis for the diagram.

8
24 hours in the city

'A temporal pattern is apparent in each and every spatial
pattern (and) space and time are separable from one another
only in abstraction.'

(Hawley, 1950, 288)

8.1 Preliminaries

Temporal factors

Human ecology

**The clock as
symbol of order**

In this chapter the city is set in a time period of 24 hours duration
and the focus is on small time variations. The chapter has six main
sections. In the first section, temporal factors in human ecological
and neo-ecological studies of urban society will be introduced. In
nearly every introduction to urban geography there is some
reference to the models and empirical work of the Chicago
ecologists: to Burgess's concentric zones and the natural areas of
the city: to competition and co-operation; to processes of
dominance, invasion and succession and to the idea of the city as a
living organism. Inevitably, however, urban temporalities tend to
be overlooked. The sector 'theory' of the economist Hoyt (1933)
and the later sociological viewpoints of Wirth (1938) and Firey
(1947) concerning the importance of consensual behaviour,
sentiment, and symbolism in the growth of urbanism as a way of
life, inevitably also have their place. The 'new' ecology of Hawley
(1950) and others, has received less attention and, significantly
from our point of view, so has their explicit acknowledgement of
the importance of temporal factors in urban environments. Indeed,
in 1938 Wirth had drawn attention to the clock as the symbol of
order in the city. So time and the ecology of the city are not new
relations; just rather distant ones that don't seem to get together
very often.

As the city becomes more complex, as its scale increases and greater scope for interaction develops (Shevky and Bell, 1955), urban places consume more space at an ever greater rate than we know to have occurred in the past. Coincidentally they also consume more time. This increase in the consumption of time, known as time colonization, is the major topic for the second section (8.3).

In the third section, 8.4, the results of an urban images study undertaken in Australia are summarized. Unlike 'conventional' urban images studies in this study the imaged spaces are set in different time frames, including the night. Recent work by Lynch (1976) and Rapaport (1977) has addressed rather similar issues.

The fourth section (8.5) reintroduces the concept of rhythms (from Chapters 1, 2, 3) and through two examples, one from the USA and one from Australia, the temporal characteristics of urban social pathologies (or ecopathologies) will be illustrated. Ecopathology (or social pathology)[1] simply refers to the occurrence of events which are 'unwanted' by any society. These include accidents, suicides, distress, illness, disorderly conduct, etc. They are disruptive social factors.

The fifth section (8.6) brings us back to the social and territorial spaces of the city, and to the variations in time use within them, as well as introducing differential time perspectives based on social status and the coherence of the images that social 'groups' have of the way socially distant groups spend and value time.

The final topic, presented in Section 8.6, reports the results of a space-time factor ecological experiment. An 'experimental city' is constructed and its space-time structures over a twenty-four hour period are identified (Taylor and Parkes, 1975; Chapman, 1977).

Increasing scale

Time colonization

Images and time

Urban social pathology or ecopathology

Time perspective

Time image coherence

Space-time factorial ecology

8.2 Temporalities in classical and neo-classical ecology

The classical human ecologists, working in Chicago in the first decades of this century, set the pace and direction for much of the later work undertaken in urban social geography. The Burgess concentric zone scheme still appears in most recent introductory urban geography texts (see for example Carter, 1972; Herbert, 1972). The combined works of Park, McKenzie, and Burgess remain to this day as an essential framework for an introduction to the social geography of the western city.

Perhaps the first extension of any consequence to the work of Park, McKenzie, and Burgess came with the publication of Hoyt's empirical studies of housing rental markets in American cities, undertaken for the US Federal Housing Department in the 1930s. At about the same time, social theorists like Wirth were introducing lifestyle and behavioural concepts into the study of urban social structure. By the 1940s a neo-classical ecology had developed under the stimulus provided by the work of Firey (1947) in Boston and more generally through the work of Hawley (1950). This fervent expression of intellectual concern for understanding the nature of the city and the social problems it 'contained' was brought together in a collection of essays edited by Theodorson in 1961. Included in this collection were essays by a group of sociologists working in California, the social area analysis group under the leadership of Shevky. Their aim had been to stimulate a new theoretical thrust into the study of 'mass society' and to break the 'habit' of classical and even neo-classical ecology.

By the early 1960s, with the rapid growth in computer-based analysis, the classical social area method (Shevky and Bell, 1955) was given a mathematical-statistical facelift. Wearing this new face and called factorial ecology, it became perhaps the most frequently used device for identifying underlying dimensions, latent structures of urban social geographic space.[2]

A careful reading of each and every one of these works reveals awareness of the significance of time in the structuring of the city. Unfortunately nowhere, (with the exceptions of Wirth (1938), Engel-Frisch (1948), and Hawley (1950)) was the temporal component given an extended treatment. It does, however, appear in the Shevky-Bell social area analysis paper of 1955 through a reference to Wirth's (1938) observation that the clock was the symbol of order in the city; a theme which Moore (1963) also takes up.

Routine and erratic behaviour

Rhythm tempo timing

Time concentration and segregation

Those ecologists like Hawley who did give explicit attention to temporality, saw routine and erratic behaviours giving the city a pulse. Rhythm, tempo, and timing became not simply additional attributes to the spatial characteristics of human ecology; rather with space they were jointly responsible for the maintenance of a living community. Everyday activities concentrated and segregated in time just as they did in space and were essential processes in the adaptation system adopted by 'man' in the development of 'his' urban environment. The resultant structures of the urban community were seen as being identifiable only when temporal aspects were considered coincidentally with the spatial aspects (Hawley, 1950).

'How do aggregates of individuals adapt to that common environment which is the root of community?' This was the fundamental question put by the human ecologists. But in spite of the obvious fact that human activities occur in space and times, it is their location in space which has received the most attention. Attempts to put the ecological perspective into a formalized space-time framework still remain to be made. The simplest ecological space-time question one might ask is, 'How is the space which contains a community occupied during various hours of the day?' From this starting point more complex issues can be raised.

Rhythm, tempo, and timing as initially conceived by McKenzie and further formalized as ecological adaptive factors by Engel-Frisch (1948), were seen as the basic ecological phenomena. The urban community and the sub-communities or neighbourhoods within, worked with greater or lesser ecological efficiency (i.e. experience less stress from the adaptation process) according to the extent that each of these three temporal factors could be coherently harnessed. **Rhythm** refers to recurrent cultural fluctuation or movement in accommodation to the physical, physiological, and cultural environment. (The associated times are universal, life, and social times and the metre of the rhythm will differ according to the time which is used, but as universe time is the only time for which a metric exists, a precise understanding of rhythms, which have para-times as their natural time, has yet to be achieved.) Rhythm is a temporal approximation to spatial pattern, a more complex parallel to our earlier association of interval and distance. **Tempo** is the number of events occuring per unit of time (as in time density, Chapter 7) and an increase in tempo is most likely to be a function of technological developments. It is the temporal equivalent to spatial density. **Timing** refers to the synchronization of rhythms and is equivalent to the pattern of spatial linkages or synchronizations among competitive and cooperative activities.

According to the classical ecologists the process of spatial invasion was an important aspect of adaptation. But there is also a process of temporal invasion involved in adaptive behaviour, for instance when 'a new rhythm is introduced upsetting the synchronized rhythmic pattern or timing system and perhaps its tempo' (Engel-Frisch, 1948, 45). Temporal invasion may result, for example, because of the introduction of a new public transport timetable or as the result of newly implemented work hours for members of a trade union or as a result of new opening and closing times for service premises. Such an invasion also occurs as the result of socio-technological innovation in which capital is substituted for labour, thus changing in a fundamental way the time environment of work places. With the invention of gas-light and later of electric light the most dramatic temporal invasion to have been experienced in human settlements occurred. The night frontier could now be crossed and new 'regions' of time could be colonized.

When, as a result of time invasion, a new timing of activities becomes a persistent attribute of everyday life, temporal succession is achieved. It is usually as difficult to reverse this process as it is to reverse and arrest the processes of spatial invasion and succession. Consider for just one moment the enormity of the investment which has been required, in personnel, capital, legislation and attendant political organization to arrest the 'natural' ecological processes of competition and co-operation which give rise to spatial

Rhythm

Tempo

Timing

Synchronization and synchorization

Temporal invasion

Persistence and temporal succession

invasion and succession, and the difficulties associated with control (planning) of the temporal environment will be put into perspective. This is underlined by the sentimental, symbolic, and even sacred values still attached to the time domain (as social times) in a manner which no longer really exists in the space domain.

Temporal segregation

Temporal segregation occurs where one pattern of timing exists within a longer pattern and it is a temporal equivalent to the notion of spatial segregation. In Chapter 3 we used the term synchronous timing in relation to the separation of item systems in time. However, while temporal segregation and synchronous timing both refer to separation, there is this difference: synchronous timing refers to activities occupying the same time location but they have no direct need of each other in the sense of being linked in some functional manner: they are functionally separate but located in the same time. Temporal segregation, on the other hand, refers to separation in time in which functional linkage, however weak, is a necessary component. Temporal segregation is an inclusive notion, synchronous timing is essentially exclusive.

Territorial subareas in the city may be segregated in time. For example, residential areas with many industrial workers resident in them will start their active occupation of the day before residential areas which have many resident office and other white collar workers. Areas with many young children will be active before areas without young children. Subareal differences in occupational and family structure of this sort contribute to the ecology of the city as dimensions of social space and it appears from factorial ecology research that social space differences are reflected in territorial patterns (Murdie, 1969; Timms, 1971; Herbert and Johnston, 1978). Thus it seems likely that it is only in conjunction with associated temporalities that ecological perspectives on the city can be given meaning.

Temporal dominance

Temporal dominance is assigned to any timing systems which 'significantly influences all other patterns and all other tempos' (Engel-Frisch, 1948). But Engel-Frisch's requirement that 'all other patterns and all other tempos' must be influenced seems as unnecessary as the requirement that a spatial ecological dominant like the CBD influences all other spatial patterns and all other spatial densities in the city. Temporal dominants include the 'working week' the '8-hour day', the 'school term and school day'. In this less restricted sense we can readily see that a temporal dominant of one structural dimension of social order (say social rank) is not necessarily a dominant for some other dimension (say family status). The school term and its day-to-day timing and tempo is clearly a dominant of high family status space but not so obviously of social rank or ethnic status.

Social rank

Family status

If different 'levels' of family status space have territorial pattern, for instance highest family status space on the periphery of the city (Murdie, 1969; Parkes, 1974; and Sections 8.6, 8.7) then the temporal dominants will differentially time territorial space, through social space. The social geography of the city is therefore inherently space-timed. Thus any statement in geography is incomplete unless it takes cognizance of time (Whittlesey, 1945, 33).

Before returning to some of these notions in Sections 8.6 and 8.7, two temporal features of contemporary urban society will be considered. The first of these is the process called the colonization of time and the second is the circadian temporality of certain social pathologies or socially undesirable events. Time budgets have provided us with a lot of information about the timing of 'normal' everyday activities in the city, but less about social problems of urban society.

It is perhaps worth recalling that the early work undertaken by the Chicago ecologists was in fact rather strongly oriented to similar problems, but emphasis was almost entirely on their spatial characteristics. The aim was to 'influence' the city administrators of Chicago to improve the lot of the urban community.

8.3   The colonization of time in the city

'As the city moves through phases of the day it switches from (essentially) coordinated actions to unconnected ones' (Melbin, 1978(b), 9). It moves from synchronized and synchorized actions to synchronous and segregated action. But, as time is colonized, especially beyond the frontier of the night this becomes less true perhaps because of an extension of the interdependence among urban items: in effect an extension of interactive urban ecosystems into the New Territories of the night.[3] The periodic night markets of Singapore, reported in Chapter 7, are an example (Yeung, 1974).

> 'Time, like space, is part of the ecological niche occupied by a species. Although every type exists throughout the 24-hour cycle, to reflect the way a species uses its niche we label it by the timing of its wakeful life ... diurnal or nocturnal. The same area of a forest or meadow or coral reef is used incessantly, with diurnal and nocturnal creatures taking their active turns. We make geographic (territorial) references to humans in a similar way ... to island people or desert people, or the people of the Arctic lands, as a means of pointing out salient features of their habitats ... time and space are containers for living'
>
> (Melbin, 1978(b), 5).

Towards the end of the nineteenth century the dominant realm of human migration began to shift from space to time. This process seems to be continuing. For the first time, in 1976, the US Bureau of Labour Statistics asked questions about the time of day that people worked. 12 million out of the US working population of 75 million worked at night and this did not include the clientele in hospitals, gambling houses, restaurants, etc.

Table 8.1. Number of radio stations and their hours of broadcasting in Boston, USA (Melbin, 1978(b), 4. Reproduced by permission of Murray Melbin and the *American Sociological Review*)

	The Span of Commercial Broadcasting									
	April 1929	April 1934	April 1939	April 1944	April 1949	April 1954	April 1959	April 1964	April 1969	April 1974
	Radio									
Number of stations	7	7	8	7	8	14	15	20	26	27
Number of 24-hour stations	0	0	0	0	0	1	3	8	12	15
Percent of 24-hour stations						7%	20%	40%	46%	57%

In Table 8.1 some evidence for contemporary time colonization can be found in the temporal growth of radio broadcasting in the Boston region (USA) between 1929 and 1974 (Melbin, 1978b). Similar evidence is easily found in Britain and elsewhere.

The Daily Telegraph, Friday, September 15, 1978

BBC SWINGS INTO ROUND THE CLOCK RADIO SERVICE

BY RICHARD LAST, TV and radio staff

FOR the first time in its 56-year history, BBC domestic radio is to go "round the clock" when the corporation's "middle of the road" music channel, Radio 2 begins a 24-hour service on November 27.

Settlement and frontier

The 'settlement of a frontier' is usually associated with acquisition of physical territory of some sort. But frontiers of time can also be settled. The more or less continuous occupation of space and time, by people and their activities, constitutes a settlement frontier where time is still relatively sparsely occupied by activities, but where occupation is increasing. A frontier exists where and when there is a consciousness within a society that there is something beyond the 'here' and the 'now'.

'The last great frontier of human immigration is occurring in time: a spreading of wakeful activity throughout the twenty-four hours of the day. There is more multiple shift factory work, more police coverage, more use of the telephone at all hours. There are more hospitals, pharmacies, aeroplane flights, hostels, always-open restaurants, car rental and gasoline and auto repair stations, bowling alleys, and radio stations always active. There are more emergency services such as auto-towing, locksmiths, bail bondsmen, drug and poison and suicide and gambling 'hot lines' available incessantly. Although different individuals participate in these events in shifts, the organizations involved are continually active' (Melbin, 1978a, 100). The remarkable composite satellite photograph of the USA at night illustrates the spatial extent of this time colonization rather well (Figure 8.1).

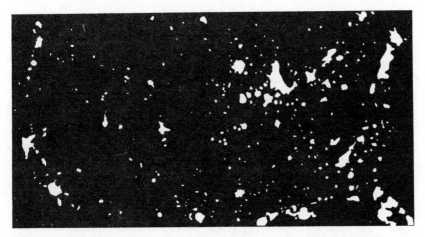

Figure 8.1. Satellite image of America by night. (Reprinted with permission from Chicago: Transformations of an Urban System, Copyright 1976, Ballinger Publishing Company)

The colonization of time parallels human occupation in space in many respects and the four-dimensional space-time container or context of social life is better understood when this parallel is kept in mind. Time is part of the container because it is occupied, along with three-dimensional space, by people and their activities. The total amount of time available in a day is finite (24 hours) but various episodes are also susceptible to colonization by expansion; by the use of new times by specific groups; by invasion. As with territorial expansion or migration this is usually initiated by a few bold pioneers. Thus Meier has observed that Singapore is becoming an intensive 24-hour city, with expansion into time well advanced. It is perhaps the best model for distributing activities round the clock, and it will become even more efficient when it is able to make full use of world market situations transmitted to it through communications satellites (Meier, 1976). The stable urban eco-system which Meier has recently discussed, requires that space and time are in balance and inducement into time colonization is unlikely to be independent of the particular spatial conditions of the system in terms of both territorial and social space.

Such colonization of time (especially into the night) suggests that two cities might co-exist within the same territorial space. Apart from the differences between the socio-demographic structure of the daytime active city and the night time city, Melbin has noted that at night time there is a decentralization of authority, a movement down the hierarchy of the equivalent day time authority rankings. The result is that at night time there are people making executive decisions (often in emergency situations) which would be referred to a higher authority during the day; for instance, nurses on night duty, foremen and assistant works managers in charge of night time production, etc.

Decentralization of authority

The ebb and flow of people involved in daily activities occurs within a four dimensional environment of time and space and all actions within it can be evaluated in terms of a newly named metric, the spant, an acronym for SPace ANd Time (Melbin, 1978a). 'Measuring people-per-spant may help explain why some recent studies have found no link between spatial crowding and social pathology, while earlier studies and some hypotheses say that they are linked' (Melbin, 1978a, 102). Melbin's term is useful and is no mere neologism. It allows us to put a name to the otherwise rather clumsy term 'four dimensional space-time'. The space-time metrics considered in the previous chapter might be measured in spants.

Spant

With the colonization of time, urban growth takes on new meaning. It requires a redefinition which incorporates the spant or something like it, so that increases in population have their full meaning conveyed, weighted in terms of space-time (spant) changes in the same period. Carlstein (1978) has adopted the term and relates it to the concepts of packing, time supply, time demand, and population flow.

What might some of the causes of the extension of the active day into the night be? Four factors have been suggested by Melbin (1978a) as:

(i) Enabling factors
(ii) Demand push factors
(iii) Supply pull factors
(iv) Stabilizing feedback factors

These operate as follows:

Causal factors in time colonization

(i) Enabling factors will usually be innovations of one sort or another. They enable the expansion of the active day. We referred in Chapter 7 to the innovation of a constant gas supply for light. This not only enabled houses and workplaces to be better lit, but the streets as well. In turn this allowed safer passage which enabled people to stretch the distance between sequential night time activities. Continuous gas supply also allowed heating. The workplace spant increased markedly. A whole new industry grew up around gas-light; the night time entertainment industry (Schlessinger, 1933, 105). Just as space colonization has relied on

Enabling

improvements in communication technology, time colonization has also enabled developments in communication technology and *vice versa*. International telephone calls across many time zones, necessary in order to maintain commercial market contact, are an obvious example. Then again large ports, air and bus terminals, operate incessantly enabling opportunities for communication which are the lifeblood of urban social and economic life. But the night time colony needs support, in much the same way as any 'colony' does, and so an available labour supply is a necessary enabling factor.

Incessance

(ii) Demand pushes many services into the night. Because most settlements are now part of a complex of urban systems depending on each other for inputs and outputs, there is a growth in time colonization as the daytime outputs of one place (space-time unit) become the night time inputs of another place. Night arrivals at airports, seaports, railway stations, and bus stations for instance, are only one manifestation of this demand push time colonization, fuelled by the global interdependencies of the settlement system. The demands of one activity are often chained to an antecedent event: a single disembarking air passenger demands a multimillion dollar support system; lighting, heating, safety (police), transport and communication, etc. Demand push of this sort is set up by chained activity phases, 'a series of procedures each of which depends on an event preceding it and in turn prompts another that follows' (Melbin, 1978a, 103-4).

Demand push

Large populations, especially in advanced scale societies, demand such a wide range of goods and services during the daylight period that it is necessary for numerous incessant utilities to be established. Hospitals, laboratories, police, and fire services, food processing and marketing establishments, are all examples of services which have been pushed by demand into the night hours.

There is also a density component in the demand push factor. The rush hour of many cities has become a crush hour on the public transport system. Staggered work hours or flextime systems which colonize new work times, are a common response to this excess demand for transport time.

(iii) The availability (or supply) of certain resources during the night has a self-reinforcing effect since it acts to attract people. Three strategies at least can be singled out in relation to this supply pull factor. The first of these is a maximizing strategy in which deliberate use is made of the facilities at hand in order to reduce overall costs. The services of large computing systems are an example, so are production systems with high shutdown and start-up costs. The second is a smoothing strategy aimed to make off-peak usage more attractive, thereby reducing peak demand. Electricity supply, transportation and telephone charges are examples. Such smoothing strategies impose a certain elasticity on the meaning of distance. Thirdly, there is the inducement

Supply pull

Maximizing smoothing inducement

component in supply pull: availability induces use. Galbraith (1958, 128-129) has called this the dependence effect, in which wants are increasingly created through the process by which they are satisfied.

(iv) Enabling factors allow the growth in active city time, but expansion is not unlimited — there are limiting factors or stabilizing feedback mechanisms in operation. There are at least three kinds of limiting factor. Social barriers such as transport systems closing down or operating at a reduced intensity 'send' people home before they might otherwise go. Zoning ordinances limit activities to specified zones and so wakeful activity is suppressed in some areas and stimulated in others. Licencing laws, social values and customs vary among the socio-demographic and territorial spaces of the city and the temporal invasion involved in colonization of the night is consequently differentially limited.

A further limiting factor is the scale of the territorial and temporal sustaining base in which any activity is located. The containing (space-time) spant must be larger than the boundaries of the specific spant. Within any city the existence of incessantly active centres may require the support of cyclic hinterlands

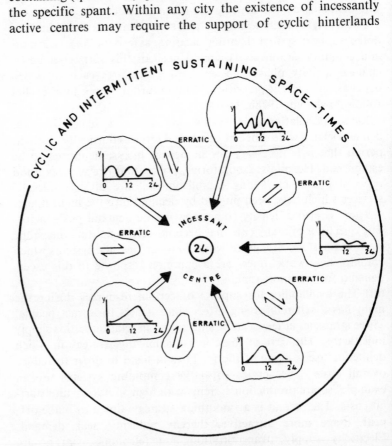

Figure 8.2. Incessantly active centres and their sustaining cyclic and intermittent space-times

(Melbin, 1978a). Thus, the strongest incessant centres will be located in the space-time centre of cyclic space-time regions. These cyclic regions in turn have zones of erratic space-time which are out of phase with each other. We have tried to illustrate this difficult idea by a simple sketch, shown in Figure 8.2. A combination of cyclic and intermittent, erratic space-times sustains the incessant centre and limits unconstrained colonization of time and expansion in space.

Cyclic and erratic space-time

A third limiting factor to time colonization acts as a negative feedback and is itself a function of the growth in active time which is available due to enabling factors such as lighting, telephones, transport and so on. This is the growth or (more accurately) perceived growth in night-time crime, which acts back upon the active growth to quell it (Melbin, 1978a). Small businesses like news stands, drugstores, garages, theatre box offices, etc. are especially vulnerable to the night-time increase in criminal late-phased activity. In Greenwich village, New York, news stand dealers, are now closing earlier, scared of hold-ups and harassment and their one-time after-dark customers are not venturing out either, thereby compounding the damage to their business. And there have been reports that some churches in a section of Boston have abandoned evening Mass because elderly worshippers were being mugged!

When an enabling resource like energy is cut off or reduced in supply such a situation can act as a limiting factor to time colonization. Melbin (1974) investigated this relation during and following the 1973-74 oil crisis, studying Boston's incessant centres and their adaptive response to a change in the availability of energy; a critical enabling factor.

As a rule 'we should review how city time (social time) might be used, with due consideration for the physiology of individuals (life times) and for impacts on families' (social time) (Melbin, 1978a). There is, for instance, evidence that price smoothing strategies can redistribute activities but care must be taken not to drive certain groups, especially the poor, farther into the night (Melbin, 1978a). Such smoothing strategies run the risk of *impelling* space-time ghettoes, increasing existing social and territorial space and time inequalities. Human geographers have yet to study these essential temporal inequalities (Smith, 1977; Bunge, 1974, 1975).

With time colonization there is a movement of physical and social items (in relation they define spaces and times) into new territories which then become the basis for the development of new ecologies. The niches in the city's eco-system, which an individual occupies, are always niches in space and time. Unless we assume that individuals adapt to their environment simply by some mindless mechanistic response, then it seems reasonable to assume that they hold pictures, images, or codes of some sort in mind, which are the basis for their behaviour in relation to new territories.

These images are constructed from information (Chapter 1.4) which arises from personal experience as well as from the generally available sources like radio, television, newspapers, friends and family, acquaintances, and strangers. Movement from one niche (a space-time unit) to another is a response to this information encoded according to a set of rules. Those who break the rules assign an additional dimension of meaning to ecological niches, encouraging or discouraging certain behaviours by others in relation to it. In 1971 intending French visitors to New York were able to obtain a map of the city, published by the newspaper *L'Aurore* (Rapaport, 1977, 154) which showed the areas which were unsafe at all times, (i.e. those which were incessantly colonized by pathological social behaviour) as well as areas which were cyclically unsafe at night (Figure 8.3) and areas which were intermittently or relatively safe. Presumably the visitors to New York, who had seen this map, adjusted the timing and spacing of their behaviour according to the information in it. Unfortunately there are not too many maps of this sort around and most people have to be content with compiling their own information and translating it into space-time pictures which will enable them to remain 'safe' in the city.

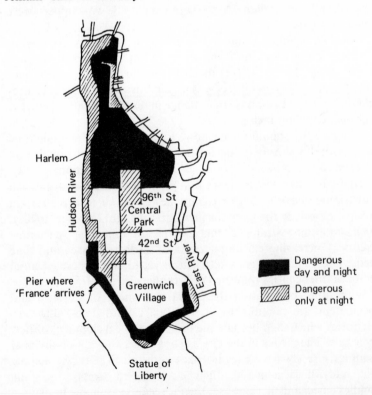

Figure 8.3. 'Safety' map of Manhatten. (From L'Aurore in *New York Times,* adapted from Rapoport, 1977, 1971)

The dominant circadian rhythm of most urban areas, together with certain higher frequency ultradian rhythms (Chapter 2), induces a temporal image which is likely to be sympathetically structured to incorporate this kind of experiential space and time information. Resultant behaviour will be a response to both. For many actions the image and the reality are so precisely congruent (as Golledge has shown in regard to learning behaviour, 1978), that the behaviour becomes almost instinctive, or perhaps more precisely habitual and undeliberated (cf. Chapter 5 and Cullen's model). Until very recently, however, urban geographers and city planners do not seem to have been very interested in trying to identify these images of the way different spaces in the city are timed nor do they seem to have been much interested in understanding the sense (Lynch, 1976) or meaning attached to them.

Before considering two empirical studies which have identified some of the cyclic, rhythmic characteristics in certain pathological urban conditions, we will describe the results of a recent study of some individual images of timed spaces in an Australian industrial city and the different meanings which were applied to the same space at different times of the day and night.

8.4 Images of timed space in the city

Over the past twenty years or so, and especially since the publication of Kevin Lynch's *Image of the City* (1960), there has been interest in the images that people hold of their city; images which, it has been suggested, are an important factor in structuring individual behaviour in the city (see, for instance, Pocock and Hudson, 1978). As has been the case with most of the ecological and general 'aggregate' survey based studies, time has tended to be absent as an integral factor in the study of the structuring of urban images except as it applies as a dimension in cognitive, learning process studies (see, for instance, Golledge, 1978, and Moore and Golledge, 1976).

It is not possible, nor is it necessary, to summarize the theory of urban image studies here. Two points only need to be made, one of which we will take up and illustrate with an example from some recent empirical work in Newcastle, NSW, Australia. First, the image that an individual has of any spatial context can be expected to change as he or she learns more about the situation. It can be expected to change with age (life-times) and with the experience gained from the utilization of different activity spaces and changing time horizons. In other words it can change over (universe) time with respect to individual and essentially personal factors, mediated by information from the social environment.

Gaining an understanding of these aspects of urban images is itself a considerable research objective. Second, we might consider the image that an individual has of a city at a particular point in time. Now this is where some troubles seem to arise; troubles which do not seem to have been investigated systematically. As we have seen, in this chapter particularly, the city has circadian rhythmic characteristics. At different times of the day there are in effect kaleidoscopic arrangements of different places; each time sees a different city. Do people who live in cities (knowing nothing of behavioural geography, urban ecology, social area analysis, and urban image structures!) seem to support this view by holding time-specific images of the city? Are there in effect images of timed space?[4] (Tranter and Parkes, 1977, 1979).

Lynch (1976) has drawn attention to 'the perception of parts of the city' and the variations in their level of activity over the day, Figure 8.4. (Also refer to Figure 8.7).

But we can go further than this by reporting the results of an experiment in Newcastle (Australia) which involved using five groups of twenty subjects and the methods of Kelly's personal construct theory (Kelly, 1955) and the repertory grid test (Kelly,

Personal construct theory

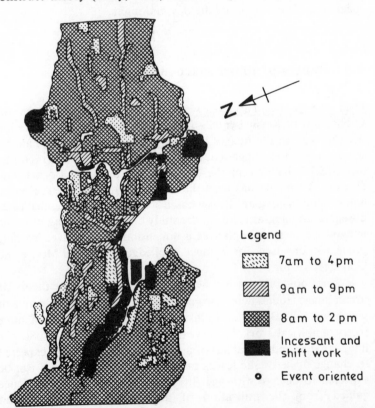

Legend

▨ 7am to 4pm

▨ 9am to 9pm

▨ 8am to 2pm

■ Incessant and shift work

o Event oriented

Figure 8.4. Time envelopes of Seattle, USA. (Adapted from Lynch 1976, 133. Reproduced by permission of the MIT Press. © 1976)

1963; Bannister, 1962; Sechrest, 1962; Bannister and Mair, 1968). Lynch (1976) has suggested the potential in this latter method for evaluating the 'sense of a region' with special regard to time, but has not published any empirical results to date. In the Australian study an evaluative approach to this problem was adopted where the objective was to seek out the meanings of the elements in the image.

One hundred and twenty subjects were assigned to one of five groups of twenty subjects. Four of the groups were given a particular time on which to structure their image, and the remaining group was treated as a control in order to produce a group with time free images, similar to those identified by Harrison and Sarre (1975) and all the other urban image studies with which we are familiar. The time structured groups were assigned as follows:

Group 1: The mid-point of the morning period of the normal working day: 10.30 a.m.
Group 2: The mid-point of the afternoon period of the normal working day: 3.30 p.m.
Group 3: The early evening: 8.30 p.m.
Group 4: The deep night: 2.00 a.m.

The control group was given no indication that time of day was a factor in the experiment and the experimental groups were not given any reason as to why they should only consider the study area at the time specified. Each subject was presented with a list of 25 standard elements obtained from pilot interviews. These standard **Elements** elements referred to 'well known' points in the city centre and the immediately surrounding zone of transition, for instance big department stores, the railway station, and the cathedral. A list of **Constraints**

Table 8.2. The standard constructs used to image timed spaces (Tranter and Parkes, 1977)

1	Feel at home in	versus	Feel strange in
2	Like	versus	Dislike
3	Beautiful	versus	Ugly
4	Visit often	versus	Visit seldom
5	New	versus	Old
6	Work	versus	Recreation
7	Cultural	versus	Commercial
8	Quiet	versus	Noisy
9	Busy	versus	Idle
10	Safe	versus	Unsafe

Table 8.3. Rotated components from the consensus matrices (Tranter and Parkes, 1977.)

	(a) Control group	I	II	III
		39·8	27·2	14·3
1	Feel at home in vs. feel strange in		0·86	
2	Like vs. dislike	0·62	0·72	
3	Beautiful vs. ugly	0·79	0·53	
4	Visit often vs. visit seldom		0·79	
5	New vs. old			0·92
6	Work vs. recreation		-0·67	
7	Cultural vs. commercial	0·70		-0·61
8	Quiet vs. noisy	0·95		
9	Busy vs. idle	-0·85		
10	Safe vs. unsafe	0·82		

Interpretations I Aesthetic/activity safety; II Evaluative/familiarity; III New vs. old

(b)	8.30 p.m. group	I	II	III	IV
		27·4	26·2	22·8	14·1
1	Feel at home in vs. feel strange in	0·48	-0·65	-0·49	
2	Like vs. dislike	0·71		-0·57	
3	Beautiful vs. ugly	0·88			
4	Visit often vs. visit seldom			-0·77	
5	New vs. old	0·05			0·95
6	Work vs. recreation			0·86	
7	Cultural vs. commercial	0·85			
8	Quiet vs. noisy		0·87		
9	Busy vs. idle		-0·94		
10	Safe vs. unsafe	0·63	-0·56		

Interpretations I Aesthetic/safety; II Activity/feel at home vs. strange; III Work vs. recreation/visit often-seldom; IV New vs. old

(c)	10.30 a.m. group	I	II	III
		38·7	24·5	20·1
1	Feel at home in vs. feel strange in		0·92	
2	Like vs. dislike	0·57	0·53	
3	Beautiful vs. ugly	0·67		0·58
4	Visit often vs. visit seldom		0·95	
5	New vs. old			0·51
6	Work vs. recreation	-0·81		
7	Cultural vs. commercial	0·89		
8	Quiet vs. noisy	0·86		
9	Busy vs. idle	-0·83		
10	Safe vs. unsafe	0·42		0·74

Interpretations I Evaluative/functional/activity; II Familiarity; III New vs. Old/Safety

(d)

3.30 p.m. group	Dimensions		
	I	II	III
	46·2	25·8	12·8
1 Feel at home in *vs.* feel strange in		0·96	
2 Like *vs.* dislike	0·73	0·58	
3 Beautiful *vs.* ugly	0·88		
4 Visit often *vs.* visit seldom		0·97	
5 New *vs.* old			0·96
6 Work *vs.* recreation	−0·68		
7 Cultural *vs.* commercial	0·84		
8 Quiet *vs.* noisy	0·91		
9 Busy *vs.* idle	−0·84	0·45	
10 Safe *vs.* unsafe	0·75		

Interpretations I Evaluative/functional/activity; II Familiarity; III New *vs.* old

(e)

2.00 a.m. group	Dimensions			
	I	II	III	IV
	28·8	17·7	27·1	13·9
1 Feel at home in *vs.* feel strange in	0·82		0·47	
2 Like *vs.* dislike	0·79			
3 Beautiful *vs.* ugly	0·47	0·80		
4 Visit often *vs.* visit seldom	0·83			
5 New *vs.* old				−0·94
6 Work *vs.* recreation	−0·61			0·48
7 Cultural *vs.* commercial		−0·94		
8 Quiet *vs.* noisy			−0·88	
9 Busy *vs.* idle	0·43		0·85	
10 Safe *vs.* unsafe			0·87	−0·41

Interpretations I Evaluative/familiarity/work *vs.* recreation; II Aesthetic; III Activity/Safety; IV New *vs.* old

10 standard constructs was also presented to each subject (described in Table 8.2) and in addition he or she used a set of personal constructs to describe the personal image of an element. For each of the five groups each subject scored the standard elements on a nine point scale and they also scored personal elements and constructs.

The basic tool in the analysis of the matrices for each group, in order to identify the factors in the underlying image, was principal components analysis.[5]

An important aspect of the repertory grid method, using personal construct theory, is that it allows the subject to construct his or her own 'personal' description of the elements presented. However, for our purposes here we will illustrate only the so-called consensus matrix results, in which the standard elements and constructs are used. With five groups there were five consensus matrices of scores with 25 elements and 10 'ways' of describing each one with a score with a range from 1-9. The principal components analysis of each group of 20 subjects produced components and loadings as shown in Table 8.3. The component structure between the analyses was tested for stability using a factor comparison test (Veldman, 1967) and the only two groups that showed a similar image structure were the 10.30 a.m. and 3.30 p.m. groups. In other words the images of the city centre and its transition zone were alike during the working day, but in the evening and again at deep night they were quite different. The control group held an image structure which was different to all others and was also most varied in its internal variability.

To provide a way to describe some of the variation in the component scores, for each of the spatial elements through the four times of the day, the scores can be mapped as shown in Figure 8.5. These graphs illustrate the relative locations of the standard 25 elements at two time locations, as they existed in the 'image-space' defined by the first two dimensions of a multiple consensus matrix (not shown here). This multiple consensus matrix extended the analysis that Harrison and Sarre (1975) undertook for Bath, by combining the consensus matrices of the four experimental groups into a single matrix of 10 standard constructs and 100 timed elements or spaces, i.e. the product of the original 25 elements × four groups.

In Figure 8.5 the lines link the same spatial elements at two temporal locations. In 8.5(a) the times are 10.30 a.m. and 3.30 p.m. The lines are much shorter than in the comparison of 3.30 p.m. with 2.00 a.m. (Figure 8.5(b). When the lines are 'short' it means that there is little or no change in the image construction of the set of spatial elements at different times. In essence this can be interpreted as meaning that 'the same place' is in mind. Thus interpreting Figure 8.5(a) it appears that between 10.30 a.m. and 3.30 p.m. there is no consistent change in the meaning of the elements, some become more active and familiar and others less so. This is indicated by the direction of the lines and whether the open circle is at the top or the bottom of them. If it is at the top and the line slopes down and to the left, then we have a spatial element which is becoming more active and more familiar as well as becoming more pleasant and associated with a cultural and recreational image rather than a commercial one. However, when the component scores for the same spatial elements are compared for the time locations 3.30 p.m. and 2.00 a.m. (8.5(b)), a very different picture emerges. Now the space-time drift of the elements is rather highly structured and almost without exception is from left to right and downward. The drift from left to right suggests a change of meaning from active and familiar to inactive and unfamiliar, implying the onset of a night frontier for these subjects rather than a time colonization or 'settlement', especially since the downward drift is from a relatively unpleasant commercial state to

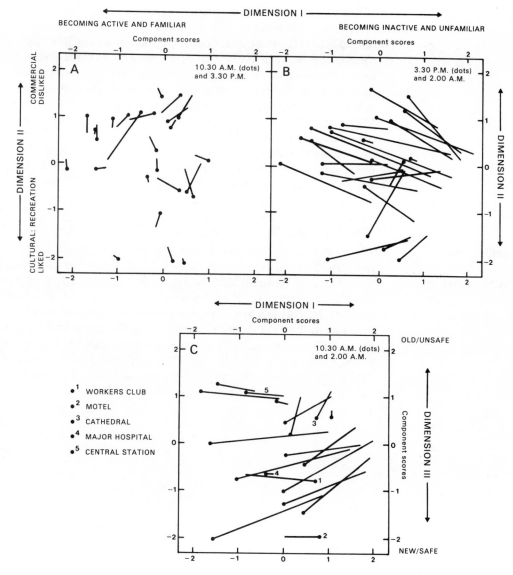

Figure 8.5. Shifts in image structure. (Tranter and Parkes, 1977)

a more pleasant cultural-recreational state. This seems to make some sense, especially since the subjects of the experiment were all in the age group 19 to 22. Note, however, that the drift in the vertical direction is not always very steep and also that the scores for many of the spatial elements on the vertical axis are around the middle of the scale, implying that intensity of the image is not very strong. On the horizontal dimension, *active and familiar to inactive and unfamiliar,* there is a very wide separation of the scores. The lines are long, implying that there is a lot of difference between the active and familiar afternoon and the inactive unfamiliar deep night.

Just as one dimension is usually not as informative as two, so it helps us to gain a better understanding of the city over a day if a third dimension can be introduced. Consider Figure 8.5(c) where a comparison between 10.30 a.m. and 2.00 a.m. is made. Now the first dimension is still the *active-familiar to inactive-unfamiliar* dimension but the one to which it is compared concerns newness, safety, and pleasing, as opposed to age, dangerous or unsafe and frightening. This is the dimension which the newspaper map of New York (Figure 8.3) described. With the approach used here it is possible to identify the drift of individual spatial elements in two dimensions. This is what the graph means: most elements become less familiar in deep night and are considered inactive, but a few also drift from being considered as generally safe and pleasing to being considered unsafe, dangerous and generally frightening. In effect they identify space-times (or *places* in the strictest sense) which are likely to be avoided. These are places which will not be in the individual's activity space and they exist in the action space only as a deterrent to behaviour. In Figure 8.6 a map of the safety-familiarity relation for the central business district and transition zone of Newcastle, Australia, illustrates the changing meanings attached to the 25 standard elements at the four times under study in the 24-hours of the experiment.(The solid 'black flags' indicate timed space when there is familiarity and safety and a general 'liking' of the place, the 'white flags' indicate lack of familiarity, danger, and general dislike).

Activity space

Action space

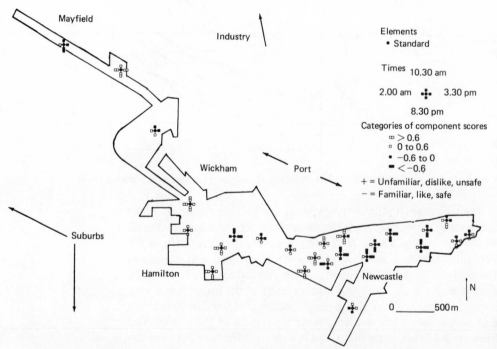

Figure 8.6. Familiarity and safety in the city centre: Newcastle (Australia)

Personal construct theory and, in particular, the repertory grid test method which has been used in this study allow subjects to define their own 'personal' constructs and elements. It is therefore of interest in the analysis of the element 'supergrid', which combines personal and standard elements with standard constructs, that a personal element was imaged either at a single time or, if in more than one time, then usually with a combination of the 10.30 a.m. and 3.30 p.m. times. Daytime, evening and deep night were well separated, distant environments. Why some space-times are 'thought' to be dangerous for particular elements and safe for others is not a question we can answer here, but some of the reasons will lie within the set of enabling, demand push, supply pull and stabilizing feedback factors discussed above (8.3). If it is thought that a city should always be 'safe' then immediate improvements in the physical elements of the environment suggest themselves, like better street-lighting, lighting or locking parks, improving police surveillance, or activity retiming and improved information.

It should now be clear that a single functional space in the city such as the central business district cannot be treated as homogeneous in space-time terms. At different times of the day and night there will be cyclic recurrence of normal and pathological incidents and the individuals who make up the city seem to have an image of this variability which is a determinant of some aspects of their conduct.

8.5 Some city rhythms

Generally speaking we live in an era of urban time growth, of growth towards incessant spaces. Incessant areas already exist in many of the world's cities, but taken as a whole, the dominant timing pattern is still one of cyclic variability rather than incessance. In such a situation a degree of symmetry from one cycle to the next is still necessary if residents of the city are to be able to programme their behaviour and meet others for satisfaction of needs, wants, and whims.

From one day to another urban spaces, whether as small as a hamburger bar or as large as the central business district, display a pattern of usage which is temporally structured, that is, they have a time shape or morphology. These patterns of utilization are manifested in the changing numbers of people 'contained' in a particular space-time (spant) unit; no-one at the hamburger bar at 0330, twenty people at 0730, etc. Such variations within 24 hours, as well as the symmetry from one day to another, when aggregated over activities, are the source of a city pulse; a symptom of city 'well-being'. However, this rhythmic order is interrupted by intermittent irregularities and disruptions from group behaviour like strikes, work-to-rule routines, riots, and hooliganism. Some of these erratic behaviours are confined to rather specific territories and times (e.g. gang 'turfs'). If a cyclic order is identifiable then there will always be a zeitgeber, an impelling factor of some sort which causes this pathology. One example is provided by the boisterous behaviour which follows large football matches in UK (and perhaps elsewhere). Another is provided by an economic crisis which inevitably leads to an increase in social unrest and movement.

Umwelt

City rhythms occur as a response to the time demands of the specific worlds (umwelt, refer back to Chapter 2), which make up its social mosaic. These specific worlds have their own appropriate times but are often in conflict with each other so that their manifest timing results from a combination of entrainment effects from an external zeitgeber and the particular timing needs of the function concerned. But the temporal dominant for even the most urban place is still the activity-rest relation: the wakefulness-sleep cycle.

As social structures, cities rely increasingly on the efficient performance of specialized roles. There is a need for finely tuned synchronization between roles as well as synchorization in the spatial location of activities. It is useful to identify the shape of the

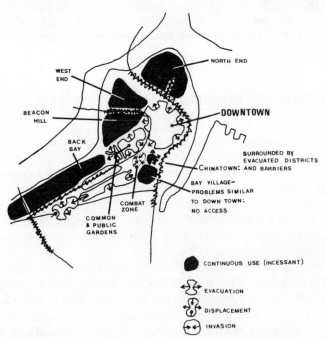

Figure 8.7. Cycles of use in central Boston, USA. (Lynch, 1976. Reproduced by permission of the MIT Press. © 1976)

'Evacuation'	: empty at night
'Invasion'	: active at night especially
'Displacement'	: shifting from day to night activity

time container which is occupied by various activities and to categorize urban subareas according to their daily temporal characteristics.

Lynch has done this for subareas in central Boston, Figure 8.7. Note the ecological emphasis: the utilization of terms which have usually been associated with purely spatial processes (Chapter 8.2 above). The cycles of daily land use in central Boston seem to be associated with four processes;

(i)	Continuous use	:	incessant areas
(ii)	Evacuation	:	empty at night
(iii)	Invasion	:	active expecially at night
(iv)	Displacement	:	shifting from day to night

Boston time

These four cycles describe particular aspects of Boston Time; its place-times, or ecological time in Gurvitch's terms (Chapter 2), because the changing relations which are observed occur, in the most precise sense, among social items and not amongst purely universe items. Although the changes amongst these items will be highly correlated with clock time, indeed many will be explicitly signalled by clock time, over the years that these activities have been spatially located in central Boston these social item relations have changed independently of the basis for clock time. Although clock time is a good basis from which to signal and monitor activity in the various subareas of central Boston, this should not preclude us from researching into other times which may be more enlightening.

Ecological time

As we have noted, time budget studies and space-time budgets are concerned with everyday behaviour, revealed through the 'normal' everyday activities that people participate in. Through them and the models upon which many of them are based, it has been possible to identify characteristics of the spaces and the times which contain recurrent behaviours. Time budget information also enables us to construct a description of the rhythmic characteristics of everyday life, with the aid of certain mathematical and statistical models, as we saw in Chapter 5. However, people living in cities, also experience disruptions to their routine everyday life. Unemployment, illness, accident, or the aggressive behaviour of others, combine with emotional disturbances (extremes of mood, loneliness, depression, and anxiety) to produce social problems which are in fact only the obverse face of the set of synchronized recurrent behaviours of 'normal' everyday life which time budgets capture.

Various public and private agencies exist to help people who have been trapped and to protect those who are in danger of becoming trapped, in one or more of the contemporary 'chronic' urban ecopathologies. A pathology is a condition which is 'harmful' to individuals and groups. In the growingly incessant urban centres of the world there is a heightened susceptibility to harmful circumstances. The agencies which maintain the well-being of the city's residents (and visitors) are costly. The maintenance of incessantly available police forces, hospitals, telephone services, fire brigades and other utilities is a considerable burden on the city's finances. Additionally, many private agencies also seek to provide a service to individuals and families in need. An improvement in the cost efficiency and the effectiveness of these services seems likely to result from a better understanding of the timing of the pathologies which they seek to intercept and to treat once they have occurred.

The city ecopathology and time

In order to illustrate this aspect of time and ecopathological features in the city, some results from two studies will be outlined. The first is based on Boston (USA) and the second on Newcastle (Australia).

(a) Boston's social pathological rhythms

Social pathology in Boston

The records of the Police Department and of emergency calls to telephone companies capture a wide range of the troubles that beset a community. Data gathered in Boston from the records of the police and telephone companies, for seven weeks during 1974 allowed Melbin (1976) to investigate:

(i) Whether there were noticeable fluctuations in the occurrence of such problems as drunkenness, public disorder, attempted suicide, stress, and mental illness over a 24 hour period?
(ii) At what time of day or night the highest frequencies occurred?
(iii) Whether some patterns were regularly in step with one another, always out of phase, or oscillating between the two states?
(iv) What is the possible significance of the results obtained for people who lived in the cities and for city administrators and agencies?

Sickness, drunkenness disorder

In Figure 8.8 the patterns for three types of pathology, based on police investigations into sick persons, drunkenness, and violations of public order are shown. An example of analysis of this kind of urban rhythm is given by comparison of these three series. Comparing reports of people being sick with the time series of reports of drunkenness showed an unlagged correlation of 0.52 [6] whereas between drunkenness and disorder the correlation was only 0.14.

But when drunkenness was allowed to lead disorder by about eleven hours in a lagged correlation analysis this relationship rose to 0.41. Similarly when disorder was allowed to lead drunkenness by two hours the correlation rose. For both conditions it became higher than when no lag was included at 0.35. This suggests that the cycles of these two pathologies are related but that there is a phase shift, possibly of up to ten or eleven hours between the onset of

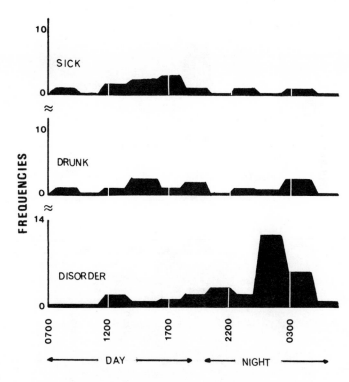

Figure 8.8. Police investigations into three categories of pathology: Boston, USA. (After Melbin, 1976)

drunkenness and outbreaks of public disorder. Possibly drunkenness and violation of public order are substitutes for one another (Melbin, 1976); caused by the same zeitgeber.

In Figure 8.9 the time series of four kinds of emergency calls to Boston telephone operators are shown. Each of these categories of individual suffering, whether self-focussed or self-inflicted, shows a peak in 'cries for help' in mid-afternoon or at about 1500.

For self-focussed suffering the peak seems to occur at about 1500. Distress is signalled at just the time when most people are up and about. Melbin's study covers a very short period and the frequencies involved are very small, but it provides a useful basis for illustration. The illustration may be extended to include the incidence of threats and fights.

In Figure 8.10 a record of the occurrence of threats and fights through the day and night is graphed. The correlation between the two pathologies is very high at 0.94 when a lag of two hours is used. This implies that threats precede fights by about two hours.

But apart from the kind of phase shifted relationships, which are implied by the lagged correlations, self-focussed problems, threats, fights, and public disorders are pathologies which exhibit a circadian rhythm. 'The rumblings begin in the morning and increase all day long, culminating in a blow-off of violence about

Circadian ecopathological rhythms

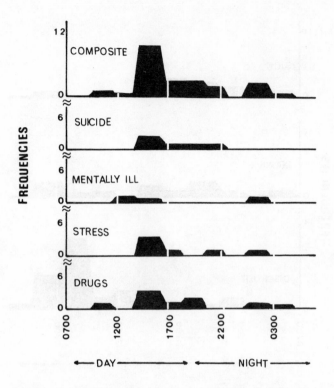

Figure 8.9. Self-focussed problems reported in emergency calls to Boston telephone operators. (After Melbin, 1976)

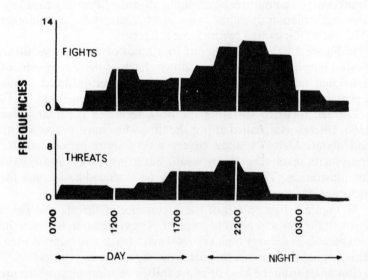

Figure 8.10. Threats and fights reported in emergency calls to Boston telephone operators. (After Melbin, 1976)

midnight' (Melbin, 1976, 6). The shape of this time is rather like that of a relaxation oscillator, described in Chapter 1. Each of the Figures (8.7-8.9) suggests tension being managed or released in the city over the course of the day. Each dominant frequency and time location has a particular phase relation to other pathologies and to 'normal' events in the life of the city. For instance, composite self-focussed problems and threats correlate at 0.62 when there is a six hour lead time for the self-focussed problem, suggesting a pattern of worrying and contemplating problems for about 6 hours and then a release of this tension as a threat. As we have already noted, the correlation between threats and fights is 0.94 with a lead time of about two hours. Reports of public disorder and fights also have a high correlation at 0.69, fights leading reports of more general disorder, noisiness and boisterous behaviour, by about two hours. Thus some people will begin the day by grappling with their own tensions inwardly, they then threaten others as the day wears on, commit violence and 'follow it with general boisterousness ... all over town' (Melbin, 1976, 7). What are some of the implications of findings of this sort? 'Isolated' individuals obviously do not realize the process which they set in motion by their conduct; but their individual problems, originating in scattered places are centralized by telephone appeals. Police and other agencies which are incessantly available are called up to deal with emergencies, 'the result is more interdependence and more reliance on standardized agencies rather than on the kinfolk equivalent in pre-urban communities' (Melbin, 1976, 8).

In large cities, people and activities have now spread out over all the twenty-four hours of the day. As a result of this colonization of time, there are forces at work in the urban environment that selectively disperse members of the population among the various hours giving a sort of time territoriality. Each phase of the day therefore has a somewhat different cultural atmosphere, forming an urban clock (Parkes, 1971). The day time, especially the afternoon, is the well-established settlement, the centre of society. It is a time neighbourhood in which formal and informal controls are plentiful and surveillance by other persons is everywhere widespread. Indeed these conditions weigh so heavily that they stifle the outward expression of tension, personal adjustment is achieved by bending one's raw edges inward. This holding back is also characteristic of the population up and about then. It has larger proportions of females and older persons. The innerfocussed, depressive forms of tension management show up at those times.

Time neighbourhood

As city life moves on into evening and the dark hours it is travelling towards its time frontier. The frontier is populated by younger people, most of whom are males. At the frontier it becomes customary to deal with problems by ignoring the law and taking matters into one's own hands. Some males step out of bars at closing time as they stepped out of saloons in the wild west!

Social life becomes more violent. Beyond that time, the deep night is an uncolonized region, a wilderness that knows no restraint except that which those who venture into it bring with them. In these population-sparse hours there are few others to observe one's conduct. Anonymity is conferred by the darkness and there is more uninhibited venting of tensions (Melbin, 1976).

(b) Newcastle, NSW, Australia

For our second example, the source of the data was a nationwide, private, telephone-based welfare agency, (Hotline) to which over 3000 recorded distress calls were made from within the Newcastle urban area during 1976. In addition, for the same year, over 12,000 records of accidents, emergency illness and other suffering were made available by the New South Wales Ambulance Service for the Newcastle urban area.[7]

Table 8.4. Hotline data record format: Newcastle, NSW

A	B	C	D	E	F	G	H	I	J
1	3	31	12	75	1055	2	08	05	04
2	3	31	12	75	1120	2	11	65	10
3	3	31	12	75	1122	1	13	03	10
4	3	31	12	75	1400	1	18	10	07
5	3	31	12	75	1445	4	21	08	10
6	3	31	12	75	1500	4	21	05	10
7	3	31	12	75	1530	3	11	10	
8	3	31	12	75	1800	3	13	15	
9	3	31	12	75	1755	1	01	05	
10	3	31	12	75	2340	1	20	10	
11	4	01	01	76	0055	3	13	10	
12	4	01	01	76	0108	5	23	02	
13	4	01	01	76	0255	2	09	20	
14	4	01	01	76	1525	2	08	20	10
15	4	01	01	76	1745	2	11	20	13
16	4	01	01	76	2215	3	11	02	02
17	4	01	01	76	2230	2	15	15	10
18	5	2	1	76	1615	3	12	07	
19	5	2	1	76	1705	1	11	10	12

A: counter
B: day of week, 3 = Wednesday
C: date of month
D: month
E: year
F: time by 24-hour clock
G: major pathology-category
H: minor pathology-category
I: duration of the call in mins
J: agency territorial code

349

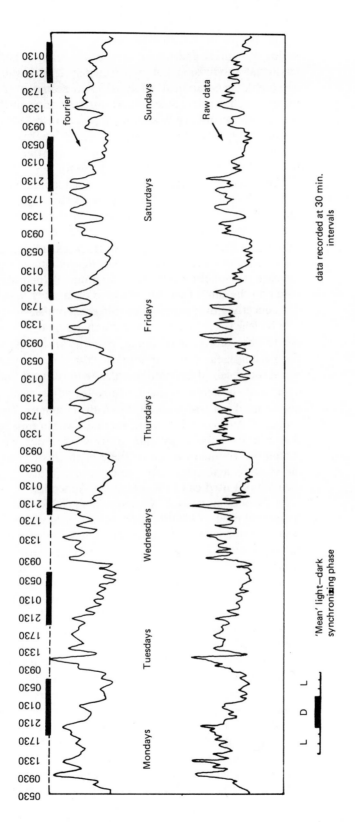

Figure 8.11. A first look at the pattern of distress Newcastle, NSW, 1976

The telephone counselling agency classified its calls into five major categories and about 25 subcategories. (There was also a territorial subdivision but unfortunately it is unsuitable for any acceptable geographical analysis). An example of the distress records is shown in Table 8.4. An interpretation of the hypercode for the first record is given below. But even with 3000 records and a period of nearly thirteen months some analyses are still not possible. Circannual rhythms cannot be investigated and nor can the effect of seasonal variation be assessed.

The interest therefore centred on circadian variability. For instance it is possible to treat data for all Mondays, as indicative of the variability within a typical single Monday. The same sort of aggregation can be applied to the working week for all days from Monday to Friday inclusively or for the weekend. Figure 8.11 shows the number of calls recorded at half hourly intervals (vertical axis) for each day of the week. The amplitude of the distress signals is based on the sum of all calls made to the agency, on the particular day concerned over a period of thirteen months.

It is clear from a glance at the graph in Figure 8.11 that there is a day to day recurrence of similar patterns of distress signals which can be analysed. (For our purpose, the time at which a distress signal was received has been rounded to the nearest 'hour' and 'half hour' and so there is some uncontrolled adjustment in the graph, but we need not worry about this at the moment). There are many statistical and control theoretic methods within time series analysis that can be applied to data of this sort, but we will simply summarize results of a few of these methods in order to illustrate the rhythms associated with these signals; (a brief explanation of the methods used can be found in the Appendix).

In chronobiology (cf. Chapter 2 above) a technique that is frequently used to identify a circadian rhythm is cosinor analysis. Here a cosine curve is fitted to the time series of the original data and represented in circular graphs of varying degrees of complexity along with certain other parameters which indicate statistical significance. You will recall that in Chapter 2.3 graphs of this type were used to illustrate circadian and circannual rhythms. In Figure 8.12 a cosine curve has been fitted to data for three of the five major pathology categories, covering the 13 months of 'continuous' records.

In Figure 8.13 the circular graphs show the distribution of the frequency of distress calls, again by half-hourly intervals, as well as the peak amplitude of the fitted cosine curve, known as the computed acrophase. The first circular graph, Figure 8.13, is equivalent to Figure 8.12, the others refer to the distress signals for the appropriate day or days, over the course of thirteen months.

Each of these pictures may be treated rather like a radar scan of the changing levels of distress in the city for the day or set of days concerned. The small 'hand' points to the peak which the best-fit

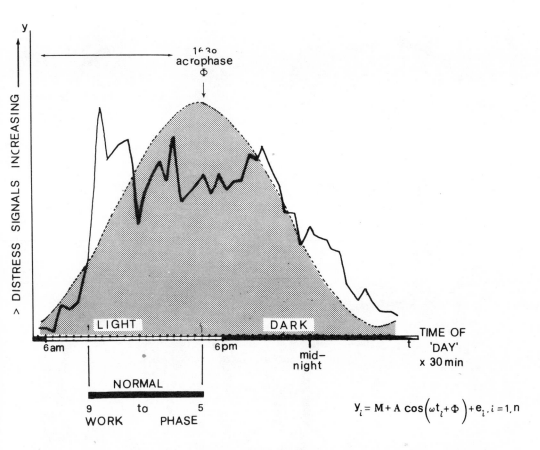

Figure 8.12. Fit and mis-fit of a simple cosine curve to a year-in-a-day of distress signals: Newcastle NSW

cosine curve identifies. Referring back to Figure 8.11 it can be seen that the signals fluctuate considerably. The cosine curve is by no means a perfect fit, especially insofar as small (or high frequency) oscillations are concerned. Other methods are clearly called for to identify the variability within the dominant circadian harmonic. To this end harmonic and spectral analyses become useful in identifying the underlying higher frequencies. From the preliminary analyses undertaken so far it appears that, apart from the rather obvious 24-hourly cycle, there are higher frequency cycles at about 12, 8, 5, 3.5, and 2.5 hourly intervals. These urban distress signals appear likely to have a frequency shape something like the generalization shown in Figure 8.14.

Autocorrelations and cross-lagged correlations enable further time domain characteristics of the pathology data to be studied. From the autocorrelations plotted as a correlogram (see Appendix) it is possible to get an indication of periodicity in ecopathological events. And the cross-lag correlations of one day's records against another day's records allows an indication of possible phase shifts.

NEWCASTLE SOCIAL PATHOLOGY STUDY 1976:FULL WEEK : RAW DATA

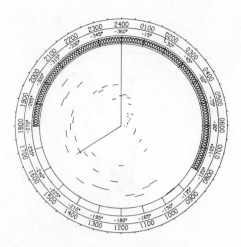

NEWCASTLE SOCIAL PATHOLOGY STUDY 1976:WEEKDAYS : RAW DATA

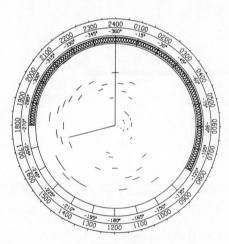

NEWCASTLE SOCIAL PATHOLOGY STUDY 1976:WEEKENDS : RAW DATA

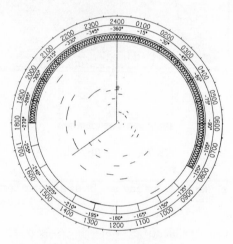

NEWCASTLE SOCIAL PATHOLOGY STUDY 1976:MONDAY : RAW DATA

NEWCASTLE SOCIAL PATHOLOGY STUDY 1976:TUESDAY : RAW DATA

NEWCASTLE SOCIAL PATHOLOGY STUDY 1976:WEDNESDAY : RAW DATA

353

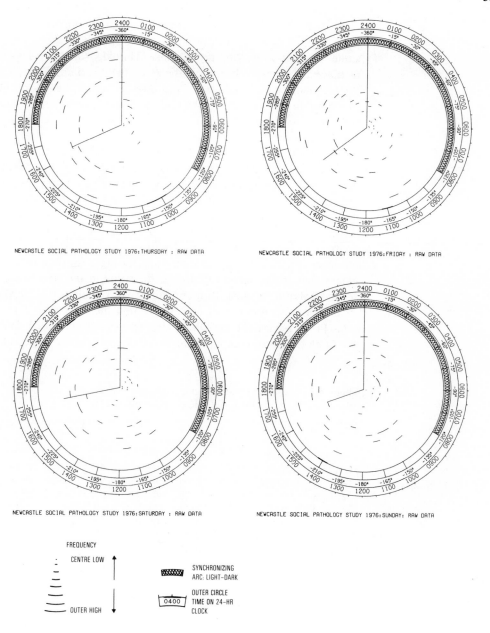

NEWCASTLE SOCIAL PATHOLOGY STUDY 1976:THURSDAY : RAW DATA

NEWCASTLE SOCIAL PATHOLOGY STUDY 1976:FRIDAY : RAW DATA

NEWCASTLE SOCIAL PATHOLOGY STUDY 1976:SATURDAY : RAW DATA

NEWCASTLE SOCIAL PATHOLOGY STUDY 1976:SUNDAY: RAW DATA

FREQUENCY

CENTRE LOW

OUTER HIGH

SYNCHRONIZING
ARC: LIGHT-DARK

OUTER CIRCLE
0400 TIME ON 24-HR
CLOCK

Figure 8.13. Frequency distribution and computed acrophase for distress signals: Newcastle NSW 1976, by Day of Week, Weekend, Working Week, Full Week

Comparing the correlation coefficients in the two matrices in Table 8.5 it appears that a half-hour lag does adjust the values slightly for some days. (Note Monday against Thursday and Friday for instances.) If these values are indicative of the correlation between the frequency of distress calls from one day to another then when the Monday times are lagged behind the Thursday and Friday times they become more alike than when the same times are correlated. Pathologies are signalled earlier on Mondays than on Thursdays and Fridays. At higher lags, for instance of four and four and a half hours, we find that Saturday lags behind a Friday and Sunday lags even further behind a Saturday. Whether the times involved are actually four and four and a half hours precisely cannot be established at the moment from the grouped data used but the direction of the phase shift seems appropriate and more rigorous analysis will provide better information. In the Tables in 8.5 the autocorrelations are in the diagonal and the lag is for the rows. This means that in the second table (b), the value in the first row and second column, against Monday and under Tuesday, is 0.59 when each Monday value is half an hour behind the Tuesday value. In table (a) however, for the same pair, the correlation is 0.63. When the same times are compared the two days are slightly more alike.

Table 8.5. Autocorrelations and cross-lag correlation coefficients in social pathology: Newcastle NSW

LAG = 0 for rows	$N = 48$							(a)
		Mon	Tues	Wed	Thurs	Fri	Sat	Sun
Mon	1	1·00	0·63	0·68	0·72	0·67	0·61	0·46
Tues	2	0·63	1·00	0·45	0·76	0·72	0·87	0·68
Wed	3	0·65	0·45	1·00	0·64	0·62	0·52	0·46
Thurs	4	0·72	0·78	0·64	1·00	0·69	0·65	0·57
Fri	5	0·67	0·72	0·63	0·60	1·00	0·52	0·58
Sat	6	0·60	0·57	0·53	0·65	0·52	1·00	0·55
Sun	7	0·46	0·68	0·47	0·58	0·58	0·55	1·00

LAG = 1 for rows	$N = 47$							(b)
		Mon	Tues	Wed	Thurs	Fri	Sat	Sun
Mon	1	0·67	0·59	0·66	*0·78*	*0·71*	0·55	*0·49*
Tues	2	0·48	0·60	*0·56*	0·57	0·55	0·61	0·55
Wed	3	*0·67*	0·44	0·50	0·58	0·52	0·63	*0·59*
Thurs	4	0·62	0·61	0·48	0·74	0·60	0·65	*0·59*
Fri	5	0·58	0·66	0·54	0·60	0·53	*0·56*	*0·56*
Sat	6	0·52	0·46	0·42	0·65	0·43	0·74	0·38
Sun	7	0·46	0·57	0·46	0·55	0·46	*0·72*	0·55

Correlations which increase after a lag of 1 unit have been italicized to assist the reading of the table. This does not mean any importance can be attached to the coefficients *as such*.

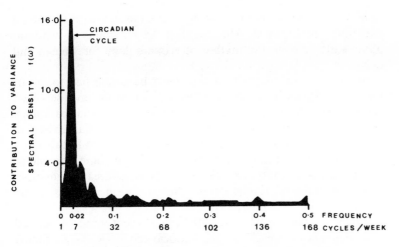

Figure 8.14. The time shape of social distress?

How do the patterns vary between the days? A summary of the difference between the days obtained from chi squared tests suggests that:

(i) There is a strong circadian variation in the frequency of distress signals.
(ii) The pattern of these variations is similar for all weekdays.
(iii) The pattern is similar for Saturday and Sunday.
(iv) The pattern differs from weekdays to weekends.[8]

From the first stages of the cosinor analysis it appears that the computed acrophase (see the Appendix) or peak of the best fit curve, appears somewhere between 1530 and 1630 on weekdays, but on Saturday and Sunday it occurs rather later at around 1700. On Mondays, Tuesdays and Fridays the calculated peak showed only three minutes of difference between the days, at 1535 on Fridays, 1538 on Mondays and 1536 on Tuesdays. In other words, if the circadian ecopathological rhythms of Newcastle do in fact have a shape which is reasonably approximated by a cosine curve (which has been found to have a good fit to many chronobiological data) then it is in the afternoon (as in Boston) that self-focussed problems, which most of these are, will reach their peak.

Apart from the 'Hotline' data discussed above, over 12,000 records of accidents and other emergency calls to the Ambulance Service were also collated from the original log sheets for the last month of 1975 and the whole of 1976. Analyses to date again suggest highly significant circadian fluctuations. Table 8.6 summarizes the acrophase and the incidence rates per thousand people for the events shown. It appears that females have accidents about 45 minutes earlier than males when all forms of accident are taken into account. However, when accidents excluding road accidents are taken into account females seem to have accidents (or to be in one way or another in need of emergency treatment) about two hours before men. This suggests the possibility a two-hour phase shift between the rhythm of ecopathology for women and men.

Because precise spatial locations could be established for all of the emergencies which were dealt with by the ambulance service it became possible to investigate certain aspects of social pathology rhythms in relation to the mosaic of urban social spaces in Newcastle, and also according to territorial categories like distance from home and distance from CBD. It also became possible to match the location of the pathologies with Census area information. Some preliminary results from this study in Newcastle are illustrated below (Tranter, 1978).

When the space in which a pathological event occurs is defined by social space dimensions such as socio-economic status, family status and ethnic status, then for all accidents and for road accidents treated separately the incidence rates per thousand people and the acrophase for each space appear to be something like that shown in Table 8.7. The social space categories are generalizations from a factorial ecology of the Newcastle urban area using 1971

Table 8.6. Acrophase from cosinor analysis and incidence rates per thousand people for various categories of social pathology: Newcastle NSW. (Tranter, 1978)

Category	Acrophase	Rate
All accidents		
— full week	15·27	14·25
— weekdays	15·05	9·87
— weekends	16·25	4·3
Road accidents	17·02	5·1
Industrial accidents	12·05	2·3
Domestic accidents	15·39	4·0
Sporting accidents	15·20	0·9
School accidents	12·26	0·58
Collapses	14·49	5·9
Heart attacks	14·05	4·2
Fits	15·00	0·7
Overdoses	18·25	0·8
Suicides	23·39	0·05
Brawls	21·54	0·27
Males		
— all accidents	15·42	—
— road accidents	17·39	—
Females		
— all accidents	14·58	—
— road accidents	15·37	—

Census data (Parkes, 1975). High socio-economic status areas are the residential areas of professional, managerial, and employer classes. High family status areas contain young families with school age children and relatively few working mothers. High ethnic status areas contain relatively large numbers of non-Australian, non-British residents. The results obtained so far suggest that, on the whole, the social space differences in the timing of pathologies are rather small. However, it is difficult to generalize because the data

Table 8.7. Social space and the incidence (I) and timing of accidents (A): Newcastle, NSW, 1976 (Tranter, 1978)

	Socio-economic status							
	High 1		2		3		4	Low
	(I)	(A)	(I)	(A)	(I)	(A)	(I)	(A)
All accidents								
— full week	1·16	14·44	1·33	15·25	1·41	15·19	1·56	15·52
— weekdays	1·11	14·21	1·30	15·11	1·44	14·34	1·51	15·16
— weekends	1·30	15·58	1·39	15·57	1·33	17·38	1·70	17·25
Road accidents								
— full week	0·38	16·40	0·48	17·20	0·47	16·04	0·48	17·40
— weekdays	0·23	16·48	0·48	18·10	0·43	15·15	0·47	16·48
— weekends	0·53	16·57	0·48	15·22	0·56	18·35	0·57	20·08

	Family status					
	High		Medium		Low	
	(I)	(A)	(I)	(A)	(I)	(A)
All accidents						
— full week	1·30	15·25	1·34	15·24	1·35	15·16
— weekdays	1·30	15·04	1·32	14·40	1·34	14·38
— weekends	1·40	16·21	1·41	16·54	1·40	17·22
Road accidents						
— full week	0·40	17·04	0·46	17·29	0·41	16·10
— weekdays	0·40	17·14	0·44	17·21	0·38	15·25
— weekends	0·57	16·34	0·48	17·45	0·50	19·12

	Ethnic status							
	High 1		2		3		4	Low
	(I)	(A)	(I)	(A)	(I)	(A)	(I)	(A)
All accidents								
— full week	1·46	15·21	1·20	15·04	1·32	15·27	1·48	15·44
— weekdays	1·48	14·41	0·92	15·05	1·31	14·40	1·37	14·57
— weekends	1·42	16·53	1·32	14·57	1·37	17·15	1·70	17·52
Road accidents								
— full week	0·40	15·36	0·43	16·54	0·44	17·51	0·5	17·37
— weekdays	0·45	15·06	0·42	17·08	0·41	17·38	0·40	16·32
— weekends	0·36	16·42	0·47	15·52	0·54	17·42	0·80	20·03

used in this analysis refer only to ambulance records, and as we have seen from the Hotline data many distress calls are transmitted in an urban area which never reach the ambulance service.

It is therefore not only the normal, everyday activities which have a circadian rhythm, and nor are these the only activities which are changing their temporal location, influenced by enabling factors such as improved transportation and communication. From the two examples above we have caught a glimpse of at least some of the circadian rhythmic characteristics associated with social pathologies. As more and more information is needed to maintain every single individual in the modern urban society, where specialized roles demand wide and efficient opportunity for interaction, knowledge about the *times* that events and activities are located in becomes increasingly important. Because this information is so important to the maintenance of an urban community many of the most recently introduced *rules* which control the flow of information and individuals relate to time use. However, as is the case for many aspects of social behaviour, it is not only on the basis of formally constituted and statutory rules that behaviour is predicated, it is also conducted in terms of subcultural value-systems and sign-systems. Thus one social category, for instance high status professional and managerial workers, may have a set of time-related values and semiologies which would not guide behaviour among unskilled workers. If differences of this sort do exist, and it seems certain that they do, then it would seem to be important to the ecological basis of the city that, where there are communitites of people with social and demographic characteristics in common, there should also be common timings and an awareness of the location of the community *in terms of the city's timing scheme,* just as there appears to be a reciprocal awareness of the more obvious variation in the location of one community in relation to others in terms of social space and geographic space. It is to some of the relations among social space, geographic space and these times to which we now turn. Once again the context is ecological and discussion is about group conditions rather than individual behaviour. The situation of the city which we will now discuss is within the 24-hours of a 'normal day' where;

> 'Different groups have different rhythms and tempos their synchronization may be a problem .... Given the extreme variability of time concepts and structuring, the sharing of temporal rules can play a major role in reducing the need to process information and make behaviour and reaction to the environment easier and more automatic ... activity systems in urban areas are intimately related to temporal rules ... it becomes important to live in areas where rule systems are shared and hence one understands when and where to do what ... and how spaces, activities and *times* are organized'
> (Rapoport, 1977, 343).

8.6 Aspects of time in social and physical space: an example

From a factorial ecology of the 1966 Australian Census and other field based data (Parkes, 1971, 1973) the formal social geographic mosaic of the Newcastle urban area could be determined. Resultant subareas and social spaces were used to structure a simple sampling frame so that 100 housewives, resident in high socio-economic status areas and 100 housewives resident in low socio-economic status areas could be interviewed about various time related characteristics of their daily lives. The temporal items of interest included time use, time perspectives and images of how time was used and valued in other social status neighbourhoods. The interviewing was carried out in the Australian summer of 1969-1970 (A more detailed account of the study which forms the basis of this section will be found in Parkes (1974). Figure 8.15 shows the location of the sampled areas. Three high status subareas and three low status subareas were selected. The high status subareas had a mean rank of 30 and the low status subareas a mean rank of 277, out of a possible 324 Census enumeration districts. It was also necessary to control for family status giving a mean rank for the high social status space of 155 out of 324 and for the low social status space a mean rank of 157 out of 324. Distance from the CBD was also controlled for and shortest road distances were 2, 5, and 11 miles for the three high status subareas and 2, 4, and 10 miles for the three low status subareas. The high status space and the low status space were both surveyed as a single social space without regard to territorial location. Due to errors, hostile respondents or incomplete schedules, only 90 records were obtained from the high status space and 84 from the low status space. But a check on the rejected subjects did not reveal any systematic bias.

Figure 8.16 suggests the association between territorial or physical space, social space and time factors: here time effectively sews together the disjoint spaces of Murdie's familiar diagram (1969). Physical space, social space, and time together act in concert as structural factors in the ecology of the city. The diagram summarizes the major components in the study, though all of them cannot be dealt with here (Parkes, 1974) (Some of them form our discussion in Chapter 2 on time and paratimes.)

An interesting point at which to begin the discussion is with the fact that each of the interviews was precisely timed. In the high status space the average duration of interview was 30.6 minutes with a coefficient of variation of 26.8 per cent. In the low status space the interview duration was 26 minutes with a coefficient of variation of 35 per cent. However, the estimated time duration of the interview of the high status group was 21.3 minutes while for the low status group it was 21.0 minutes. The coefficients of variation were 42.6 per cent and 46 per cent respectively. The

Factors in the ecology of the city

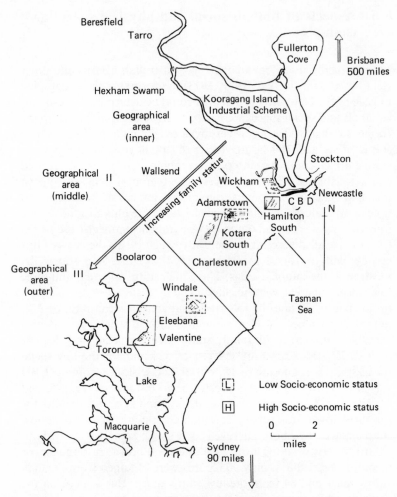

Figure 8.15. Subareas selected for sampling: Newcastle NSW (Parkes, 1974)

implication might be that the high status space subjects 'enjoyed' or were more interested in their interview session. The low status group 'got the interview over as quickly as possible' but still felt that it lasted as long as the interviews in the high status areas! Here psychological life time was involved and no doubt influenced responses.

From the recent multinational time budget study it appears that, on average, people engage in about 23 activities a day (Szalai, 1972). Cullen also found an average of 23 non-travel episodes of all sorts (Cullen and Godson, 1972) in London and in Washington, DC, Chapin found a mean of 18.4 activity episodes (Chapin, 1971, p.416 footnote 8). Are there interesting differences in the usual amount of time allocated to activities? Do the proportions of participants differ from activity to activity and between social spaces?

DIMENSIONS OF FORMAL SOCIAL–GEOGRAPHIC STRUCTURE

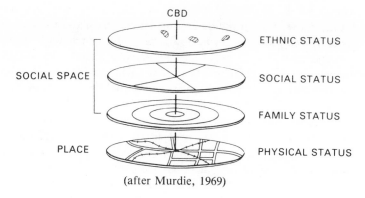

(after Murdie, 1969)

DIMENSIONS OF FUNCTIONAL SOCIAL–GEOGRAPHIC STRUCTURE

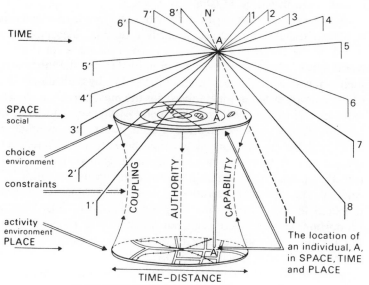

ELEMENTS OF SOCIAL TIME AND ACTIVITY

1 1' Circadian Rhythms (Bio–Psych.)	5 5' Time surplus–Time Deficit
2 2' Time Scale (UBT–UST) e.g. hour, day, year, life	6 6' Elasticity (Oblig.–Disc.)
3 3' Duration t_o–t_{o+m}	7 7' Time Perspectives / Images
4 4' Sequence	8 8' Monochronic–Polychronic

N N' Other elements of social time

Figure 8.16. Social space, physical space and time in the ecology of the city. (Parkes, 1974)

At this stage it is the inter-social space differences which are of most interest to us and territorial activity location is not considered. If a space-time budget study had been undertaken then the proportions of housewives, and the various non-residential space locations occupied *at the same time,* would provide an indicator of the divergence or convergence of social space elements into geographical space. This latter issue is taken up later in the chapter when a space-time factorial ecology design is proposed.

Many more activites were included in the Newcastle study than would fill a single day because a time diary was not used. Furthermore, not all the data considered were activity data, more general temporal information was also included. Interest concentrated on four aspects of an activity episode, its duration, the proportion of participants and the variability of times allocated as an index of activity elasticity, and the time location of the activity (See Table 8.8).

In calculating the amount of time allocated to activities the contingent N of the sample was used. That is to say, only the participants in an activity were considered. The mean time allocated to an obligatory activity is obviously the mean based on all subjects but for discretionary activity the mean is calculated simply on the number of participants and not on the total N of the sample. The notion of obligatory and discretionary activity has already been discussed in Chapters 4 and 5.

In Table 8.9 a selection of the activities and events to which time is allocated has been ranked according to their elasticity value based on the coefficient of variation. The ranking is for the high status space. (A rank order correlation of 0.69 for the thirty eight items shown here). There is no need for a blow by blow description of the data recorded in these tables; for the most part they are self-explanatory. In Table 8.8 however, items 41-46 and items 50-53 do perhaps require some explanation. Public transport is available to all city dwellers but we can ask whether there are any differences in the amount of time that housewives are prepared to allocate to travel on public transport to participate in various activities. For each of the activities considered, low status housewives were prepared to spend more time on public transport, indeed higher proportions of them were prepared to use public transport at all. For instance, only 46 per cent of the housewives resident in high status space were prepared to use public transport for daily shopping needs, and then were only willing to allocate 30 minutes to it. (This was 2½ times the length of the average shopping trip). On the other hand, 86 per cent of low status housewives were prepared to use public transport for daily shopping and they would allocate twice as much time to it. These results, along with many of the others in Table 8.7, suggest that there is a difference in the *cadence* **Activity cadence** of activity in urban social spaces, a difference which is also indicated by the elasticity values and the participation rates. Items 50-53 concern residential mobility and the temporal perspectives that housewives have about when they anticipate 'moving house' again, as well as the amount of time they might need, in terms of notice, for moves of varying length. Only 29 per cent of low status residents expected to move house again (possibly a function of the somewhat higher average age of respondents), against 49 per cent of residents in the high status group. Amongst those that did expect to move there was a 4½ year horizon in low status space and a 5½ year horizon in high status space. The elasticity was, however,

Table 8.8. Differences in time allocations, participation proportions and activity elasticity by social space: Newcastle NSW (1969-1970) (Parkes, 1974)

	Variable description	Means		% Participation Proportions			Elasticity $CV_H CV_L$
		1	2	3	4	5	6
		High	Low	High	Low	High %	Low
1.	Years of telephone ownership	7·4	4·7	76	(34)	99·0	143·5
2.	Years of residence at this house	9·2	(17·5)			93·7	80·2
3.	Age of respondent (years)	42·4	(49·6)			30·0	33·8
4.	Time husband leaves for work	7·9	7·2			13·2	32·7
5.	Time husband starts work	8·3	7·9			12·6	30·2
6.	Time husband finishes work	17·0	(15·4)			7·2	21·2
7.	Time husband arrives home	17·5	(15·7)			7·2	26·2
8.	Duration of journey to work (calculated)	0·4	(0·7)			70·4	48·5
9.	Duration of journey from work (calculated)	0·5	(0·7)			62·3	66·7
10.	Respondent's bedtime (weekday)	22·5	(21·9)			3·7	8·0
11.	Respondent's getting-up time (weekday)	6·7	(6·3)			18·6	13·7
12.	Sleep duration (weekday)	8·2	8·0			11·6	17·8
13.	Journey time to usual shops	0·2	0·2			105·0	62·5
14.	Time spent at shops	1·4	1·3			67·6	67·7
15.	Departure time to shops	11·3	10·9			21·0	23·0
16.	No. of houses visited (weekly)	2·1	1·6	82	74	79·1	75·5
17.	Duration of most regular visit	2·5	(3·9)	82	74	85·0	97·7
18.	Duration of second regular visit	2·6	(4·9)	63	54	101·6	114·1
19.	Time to furthest regular visit	0·3	0·6	82	74	61·8	218·6
20.	Time to nearest regular visit	0·2	0·3	82	74	61·9	186·7
21.	No. of telephone calls (weekly)	4·4	3·7	63	(19)	73·7	90·0
22.	Journey time library	0·1	(0·2)	46	(27)	120·0	55·0
23.	Listening to radio 5am-12 noon	2·4	2·4	69	71	77·6	75·7
24.	Listening to radio noon-5 pm	1·8	2·1	26	25	92·8	77·7
25.	Listening to radio 5pm-midnight	0·7	1·3	7	13	49·2	45·8
26.	TV 5am-noon	1·0	1·0	12	15	42·7	46·6
27.	TV noon-5pm	1·5	(2·9)	57	68	68·5	156·3
28.	TV 5pm-midnight	3·0	3·1	96	93	37·6	36·6
29.	ABC TV (National Network)	2·7	(1·5)			135·1	106·8
30.	NBN 3 TV (Local Network) (Commercial)	2·0	(2·9)			104·0	71·8
31.	Time on education (weekly)	4·9	2·6	27	25	73·6	175·9
32.	Time in a park (summer) hours	2·2	2·8	48	42	71·2	146·6
33.	Dept. time to park (summer)	12·2	11·1	48	42	16·2	38·1
34.	Journey time to park	0·3	(0·7)	48	42	179·4	145·9
35.	Time in a park (winter) hours	4·1	4·0	50	(34)	50·0	113·3
36.	Dept. time to park (winter)	10·7	10·4	50	(34)	19·1	37·2
37.	Days in sport (family)	1·4	1·2	66	56	115·3	122·7
38.	No. of visits to cinema (yearly)	6·7	6·8	56	44	88·7	153·6
39.	No. of visits to theatre (yearly)	4·1	5·5	46	(13)	117·4	129·5
40.	No. of visits to concert (yearly)	6·1	4·7	17	(3)	107·4	135·9
41.	Public transport/cinema	0·6	(1·0)	43	(76)	78·1	81·6
42.	Public transport/theatre	0·8	1·0	39	(70)	92·7	83·5
43.	Public transport/concert	0·8	1·0	27	(61)	70·7	85·3
44.	Public transport/beach	0·5	(1·1)	47	(63)	66·7	171·0
45.	Public transport/park	0·4	(0·8)	24	(55)	85·4	114·1
46.	Public transport/shops	0·5	(1·1)	49	(86)	68·5	76·7
47.	Time in hotel (hours)	1·5	1·9	3	(20)	66·7	83·6
48.	No. of evenings club (yearly)	8·2	12·8	57	58	109·4	133·5
49.	Time at club (hours)	3·8	3·8	57	58	49·6	107·3
50.	Anticipated change of residence (years)	5·7	4·6	49	(29)	93·4	108·0
51.	Days notice needed to change address within Newcastle	10·0	9·7			201·2	102·2
52.	Days of notice needed to change address within State	13·0	17·6			156·6	104·9
53.	Days notice needed to change address inter-state	17·4	26·3			142·8	110·4
54.	Estimated duration of interview (minutes)	21·3	21·0			42·6	46·0
55.	Actual duration of interview (minutes)	(30·6)	(26·0)			26·8	35·0

( ) Significant difference at 0·05 for t and z. All values are in hours unless otherwise stated in variable description; 0·1 = 6 minutes, 0·5 = 30 minutes.

slightly greater in the low status space. As the distance of the anticipated move increased, low status residents declared a need for more time (of notice) than the high status housewives. Elasticities were generally higher in the high status space but decreased as the distance increased. Newton (1972) found empirical support for these results in an independent residential mobility survey in Newcastle, Australia.

High status space housewives 'spent' more time, would be prepared to spend more time or needed more time, in only 2 out of 46 activities, compared with low status space housewives. For 13 activities the same amount of time was involved. Therefore in 68 per cent of the activities which were evaluated, low status space had greater time demands. Again, the suggestion is of quite marked temporal variability between urban social spaces.

Further insight into the temporal characteristics of social space can be obtained by considering the correlation between pairs of activities for those people who *participate*. The number of participants in pairs of activities can also be identified. In Table 8.10 the upper triangle of each matrix shows correlation coefficients which were greater than ± 0.2 whilst the lower triangle shows the number of participants in pairs of activities given. The diagonal refers to the number of participants in a single activity. In interpreting the size of the coefficients therefore, the contingency N or number of participants must be taken into account.

The upper matrix is for low status space, the lower matrix for high status space. The different values suggest that *social spaces* might be amenable to discrimination simply in terms of a range of *temporal* characteristics. Factor analyses of matrices such as those in Table 8.10 were undertaken to identify structural characteristics of the two spaces but were rejected because of instability or difficulty of interpretation. Instead, in order to pursue the possibility that the social spaces might be discriminated by temporal data, a linear multiple discriminant model was developed (Parkes, 1974). Using nine time related variables it was possible to reallocate all but 14 of the high status housewives (i.e. 76 out of 90) back to their original social space. With the low status group, where elasticities (or variance) were rather higher, all but 24 of the subjects were reallocated to the appropriate space (60 out of 84). To further test the discriminant model the 174 subjects were randomly allocated to one of two new groups and then, using the same data, the model was rerun. The reallocation changed to 47 out of 90 and 54 out of 84. The mean probability of correctly allocated subjects, in the social space discrimination, was 88 per cent. However, in the randomized grouping it was only 57 per cent. This is only marginally above the prior probability of a 50/50 chance of being right. The variables which were used in the discriminant

Table 8.9. Selected elasticities by rank order comparison of high and low social status space: Newcastle NSW (1969-70) (Parkes, 1974)

Time locations and activities by high status rank order (*low status in brackets)	$\bar{X}_{high}$	$\bar{X}_{low}$	percent C V	or elasticity	Signif. $\bar{X}(t)$	F	Low status * Most elastic
1. Bedtime (1)	22·5	21·9	3·7	8·0	Y	Y	*
2. H finishes work (4)	17·0	15·4	7·2	21·2	Y	Y	
3. H arrives home (6)	17·5	15·7	7·2	26·2	Y	Y	*
4. Sleep duration (3)	8·2	8·0	11·6	17·8		Y	*
5. H starts work (7)	8·3	7·9	12·6	30·2		Y	*
6. H leaves home (8)	7·9	7·2	13·2	32·7		Y	*
7. Get up (2)	6·7	7·3	18·6	13·7	Y		
8. Go to shops (5)	11·3	10·9	21·0	23·0			*
9. Interview length (9)	30·6	26·0	26·8	35·0	Y		*
10. T.V. 1700-2400 (10)	3·0	3·1	37·6	36·6			
11. Interview estimate (12)	21·3	21·0	42·6	37·0			*
12. T.V. 0500-1200 (13)	1·0	1·0	42·7	46·6			*
13. Radio 1700-2400 (11)	0·7	1·3	49·2	45·5			
14. Clubs (Time at) (30)	3·8	3·8	49·6	107·3		Y	*
15. T.V. Total (14)	5·5	7·0	54·3	50·8	Y		
16. Public Trans. (Cafe) (27)	0·5	0·9	60·4	93·3	Y	Y	*
17. Furthest visit (Journey) (38)	0·3	0·6	61·8	218·6		Y	*
18. Nearest visit (Journey) (37)	0·2	0·3	61·9	186·7		Y	*
19. Journey from work (16)	0·5	0·7	62·3	66·7	Y	Y	*
20. Public Transport (Beach) (30)	0·5	1·1	66·7	171·0	Y	Y	*
21. Hotels (24)	1·5	1·9	66·7	83·6			*
22. Shops (Time at) (17)	1·4	1·3	67·6	67·7			*
23. T.V. 1200-1700 (35)	1·5	2·9	68·5	156·3	Y		*
24. Public Transport (Shops) (21)	0·5	1·1	68·5	76·7	Y	Y	*
25. Public Transport (Cinema) (25)	0·6	1·0	70·7	85·3	Y	Y	*
26. Telephone Calls (Weekly) (26)	4·4	3·7	73·7	90·0			*
27. Radio 0500-1200 (20)	2·4	2·4	77·6	75·7			
28. Homes visited (weekly) (19)	2·1	1·6	79·1	75·5		Y	
29. Most regular visit duration (28)	2·5	3·9	85·0	97·7	Y	Y	*
30. Public transport (Park) (32)	0·4	0·8	85·4	114·1	Y	Y	*
31. Radio 1200-1700 (22)	1·8	2·1	92·8	77·7			
32. Change of residence anticip. (31)	5·7	4·6	93·4	108·0			*
33. Years of residence (23)	9·2	17·5	93·7	80·2	Y	Y	
34. Next regular visit duration (32)	2·6	4·9	101·6	114·1	Y	Y	*
35. Local T.V. (18)	2·0	2·9	104·0	71·8	Y		
36. Shops (Time of) (15)	11·3	10·9	105·0	62·5		Y	
37. Clubs (No. Evenings) (34)	8·2	12·8	109·4	133·5	Y	Y	*
38. ABC T.V. (National) (29)	2·7	1·5	135·1	106·8	Y	Y	

H = Husband

Significance tests for $X(t)$ and F were at the 0·05 level; Significant differences are Y.

* low status elasticity > high status elasticity and F significantly different, Elasticity increases, in general from 1-38, rho = 0·69, and r = 0·57: t tests for $\overline{X_h}/\overline{X_l}$ were based on $\sigma_1 \neq \sigma_2$.

Table 8.10. Contingency N and correlation coefficients among thirty items and activities for high and low social status space: Newcastle NSW (Parkes, 1974)

LOW

1	2	3	4	5	6	7	8	9	10	11	12	13	14	15	16	17	18	19	20	21	22	23	24	25	26	27	28	29	30
29	2	3	-2	3					-3	2			4		-2					4	9		2	-2	9		9		2
29	84	-3	-3	5	2					2					5	-2	-2			2	-2	-2	-2	-2		8	3	9	
NA	NA	NA	2	-4	-2						-2	3			-2			2		-2	-3	2	-2	3				-2	2
NA	NA	NA	NA	-4											-4	-2	4			3	-3	-2	-2	5				-3	2
29	84	84	84						2					-2	2	-2	-2			4	9	3	3	-2	-2		2	2	2
29	84	84	84	84								-8			3					3	3	3	2	-2			-2	-3	-3
29	84	84	84	84	3	-2				2	2					2			-2		2	3	2	-3	2		-2	-2	-2
29	84	84	84	84	84	2			2					3	2														
29	84	84	84	84	84	84	-3							3	2	2					-3			2	2				
29	84	84	84	84	84	84	84	84		2			-5	2	2		5	2		4	-2	3	3	2					3
22	62	62	62	62	62	62	62	62	62					3	3	2	2	2		-2		7	2		2	-2	-2		
21	59	59	59	59	59	59	59	59	59	59	59			7	5		4	2		-2	-2	7			-2	-2	3		
19	58	58	58	58	58	58	58	58	58	58	58	58			3	3	6			2	-2	-3	-3	-2		-3	-2	-2	
15	49	49	49	49	49	49	49	49	49	49	39	38	37	49	-5	-2	5	4	-2	2	2	4	2	-2	-2	-3	-2	3	-2
8	26	26	26	26	26	26	26	26	26	26	21	20	20	18	26		3	6		4	-3	-3	2		2				2
11	33	33	33	33	33	33	33	33	33	33	30	28	28	23	14	33					-3	4	3			-3	-3	-2	-3
23	60	60	60	60	60	60	60	60	60	60	47	45	44	37	22	28	60	6	5		5		2		2		-3	2	4
10	21	21	21	21	21	21	21	21	21	21	15	15	12	15	9	11	19	21				4	3	7		2		2	3
20	57	57	57	57	57	57	57	57	57	57	45	43	43	33	17	22	40	12	57	78	5	4	3		2		-2	2	2
29	78	78	78	78	78	78	78	78	78	58	55	54	45	24	28	57	19	53	78	21	-2	3	3	-4	-2	-2	-2		
4	21	21	21	21	21	21	21	21	21	16	15	16	15	8	11	14	4	16	20	21	-2	-2	3	-2	2			-2	
19	36	36	36	36	36	36	36	36	36	30	30	28	24	15	17	27	13	22	34	6	36	36	3	-4				-2	
19	38	38	38	38	38	38	38	38	38	32	32	30	25	15	18	28	13	24	36	9	36	36	2	-2	-2		4		
15	37	37	37	37	37	37	37	37	37	34	34	33	27	17	19	30	12	25	35	13	22	38	24	37	-3	-2	-2		
22	64	64	64	64	64	64	64	64	64	50	49	47	40	23	26	47	15	46	62	18	30	32	24	37	64	2	9	7	3
16	53	53	53	53	53	53	53	53	53	53	41	40	39	33	22	21	39	12	36	50	15	27	30	28	31	50	53	5	4
15	46	46	46	46	46	46	46	46	46	38	37	37	28	15	21	35	10	31	44	14	22	31	23	27	43	41	46	4	3
25	72	72	72	72	72	72	72	72	72	72	54	52	52	42	23	27	53	17	52	70	20	31	33	36	62	52	48	72	45
19	50	50	50	50	50	50	50	50	50	50	38	37	38	30	18	21	38	14	36	46	12	24	25	26	38	29	26	45	50

HIGH

1	2	3	4	5	6	7	8	9	10	11	12	13	14	15	16	17	18	19	20	21	22	23	24	25	26	27	28	29	30
69	6	-3		4		2								-2	-2			-2			-3	-2		-2	-4	-2	-2	-2	-2
69	90	-3	2	6	-2														3		-3	-2			-2		-2	-3	-2
NA	NA																												
NA	NA	NA	-6	-4	-2																								
69	90	90	90	90	90	90	90	90								2	3	2					2		2		2		
69	90	90	90	90	90	90	-7	90	-2								3	3			-3				-2		-4	-2	
69	90	90	90	90	90	90	7	90					-2								-2	-3			-4	-2	2	2	2
69	90	90	90	90	90	90	90	90																					
69	90	90	90	90	90	90	90	90		-2						3	2				-2			2				2	2
69	90	90	90	90	90	90	90	90		-2	-2					2	2				-2				2	2	2	2	2
56	74	74	74	74	74	74	74	74	74	72	72																		
54	72	72	72	72	72	72	72	72	72	72	72	72																	
51	68	68	68	68	68	68	68	68	68	68	68	68	68	59	40														
45	59	59	59	59	59	59	59	59	59	47	46	43	43																
32	40	40	40	40	40	40	40	40	40	35	34	33																	
33	46	46	46	46	46	46	46	46	46	40	40	37																	
45	61	61	61	61	61	61	61	61	61	51	51	49																	
14	23	23	23	23	23	23	23	23	23	18	18	15																	
38	51	51	51	51	51	51	51	51	51	43	42	39																	
66	86	86	86	86	86	86	86	86	86	71	69	65	57	40	44	58													
23	24	24	24	24	24	24	24	24	24	20	19	18																	
32	42	42	42	42	42	42	42	42	42	36	35	33																	
34	44	44	44	44	44	44	44	44	44	38	37	33																	
39	50	50	50	50	50	50	50	50	50	40	40	38																	
28	39	39	39	39	39	39	39	39	39	32	32	29																	
32	43	43	43	43	43	43	43	43	43	35	34	32																	
17	23	23	23	23	23	23	23	23	23	19	19	17																	
35	45	45	45	45	45	45	45	45	45	35	35	34																	
40	52	52	52	52	52	52	52	52	52	46	46	43																	

1. Years of telephone ownership
2. Years of residence at this house
3. Children under five years old
4. Children at school
5. Age of respondent
6. Respondent's bed time
7. Respondent's getting up time
8. Sleep duration
9. Journey time to usual shops
10. Time at shops
11. Departure time to shops
12. Duration of most frequent visit
13. Time to furthest regular visit
14. Time to nearest regular visit
15. Time writing letters (social)
16. Time in group activity
17. Time on novels
18. Radio 0500-1200
19. Radio 1201-1700
20. TV 1201-1700
21. TV 1701-2400
22. Time on personal education
23. Time at church
24. Time to church
25. Number of cinema visits
26. Public transport to cinema
27. Public transport to beach
28. Public transport to park
29. Public transport to shops
30. Time in clubs

Table 8.11. T-score group means and their differences for social spaces and randomized spaces

T-standard score $\overline{X} = 50\cdot0$, Sd $= 10\cdot0$

Variable	Social Space (S) HS	LS	Random Space (R) RA	RB	Difference |HS-RA|	|LS-RB|
1 UBT	46·6	53·6	50·1	49·9	3·5	3·7
2 UBT	47·7	52·5	49·9	50·1	2·2	2·4
3 UST	51·8	48·3	49·4	50·8	2·4	2·5
4 UST	50·9	49·1	50·0	48·9	0·9	0·2
5 UST	48·5	51·5	50·9	49·0	2·4	2·5
6 UST	50·3	49·8	51·3	49·0	1·0	0·8
7 UST	52·5	47·3	50·5	49·5	2·0	2·2
8 UST	46·4	54·1	49·6	50·6	3·2	3·5
9 UST	45·6	54·9	49·8	50·3	4·2	4·6
	$N=90$	$N=84$	$N=90$	$N=84$		

Description of variables
1. Years of residence at present address.
2. Age of respondent.
3. Normal time of getting up on a week day.
4. Normal time for going to shops for weekly needs.
5. Duration of most frequent visit to friends or relatives.
6. Journey time to furthest regular visit.
7. Time spent watching ABC (TV): This is a national, non-commercial channel.
8. Use of public transport for discretionary activity (cinemas).
9. Use of public transport for obligatory activity (shopping).

function are shown in Table 8.11 where the T-score group means and their differences are also given.

In selecting variables the object was to combine Small Time and Big Time characteristics of the social spaces and not simply to derive the most discriminatory set of variables, since this could easily be achieved by a step-wise multiple discriminant model.

Time discrimination of social space

The result is that the discrimination of social space by time is not as incisive (in terms of these results) as it might in fact be. The Mahalanobis D^2 statistic between the *social spaces* was 130.4 which implies a highly significant discrimination by the variables in Table 8.11, but for the *randomized spaces* the Mahalanobis D^2 value fell to 7.3. When subjects were grouped according to the three territorial/geographical subareas shown in Figure 8.14, with a high status cell and a low status cell considered together, there was again a significant discrimination but it was not as strong as when social spaces were used.

'As industrialization has meant increasing social differentiation, so also has it meant temporal differentiation' (King, 1976, 9). An urban area occupies a territorial space which it is always hard to define adequately. It is also a complex of sets of interlocking spaces and activities. Individuals wishing or needing to use these spaces and activities require high levels of accurate information about dependent elements. Satisfaction with life patterns, and perhaps the assurance of an adequate 'quality of life', will increase as the amount and accuracy of the information about the interdependence of activities in place (space and times) increases. Ability to take up the options which exist in daily prisms depends on a high level of mutual understanding by people about temporal differences in the time location and duration of activities in which they participate. The social area approach provides a useful first approximation to the social structure of an urban population, and the derived spaces appear to be differentially timed. But for the urban ecosystem to 'work' the individuals resident in these various derived spaces need to interact. A survey of the images that individuals from one social subspace have of the urban schedules of others may give some insight into the way the city coheres as a result of this interaction.

The linkage of individual activities into an urban system is based on a wide range of 'forms and complexities of collective activity' (Hawley, 1950, 207) and requires a highly developed time sense about the activities of other urban dwellers. Therefore, a means of reckoning time accurately is essential. 'Progressive involution in the broadened pattern of urban interdependences called for... refinement and standardization of time units. The timing of manifold interlocking sustenance activities, each with a rhythm peculiar to itself, depends on a system composed of minute and exact time units' (Hawley, 1950, 298).

In Chapter 2.3 (v) we referred to Frank's assertion of differences in the time perspectives of different social classes (Frank, 1939), How do status-distant groups consider each other's time allocations and values? How consistently do they describe each other's time use and time 'value' patterns? Questions of this sort are related to the intersubjectivity of temporalities as discussed by Schutz and Luckman (1973) to which we referred in 2.4 on social time. In an integrated and interdependent system, temporal images should be well focussed between social groups. Results of χ^2 tests should have high values if the images of one group complement the images of another. Results such as these would suggest that the differences among actual time allocation values recorded in Table 8.8 fit the 'perceived' time allocations and that social status differences are linked by images of different social 'times'.

Images of time use

Subjects in high status space were therefore asked to consider whether residents of low status areas (defined for the purposes of the interview by appropriate suburb or district names) undertook certain activities at times which differed from those they would expect of residents in their own 'home area' or 'neighbourhood' (Parkes, 1974). In addition they were asked if they allocated

Table 8.12. Images of elative time location of activities (Parkes, 1974)

	Earlier	Same time	Later
High Status	52	31	7
Low Status	9	40	36
(a) Getting up time:	$\chi^2 = 50\cdot909$,	df = 2,	$p = 0\cdot0000$
High Status	60	23	7
Low Status	4	35	46
(b) Leave home for work:	$\chi^2 = 80\cdot103$,	df = 2,	$p = 0\cdot0000$
High Status	51	20	10
Low Status	10	33	42
(c) Finish work:	$\chi^2 = 47\cdot404$,	df = 2,	$p = 0\cdot0000$
High Status	36	25	29
Low Status	11	27	47
(d) Arrive home from work:	$\chi^2 = 17\cdot509$,	df = 2,	$p = 0\cdot0004$
High Status	28	34	28
Low Status	8	33	44
(e) Go to bed:	$\chi^2 = 14\cdot551$,	df = 2,	$p = 0\cdot0011$

different *quantities* of time to selected activities, suggesting a different time value system and allowing some comparison with reported time allocations. The same set of questions were also asked of the low social status respondents. χ^2 tests were used as the standard method of assessment with contingency coefficients assisting the interpretation. Some of the results obtained are presented in Table 8.12 and Table 8.13.

High status residents generally considered that (a) low status residents got up at the same time or earlier than they did; (b) that they left home for work earlier than they did, and (c) that they finished work earlier than they did (Table 8.12). Possibly due to ideas about the incidence of overtime or shift work, there was also a tendency for high status housewives to think that low status workers left for work earlier than their high status counterparts, and also tended to finish work later, thus working a longer day, but the differences were very small as can be seen by comparing (b) and (c). With (d), arrival home from work, a considerable shift in the distribution of high status images about low status area work-residence journeys was revealed. Low status images remained more consistent but there was still a shift towards the 'later' category, (d). Despite leaving home earlier for work, and also finishing work earlier, low status space residents were also thought to arrive home later than high status residents. About 32 per cent of high status housewives had this image.

So part of the image that high status area housewives have of low status area work-journey schedules is of earlier departure times to work, but a somewhat protracted journey back home after work. But these views are not generally supported by the reported times — low status area residents do start work earlier than residents in high status space but whereas their journey *to* work times are longer (taking up to 20 minutes longer on average) their journey *from* work times, rather than taking longer than journey to work, took the same average time, although there was some increase in elasticity. It was the high status area residents who took longer to get home, but the difference here was only 6 minutes and this can probably be accounted for by a combination of sampling errors and the real 'delays' which occur with evening peak hour traffic. Evening peaks are likely to have a greater impact on travel times that morning peaks, because of the late afternoon spatial and temporal concentration of origins in the central core. This provides less opportunity for individual discretion about travel programmes, than occurs in the morning. Then again, the size of the population of the central area is increased in the late afternoon by a population of shoppers and other 'consumers' of urban services. These do not arrive in the central city area at the same time as workers, but tend to leave at approximately the same time.

As well as considering images of the differences in the time location at which activities occur, we can also consider images of differences in the duration of activities. Shopping is an important activity to any housewife, but as one moves up the income-status scale the obligatory proportion of all shopping trips probably falls. We might expect there to be few differences between the groups with regard to shopping duration images unless there were extreme geographical eccentricities in location of retailing outlets, thereby significantly disadvantaging one group more than the other. But turning now to Table 8.13 we see that 84 out of the 90 high status area housewives considered low status area housewives spent either more or the same amount of time than they did on necessity shopping, but 66 out of the 84 low status area housewives also considered that high status housewives spent more or the same amount of time on this activity. When such 'ecological incongruence' occurs between images, perhaps it means that the time component of shopping activity is not an important element in status differentiation. (But the amount of money spent is likely to be highly discriminatory!) Indeed, when the actual duration of activities was considered (in Table 8.8) no difference was apparent.

However, by contrast, images of the amount of time spent on personal *education* showed a very clear separation indeed. *Recreation* activities, like visits to the cinema, watching TV or visits to hotels (pubs) are obviously discretionary, but while there were some emphatic differences in images, they were not always consistently expressed by both groups. The low status image seemed quite clear; high status residents spend more time in *cinemas*. But in fact, actual differences of time allocation to cinema

Table 8.13.   Images of relative time-value of activities (Parkes, 1974)

	More time	Same time	Less time
High Status	40	44	6
Low Status	24	42	18
(a) Shopping (household necessities)	$\chi^2 = 9 \cdot 877$	$df = 2$	$p = 0 \cdot 0076$
High Status	1	3	86
Low Status	78	5	1
(b) Education	$\chi^2 = 158 \cdot 6$	$df = 2$	$p = 0 \cdot 0000$
High Status	12	59	19
Low Status	24	42	18
(c) At church	$\chi^2 = 6 \cdot 399$	$df = 2$	$p = 0 \cdot 0399$
High Status	32	31	27
Low Status	60	20	4
(d) Cinema	$\chi^2 = 27 \cdot 389$	$df = 2$	$p = 0 \cdot 0000$
High Status	8	19	68
Low Status	76	6	2
(e) Theatres	$\chi^2 = 122 \cdot 638$	$df = 2$	$p = 0 \cdot 0000$
High Status	78	10	2
Low Status	25	34	26
(f) Hotels	$\chi^2 = 60 \cdot 841$	$df = 2$	$p = 0 \cdot 0000$
High Status	69	16	5
Low Status	58	21	5
(g) Clubs	$\chi^2 = 1 \cdot 759$	$df = 2$	$p = 0 \cdot 5817$
High Status	3	22	65
Low Status	47	33	5
(h) ABC TV	$\chi^2 = 92 \cdot 281$	$df = 2$	$p = 0 \cdot 0000$

visits, in terms of evenings per year, were very small; (on average 6 evenings per year for high status and five for low status), although high status space does have a much larger proportion of regular cinemagoers.

On visits to *theatres* the images were more clearly congruent; this activity has greater *status accessibility* (Chapin and Hightower, 1966) differentials than cinema visits.

City-fast City-slow

City-fast and city-slow 'sectors' are potentially identifiable, not only in terms of the amount of time allocated to activities but also in terms of participant levels. We might consider an urban space, geographically or socially defined, as 'city-fast' if the activity participation level was high *per capita* and per unit of time. In other words, frequency of participation is an indicator of pace. Conversely we might consider an urban space as 'city-slow' where the participation levels are low. Taking the example of theatres, the central area location of theatres is in one sense, 'un-just' — high rental space is servicing only a small part of the population space-time, to which it is 'central'.[9]

As opposed to out-of-home recreation, like club and hotel visits, *TV watching* consumes a high proportion of the daily discretionary time budget... 'passive diversion activity is dominated by a single activity, television viewing' (Hammer and Chapin, 1972, 175). However, its consumption is again linked with status. Thus Hammer and Chapin also found decreasing amounts of time allocated to this single activity as status level increased. The type of television watched, whether on commercial or national channels, also varies with status (BBC, 1975). This is supported in Newcastle, Australia, where ABC TV (the national network with no advertising) was associated with high status space viewers. Both samples supported this quite clearly. Actual reported time differences (Table 8.8 variable 29) confirmed these images. High status ABC viewing averaged 2.7 hours, low status averaged 1.5 hours. Whereas evening viewing is very consistent between spaces, in total time allocated ($\simeq 3.0$ hours), the proportions (96 per cent and 93 per cent) for high and low status, and elasticity, 37.6 and 36.6, watching TV during the day does differentiate the spaces. As in Chapin's Washington study, high status area residents were found to spend less time during the day on this passive diversionary activity than low status area housewives (Hammer and Chapin, 1972).

Finally and very briefly, another aspect of temporality in urban ecology can be revealed through evaluation of the time perspectives or horizons of differing social status spaces; assuming once again that the aggregation of the images of sampled individuals reflects the attributes of the space.

Time horizons

LeShan has proposed a systematic relationship between time perspectives and social status (LeShan, 1952; Orme, 1969). Other differences might also be identifiable in the values attached to the past, present, and future by residents of socially distant spaces. As well as considering such differences according to social status, subjects were regrouped according to geographical space (Figure 8.15) and this meant that pairs of high and low status spaces were amalgamated. Differences in time perspectives were again tested using the χ^2 test. Subjects were simply required to order the past, present, and future according to how important each was in their own lives. Whilst this is obviously a very simple criterion the results shown in Table 8.14 suggest that status differences in time perspectives are more clearly established in social status space than in the simple concentric configuration of geographical space.

These results imply that the past, present, and future are rated differently by residents of different social space. However, there does not appear to be any difference when the two social space categories used here are amalgamated by a territorial grouping. If time perspectives are related to behaviour, and it is rather hard to imagine that they are not, then the day to day ecology of the city must take account of this factor. Explanations of everyday behaviour in the city will have to be made in terms of social space differences *and* associated time perspectives.

374

Table 8.14. Differences in time perspectives: Newcastle NSW (Parkes, 1974)

		χ^2	df	p	Sig. at 0·05	Interpretation
1.	Past	6·027	2	0·0481	Yes	Different
2.	Present	8·687	2	0·0131	Yes	Different
3.	Future	12·462	2	0·0025	Yes	Different

above: social space of high and low social rank
below: territorial subareas of inner, middle outer

		χ^2	df	p	Sig. at 0·05	Interpretation
1.	Past	6·298	4	0·1776	No	No difference
2.	Present	3·940	4	0·5843	No	No difference
3.	Future	2·237	4	0·6957	No	No difference

But these perspectives do not vary simply according to social status. They also vary demographically, especially by age. Whereas the image of space is probably more important to the young child, than is the image of time, in old age it is the image of time which dominates. Indeed, 'places' may recede into a three-dimensional time container, for which *when, what happened* and for *how long*, are the dimensions of reference. Time becomes the life space of the very old. And because time gives 'order to events by identifying them as co-existing or successive' (Lynch, 1972, 120) it follows that the evaluation of the relative importance of the past, present, and future to people of different ages and social status has some significance to their use of time and space in the city.

The Newcastle study underlines this point. For instance, housewives under the age of 45 (Table 8.15) showed no difference in their ranking of the past, but those over the age of 44 did show a difference. High status area housewives valued the past more highly than those resident in low status space. Whilst there was no difference in the rating of the present by either age group, in their rating of the future both age groups differed. The low status area respondents valued the future more highly. Differences thus seem to be greatest between social space for the older age groups. And where age differences are at their greatest, time perspectives show the greatest divergence. This combination of status and age differences is probably a major factor in the development of variable activity and action spaces in the city. We can now draw tentative conclusions about some of the temporalities involved in urban ecology from the examples given above.

Differences in time schedules and different time use images have been suggested. The small sample and the method of time data collection used only allow ordinal differences to be identified and the crucial problem of the size of differences remains to be studied. But assuming that these results do have some substance (and accepting statistical significance as indicative of real differences) we can now summarize the temporal dimensions of social space revealed by the study of Newcastle.

Table 8.15. Time perspectives, social status and age group: Newcastle, NSW (Parkes, 1974)

	χ^2	df	p	Sig. at $0\cdot05$	Interpretation
Past (under 45)	$0\cdot246$	2	$0\cdot8843$	No	No difference
Past (over 44)	$9\cdot714$	2	$0\cdot0081$	Yes	Different
Present (under 45)	$2\cdot994$	2	$0\cdot2226$	No	No difference
Present (over 44)	$5\cdot237$	2	$0\cdot0715$	No	No difference
Future (under 45)	$5\cdot920$	2	$0\cdot0507$	Yes	Different
Future (over 44)	$7\cdot936$	2	$0\cdot0188$	Yes	Different

(i) Many of the activities undertaken in common by residents of high and low social status space differ in terms of the time allocated to them. But, in general, it is the number of participants in an activity which shows the greatest difference, and not the amount of time spent on that activity or the frequency of engagement. Many activities are more or less time 'fixed' in terms of location and duration.

(ii) For any social status space it is possible and informative to isolate the elasticity of activities. Low status spaces generally reveal higher group elasticities. Whilst in many cases this may be a function of other dimensions like family status or ethnic status or of relatively less homogeneity in low status space, other causes ought to be sought.

But there are a number of problems associated with interpreting comparative cross-sectional data using an elasticity measure based on the coefficient of variability if there are sampling errors in the groups which are being studied. The propensity to make trade-offs in time allocations might also prove useful to study, but finding a measure of this propensity could prove difficult.

From the planners point of view, it might be useful to predict changes in activity participation, or in time allocated to activities, or indeed changes in the sequence of activities which result from a unit change in the amount of time available in some defined finite period, such as a day or a working week.

(iii) Activities which are inelastic in high status space will generally be more elastic in low status space.

(iv) If higher participation levels in out-of-home activities tend to be found in the high social status areas of the city, then these areas may be described as 'city fast' and the low status areas as 'city slow'. But such a simple dichotomy as this is complicated by consideration of a multidimensional social space. For instance, low social status, combined with high family status, will run slower than low social status combined with low family status. However, one possible exception to this is that low family status is also indicative of older age concentrations. In this case other life time factors complicate the time dimension.

(v) Socially distant groups hold temporal images of each other. This is a necessary condition for integrated urban living.

(vi) Time related variables, in a linear combination allow the allocation of subjects to social space. But there was a weak theoretical basis for the selection of these variables. Some time framework, analogous perhaps to the social area typology, needs to be developed as a basis for variable selection, and if possible, combined with the social area variables into a single composite 'model' of urban social space-times.

8.7   Urban factorial ecology in space and time

In an earlier section (8.2) we summarized aspects of urban ecology and pointed to the neglect of the role played by time in the study of adaptation processes in urban areas. In this section we illustrate one approach to the study of the ecology of urban areas aimed at rectifying this situation. This approach adopts an 'absolute' space and time framework in contrast to that adopted in the previous section where we discussed aspects of time in urban social geography using a 'relative' space and time approach. Chapman has recently supported the value of this absolute time and space approach although emphasizing the need for an appreciation of relative space and time concepts (Chapman, 1977, 37). The reader should not be misled into concretizing these categories of absolute and relative space and time. It is the notion of space and time as dimensions of a 'container' within which human action is set which is important.[10]

The city described in the factorial ecology literature is essentially static. The source of data is usually the various national population censuses. These record the characteristics of the *residents* of the city as they exist at midnight on census night. However, every workday morning people commute from their homes to other parts of the city to carry out their day time activities; usually with a regularly recurrent pattern, interrupted from time to time by erratic and innovative actions. Thus conventional factorial ecologies only tell us about one aspect of the city's social geography — as it is at midnight. But with a space-time factorial ecology the changing social geography of the city through a 24 hour period can be illustrated. The same basic procedure is used as that adopted in conventional factorial ecologies but because 'real' data of the sort required are not yet available, the study we report here (Taylor and Parkes, 1975) is *experimental*. It uses artificial data generated by the authors (For a full discussion of the method of data creation see the original article in *Environment and Planning*).

One reason why social geographers have concentrated on midnight spatial patterns has, of course, been the ready availability of data from census sources. There is no other equivalent

Table 8.16. The matrix of experimental data for a space-time ecology[11] (Taylor and Parkes, 1975. Reproduced by permission of Pion Ltd.)

Number	Zone	Sector	Period	Residential	Retail and service	Other commercial	Industrial	Entertainment	Open space	Population	Persons/dwelling	Male adults	Age 0-4	Age 5-17	Age 18-29	Age over 65	Married adults	Employed females	Upper class	Lower class	Professional services	Manual services	Private transport	Public transport	Access
1	1	0	1	10	00	00	00	03	00	08	10	45	01	04	45	20	51	07	15	60	05	18	02	00	08
2	1	0	2	10	05	00	05	00	00	65	08	35	00	02	45	10	62	22	22	50	10	25	15	15	10
3	1	0	3	01	30	30	10	00	05	80	01	65	00	02	46	05	76	65	29	42	60	23	08	06	10
4	1	0	4	06	30	25	03	06	05	78	08	60	00	02	46	05	72	63	29	43	58	23	09	06	10
5	1	0	5	01	30	30	10	07	05	85	01	50	00	02	40	08	77	60	29	41	62	24	13	14	10
6	1	0	6	07	07	20	05	07	09	05	66	07	50	01	04	45	10	64	30	28	43	42	23	11	09
7	1	0	7	06	05	00	01	20	05	15	08	50	00	06	61	05	44	15	19	58	10	20	05	02	09
8	1	0	8	06	01	00	01	14	00	10	09	55	01	04	50	10	46	09	19	58	07	18	03	01	08
9	2	A	1	50	00	00	00	01	00	11	20	48	05	10	20	20	65	00	40	21	01	01	01	00	07
10	2	A	2	40	05	00	00	00	00	08	18	45	08	12	15	30	65	03	43	19	01	15	10	05	08
11	2	A	3	25	16	08	04	00	07	08	11	40	10	15	10	40	64	18	40	14	09	12	04	02	09
12	2	A	4	35	12	08	02	02	07	08	17	44	08	11	16	31	75	13	40	24	09	12	05	02	09
13	2	A	5	25	16	08	04	02	07	08	12	40	11	14	10	40	62	14	35	29	09	12	04	02	09
14	2	A	6	38	11	02	02	03	07	09	16	43	07	13	14	30	65	08	40	22	05	06	09	04	08
15	2	A	7	42	04	00	00	04	07	10	16	47	06	11	18	25	66	04	40	21	01	02	03	01	08
16	2	A	8	45	00	00	00	03	00	11	17	48	05	10	19	22	66	03	40	21	01	01	02	01	07
17	3	A	1	72	00	00	00	00	00	10	31	49	08	15	23	12	73	00	40	20	01	01	00	00	06
18	3	A	2	64	00	00	00	00	00	07	24	38	10	17	19	18	74	09	42	19	01	13	08	04	07
19	3	A	3	56	12	04	01	00	11	06	16	25	13	20	14	23	73	11	40	21	06	11	05	02	07
20	3	A	4	60	10	04	00	01	11	06	25	30	11	17	18	18	73	09	37	25	06	11	05	02	07
21	3	A	5	57	12	04	01	00	11	06	16	20	13	20	14	24	72	06	39	31	06	10	05	02	07
22	3	A	6	65	08	00	01	01	11	09	25	36	09	18	16		73	03	40	20	03	05	070	03	07
23	3	A	7	70	02	00	00	02	11	09	26	48	09	16	21	13	77	02	40	20	01	02	02	01	06
24	3	A	8	71	00	00	00	01	00	10	29	49	07	15	22	10	76	01	40	20	01	01	01	00	06
25	4	A	1	70	00	00	00	00	00	10	40	50	10	20	25	05	80	00	41	19	01	01	00	00	06
26	4	A	2	68	02	00	00	00	00	08	30	30	12	22	23	06	83	01	42	18	01	12	04	02	07
27	4	A	3	66	10	03	01	00	21	07	25	10	15	25	18	08	81	08	39	20	03	09	01	01	07
28	4	A	4	68	07	03	00	01	21	07	32	15	14	24	19	08	80	04	39	20	03	10	01	01	07
29	4	A	5	66	10	03	01	00	21	06	20	10	16	26	17	08	81	04	38	21	03	09	02	01	07
30	4	A	6	67	06	00	01	01	21	08	35	30	12	22	23	06	82	02	41	19	01	04	03	02	07
31	4	A	7	70	01	00	00	01	21	10	34	50	11	21	24	05	87	01	41	19	01	02	02	01	06
32	4	A	8	70	00	00	00	00	00	10	39	50	10	20	25	05	85	00	41	19	01	01	01	01	06
33	2	B	1	40	00	00	04	00	00	11	19	49	05	11	22	19	65	00	15	30	01	01	00	00	05
34	2	B	2	32	03	00	12	00	00	10	18	47	05	12	21	20	67	03	11	34	01	18	08	09	06
35	2	B	3	21	11	05	20	00	04	12	11	45	06	13	20	22	64	15	10	34	02	13	03	04	06
36	2	B	4	27	09	05	18	02	04	12	16	46	06	12	21	20	64	10	10	34	02	13	04	04	06
37	2	B	5	20	11	05	20	02	04	11	12	44	06	13	19	22	64	13		32	02	12	04	05	06
38	2	B	6	31	06	01	18	02	04	10	16	46	05	13	21	20	65	03	14	31	01	05	07	07	06
39	2	B	7	35	02	00	06	04	04	11	17	46	05	12	21		67	02	15	30	01	02	02	02	05
40	2	B	8	36	00	00	04	03	00	11	18	48	05	11	22	19	57	01	15	30	01	02	02	02	05
41	3	B	1	68	00	00	02	00	00	10	30	50	08	15	23	12	72	00	15	30	01	01	00	00	04
42	3	B	2	59	05	00	06	00	00	08	27	39	09	17	22	13	74	01	12	33	01	17	07	06	05
43	3	B	3	52	13	02	10	00	08	07	19	26	11	18	20	14	71	04	11	33	02	11	03	03	05
44	3	B	4	58	11	02	09	01	08	07	23	27	11	18	21	14	71	09	11	33	02	13	02	03	05
45	3	B	5	50	13	02	10	00	08	06	17	25	12	20	17	15	71	06	12	32	02	12	06	06	05
46	3	B	6	60	09	00	06	01	08	08	26	38	10	18	23	13	72	02	14	31	01	07	02	02	04
47	3	B	7	65	03	00	03	02	08	09	27	47	10	17	22	13	75	02	14	31	01	03	01	01	04
48	3	B	8	66	00	00	02	01	00	10	29	48	10	17	23	12	75	01	15	30	01	01	01	00	04
49	4	B	1	68	00	00	01	00	00	10	41	51	11	19	24	05	79	00	14	30	01	01	00	00	03
50	4	B	2	65	02	00	03	00	00	08	33	33	13	21	23	06	82	01	12	32	01	17	03	03	04
51	4	B	3	61	11	03	04	00	09	07	26	11	15	24	10	07	80	05	12	32	03	12	01	02	04
52	4	B	4	63	10	03	04	01	09	07	33	14	14	24	20	08	80	03	12	32	03	11	01	02	04
53	4	B	5	60	11	03	04	00	09	07	22	09	16	27	16	07	80	03	11	32	03	12	02	02	04
54	4	B	6	66	07	00	03	01	09	08	36	29	13	21	24	06	80	01	13	31	01	08	03	04	04
55	4	B	7	67	02	00	01	01	09	10	36	47	11	22	23	05	83	01	13	31	01	04	01	01	04
56	4	B	8	68	00	00	01	01	00	10	40	49	11	19	24	05	82	00	14	30	01	02	01	00	03
57	2	C	1	30	00	00	10	00	00	11	21	49	06	11	20	18	64	00	05	80	01	08	00	00	03
58	2	C	2	20	02	00	30	00	00	13	17	52	05	10	22	15	72	08	06	74	01	18	06	12	05
59	2	C	3	08	06	03	40	00	01	15	08	70	04	08	25	12	78	35	15	50	02	14	03	07	05
60	2	C	4	12	06	03	35	12	01	15	17	60	05	10	22	14	76	30	10	57	02	15	03	07	05
61	2	C	5	05	06	03	40	01	01	13	08	71	04	08	26	12	78	36	17	48	02	14	03	08	05
62	2	C	6	18	06	00	30	01	01	11	16	55	05	11	21	15	71	11	09	61	01	10	05	11	05
63	2	C	7	22	04	00	15	02	01	11	19	54	05	11	21	17	68	03	05	80	01	09	02	03	04
64	2	C	8	24	00	00	10	00	00	11	20	53	06	11	20	17	66	02	05	80	01	08	02	03	03
65	3	C	1	50	00	00	05	00	00	10	31	45	10	19	20	12	72	00	05	80	01	04	00	00	02
66	3	C	2	42	03	00	15	00	00	08	25	42	11	20	17	11	77	05	07	75	01	16	05	08	04
67	3	C	3	36	14	02	18	02	03	07	19	45	12	15	21	10	79	18	12	58	02	12	03	05	04
68	3	C	4	39	10	02	17	01	03	07	26	42	12	16	20	10	78	15	10	62	02	12	04	05	04
69	3	C	5	34	14	02	18	01	03	07	19	42	12	15	23	10	75	18	14	56	02	09	02	04	04
70	3	C	6	41	08	00	14	02	03	08	27	47	12	17	20	11	74	06	08	67	01	07	04	08	04
71	3	C	7	46	02	00	10	03	03	09	30	48	11	16	19	12	75	02	05	79	01	06	02	03	03
72	3	C	8	48	00	00	05	01	00	10	31	49	11	17	20	11	74	01	05	80	01	03	01	01	02
73	4	C	1	60	00	00	02	00	00	10	42	42	15	25	20	05	81	00	04	80	01	00	00	00	01
74	4	C	2	56	02	00	04	00	00	08	33	35	18	24	19	06	81	03	04	74	01	13	03	05	03
75	4	C	3	52	10	01	07	00	00	07	28	20	21	23	17	07	80	06	09	67	03	10	01	03	03
76	4	C	4	54	08	01	06	02	07	07	32	25	19	24	18	06	81	05	09	67	03	11	01	03	03
77	4	C	5	51	10	01	07	01	07	07	28	24	20	24	19	06	81	06	10	66	03	10	01	03	03
78	4	C	6	55	06	00	05	01	07	08	37	40	18	24	10	06	81	02	07	72	01	05	02	03	03
79	4	C	7	58	01	00	03	02	07	10	40	43	17	24	18	07	82	01	04	79	01	02	01	02	02
80	4	C	8	59	00	00	02	00	00	10	40	44	16	25	19	05	82	00	04	80	01	01	00	00	01

systematic enumeration of people at other times of the day. Where researchers have wanted to consider the activities and locations of people during the day, they have had to resort to collecting their own data.

The result has been the publication of a number of specific space use studies and a number of specific time-budget studies (Chapin and Hammer, 1972; Szalai *et al.,* 1972; Chapin, 1974; see Chapter 5 above.) These studies form a useful source of background information for the present study without furnishing all the data needs. Although the experimental city is British in character, much of the guidance in determining variable values of activity profiles has been derived from American literature (Szalai, 1972; and Chapin, 1974) for the time use surveys and for the land-use pattern (Browning (1964) and Niedercorn and Hearle (1964)). The data is meant to be representative of a medium sized British city, say with 200,000 people. In terms of the time context the city is set in a typical working day, in summer time, in a period of 'full' employment. 1760 data cells were filled, the product of a raw data matrix of 80 observation units and 22 variables, given in full in Table 8.16.

The determination of the base units involved integrating a classification of the city's social-territorial space with a classification of diurnal *social* time, and therefore involved decisions about a classification of sequence. The spatial classification was guided directly by the results of conventional factorial ecologies, and the social and physical space schemes such as that of Murdie (1969) (Figure 8.16). The experimental city is divided into ten districts: a central zone (which is more extensive than a strictly defined CBD as it includes some residential space) and nine other districts. These are defined by the interaction of the three sectors, A, B, and C with three concentric rings, as in Figure 8.17. Following the findings of many factorial ecology studies, Taylor and Parkes (1975) allocated variations in terms of social rank or economic status to the sectors and variations in terms of family status to the rings or zones (refer again to Figure 8.15). Thus upper status (A), middle status (B), and lower status (C) *sectors* interact with increasing family status or suburbanism away from the *central zone.*

Time is divided into eight periods, each reflecting a dominant activity bundle or 'envelope' (see Lynch, 1976; and Figure 8.3) for the city as a whole. These times are set out in Table 8.17. The space-time base units therefore consist of districts of the city in specific periods of time. Since each of the ten districts operates through each of the eight defined periods, there are eighty space-time units in all. In Figure 8.16 these units are numbered 1-80 and the variables corresponding to them will be found in Table 8.16. The central zone constitutes the first eight space-time units and unit 80

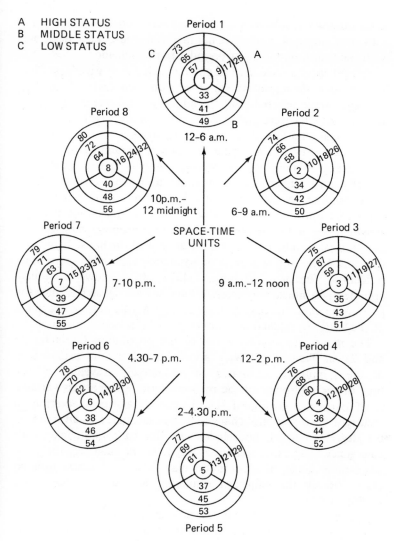

A HIGH STATUS
B MIDDLE STATUS
C LOW STATUS

Figure 8.17. The space-time base unit pattern. (Taylor and Parkes, 1975. Reproduced by permission of Pion Ltd.)

Table 8.17. The time periods and associated activity bundles (Taylor and Parkes, 1974. Reproduced by permission of Pion Ltd)

Period	Duration	Activity bundle
1	12 midnight to 6 am	sleep
2	6 am to 9 am	the transfer to worktime
3	9 am to 12 noon	morning phase of worktime
4	12 noon to 2 pm	lunchtime
5	2 pm to 4.30 pm	afternoon phase of worktime
6	4.30 pm to 7 pm	the transfer to residential-time
7	7 pm to 10 pm	early evening discretionary-time
8	10 pm to 12 midnight	late evening and bedtime

is the outermost, suburban zone in Section C, operating between 10 p.m. and midnight. Conventional factorial studies have dealt only with the space-time units of period 1 in this scheme.

Several criteria were used to guide the choice of variables. First, typical socio-demographic variables from conventional factorial ecologies were selected. But, in addition, *movement* and *access* related variables were included, as were commercial land use and recreational land uses, like public open space.

The land uses which appear on statutory planning maps were therefore translated into the potential activity spaces which they actually are. Each variable was allocated eighty values corresponding to the space-time base units. It is essential to appreciate that for each variable a value in space-time is being assigned, thus if a residential area has a high proportion of families in it, but all members of the family are out of the house between 9 a.m. and 4 p.m. this space is proportionately reduced in terms of its residential function. Chapman (1977, 48) follows a similar argument, 'A house may cease to have any spatial significance for part of the day'. In terms of their ecological role normal occupiers of the house (residents) are 'working' or 'at school' and so on.

The movement characteristics of the population (variables 20 and 21 in Table 8.18) were approximated by the proportions of the population estimated to be actually travelling at the time of the mid-point of the relevant period. The access variable is of particular ecological significance. It assists adaptation to recurrent environmental changes. It was estimated as an index of the degree to which households are able to communicate with emergency services, especially at night. As we shall see, for some space-time niches, the afternoon may be a particularly isolated place. (Recall the empirical results of Section 8.5 above. See Melbin, 1976).

Readers familiar with the factor ecological literature will note that there is no ethnic dimension and that the input units are in effect the smoothed output configuration for conventional factorial ecologies. Obviously some expediency had to guide the experimental design. In an *empirical* replication of this scheme, (a primary objective of the experiment) complicating criteria such as an arbitrary geometry to the enumeration areas, would have to be accommodated.

Acknowledging the value of Hanham's replication of this experiment using a more complex and sensitive 3-mode factor analytic model (Hanham, 1976), but also noting that he found that the original results were to all intents and purposes supported, we can now summarize some of the ecological configurations which were produced by the analysis of the experimental data using the original 2-mode common factor analysis (Taylor and Parkes, 1975).

Table 8.18. The variables and the rotated factor loadings for the experiment (Taylor and Parkes, 1975. Reproduced by permission of Pion Ltd.)

Variables	Factors						
	1	2	3	4	5	6	7
(1) Residential	−0·41		−0·42	0·42	−0·35		0·39
(2) Retail and service.	**0·84**						
(3) Other commercial	**0·96**		0·36				
(4) Industrial							−0·85
(5) Entertainment			0·83				
(6) Open space		0·38			−0·68		
(7) Population	**0·78**					0·37	
(8) Persons/dwelling	−0·43		−0·37	0·57			0·35
(9) Male adults					0·89		
(10) Age 0-4			−0·47		−0·68		
(11) Age 5-17			−0·44	0·37	−0·65		
(12) Age 18-29			0·82				
(13) Age over 65				−0·92			
(14) Married adults			−0·60	0·72			
(15) Employed females	**0·85**						
(16) Upper class		0·92					
(17) Lower class		−0·88					
(18) Professional services	**0·92**						0·12
(19) Manual services	0·45					0·65	
(20) Private transport	0·37					0·82	
(21) Public transport						0·83	
(22) Access	0·39	0·68	0·37				
Eigenvalues	9·05	3·42	2·82	1·90	1·23	0·93	0·63
Cumulative percentage of total variance	41·1	56·7	69·5	78·1	83·7	88·0	90·8

Loadings below ±0·35 not shown, in order to simplify table.

'Factorial ecology comprises the application of factor analysis to data describing the *residential* differentiations of the population, generally the urban population' (Timms, 1971, 53-54). But, as we have noted, the emphasis on the residential population seriously depletes the full power of the notion of an ecosystem. In an abridged summary of the results which now follows, based on the interpretation of Table 8.18 together with the factor score maps, the advantages of a space-time approach will be apparent. Data availability is not an insuperable problem; sampling designs provide one solution. The inclusion of time records in the census are another, a solution proposed by a number of people, especially in recent years (Parkes, 1971, 1973; Michelson, 1976; Ware, 1977).

White collar Workday cycle

Consider factors 1, 3, and 4 by way of illustration. Factor 1 is a white collar workday cycle dimension, identifying the ecological niches of tertiary and quaternary urban activity. The bold type values under factor 1, in Table 8.18 identify the variables which allow this interpretation. The correlation of 0.78 for population size in the various space-time units indicates the shifting mass of the city during a 24 hour period. (Imagine a time-lapse camera focussed over a city, the rerun condensed into say 24 minutes would summarize this factor rather well!) In Figure 8.18 the estimated factor scores for each space-time unit have been mapped for Factor 3. Figure 8.18(a) maps those scores in the conventional factorial ecology manner. Figure 8.18(b) shows an alternative and perhaps more informative method of mapping scores for space-time factorial ecologies (Parkes, 1977). Here the 24 hour ecological 'image' of the city is represented as a radar trace might show it, were it possible to scan the changing space-time of the city's major dimensions. Such a device is called a *T-P* graph.

T-P graph

Although much is written and spoken about the decline of the central zone in large cities, it nonetheless remains the principal area for evening and night time relaxation. Factor 3 of the experimental city is a bright lights dimension; and the evidence of contemporary time colonization (Section 8.3). In Figure 8.19(b) it is illustrated using the *T-P* graph method. The *T-P* graph for factor 1 is shown in 8.18(a) to emphasize the comparative values of the method. When those two graphs are superimposed on each other, the 'incessant' heart of the city is clearly revealed. The other subareas sustain the central zone by their cyclic and their erratic contacts with it. The white collar cycle is the most significant and regularly recurrent factor, the bright lights of the night time colonization being sustained by more erratic but high frequency contact from the other space-time zones. The 'blip' which appears in the *T-P* graph trace of 8.18(b) captures the 'business lunch-concert-club-pub-break' in the working day: a sort of 'mini-night' in the middle of the day; itself a social time colonization feature.

Bright lights dimension

Table 8.18 indicates the differing social and demographic elements in the structure of each experimental dimension. It will be noted that variable (5), entertainment land use, has the highest loading, 0.83. The positive relationship to variable (12), age group 18-29, brings together two urban ecological attributes of the city which are the hallmark of contemporary urbanism. Their inverse relation with variables (10), age group 0-4, (11), age group 5-17, (14) married adults and (1) residential land use, emphasizes the non-familistic (or urban) elements of the dimension. Conventional factorial ecologies could not isolate such a dimension.

Although this is a picture of an experimental city, it can be appreciated that by altering the values of the variables represented

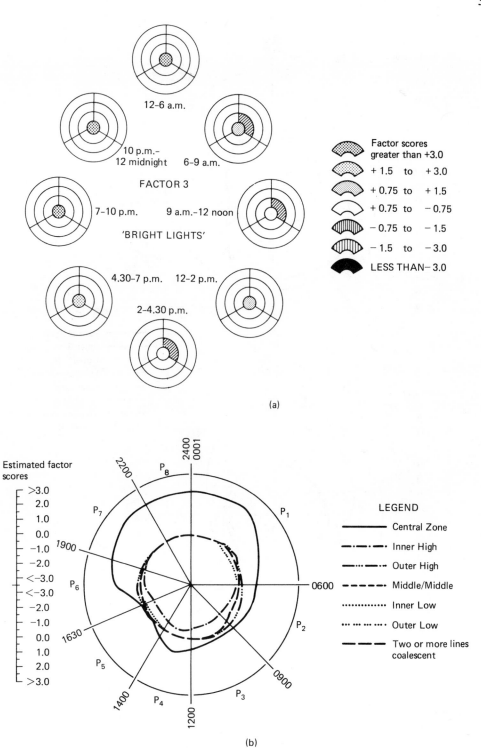

Figure 8.18. Mapping the space-time factorial ecology. (a) Conventional method. (Taylor and Parkes, 1975. Reproduced by permission of Pion Ltd.). (b) *T-P* graph. (Parkes, 1977)

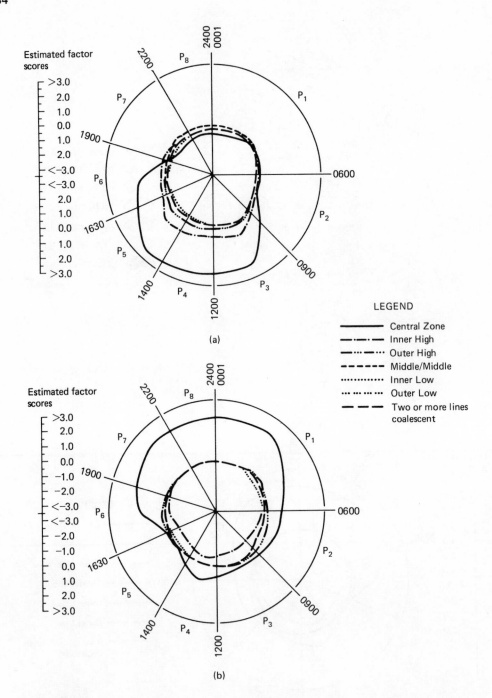

Estimated factor scores

>3.0
2.0
1.0
0.0
1.0
2.0
<-3.0
<-3.0
2.0
1.0
0.0
1.0
2.0
>3.0

(a)

LEGEND

——————— Central Zone
—·—·—·— Inner High
—···—···— Outer High
– – – – Middle/Middle
············ Inner Low
···—···—··· Outer Low
– – – Two or more lines coalescent

Estimated factor scores

>3.0
2.0
1.0
0.0
-1.0
-2.0
<-3.0
<-3.0
-2.0
-1.0
0.0
1.0
2.0
>3.0

(b)

Figure 8.19. *T-P* Graph of the city's white collar (a) and bright lights (b) cycles. (Parkes, 1977)

in the space-time base, it would be possible to simulate the impact of alterations to the space and time location of urban functions and also to simulate the impact of change in socio-demographic structure on the 24 hour ecology of the city.

Our final illustration is worthwhile because through it there is the chance to illustrate another dimension of city life, one close to the discussions in Section 8.5 on urban social pathologies. This is factor 4, in Table 8.18, the *old-age dependency* dimension. This space-time ecology dimension of the city identifies a distinctive aspect of urban social structure, one with which we are all familiar, and it also identifies the space-time envelope (Lynch, 1976) to which welfare oriented effort is so often addressed. Many planning strategies have perhaps been too concerned with the 'where' of old people rather than the 'when-where'. Variable (13), people over the age of 65, correlates —0.92 with this dimension and the inverse relation with people now married (0.72), persons per dwelling (0.57) and age group 5-17 (0.037) emphasises the spatio-temporal segregation (Section 8.2) of the old from the 'family-based' and 'young' space-time units of the suburbs. We can see that old people are segregated or 'isolated' in space time to a greater extent during the day than at night when their 'younger working neighbours' return. In Figure 8.20 the *T-P* graph captures this condition.

During period 1, midnight to 6 a.m., the scores are low because other residential space-times, even the few other residents of the central zone, are close at hand. But, by periods 2 to 6, the trace is quite different. Now we find that there are concentrations of old people, particularly in space Sector A. (The fact that the concentration is in Sector A is not by itself of substantial importance since it merely reflects the location given to old people in the original raw data generation.)

The trace clearly shows that the inner high status space is distinctly different in its profile from the other spaces. The distance along any radius, from one trace to another, is a measure of the space-time ecological separation of these 'spaces'.(12) This 'distance' is greatest in periods P3 and P5. As the central zone returns to a city-slow space in period P1, this segregation decreases. The central zone also moves closer to the inner low status zone which, during the day, is rather distant from it.

The problems associated with the use of artificial data are not unique to geography. They exist in other areas of research. In plant ecology, for example, Greig-Smith conducted scale and classification experiments on artificial data (Greig-Smith, 1952). The Lund time-geographic scheme, outlined in Chapter 6 above, also utilizes deductively derived 'artificial' data, as did the early Swedish innovation diffusion studies. But use of experimental data does not preclude the use of real data. Indeed it increases the possibilities of using real data properly, by identifying theoretical and practical difficulties in advance. It also allows others to suggest

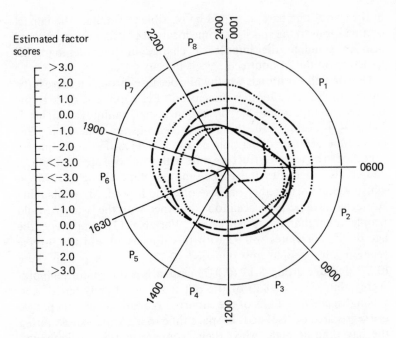

Figure 8.20. *T-P* graph trace of the old-age dependency factor. (Parkes, 1977)

improvements, as Hanham (1976) was able to do, and can provide a basis for simulation studies. There is no reason, apart from the associated hard work, why the space-time scheme outlined here should not be adapted and adopted to a *real* world study.

The results given here are obviously not substantive, but they do illustrate the possible utility of a space-time factorial ecology approach in extending geographic practice and theory.

They are somewhat less artificial than they may seem in that they indicate the space-time structure of a medium sized city as perceived by two geographers. It is their space-time 'image' of the city. What is beyond discussion is that a space-time ecology along the lines of this experiment is in every way superior to the static, *one-time* view of census information referenced to midnight.

In Chapter 8 we have been discussing chronogeographical aspects of the city at a scale of 24 hours. Chapter 9 moves into a bigger time scale and is concerned with 'long period' cyclic changes in the built fabric of the western city.

Notes

1. These terms can be used interchangeably. We will usually use the term social pathology.
2. As early as 1952, Bell (in association with Shevky) had used factor analysis to test the 'theory' of the three dimensional structure of urban

social space, see Bell (1952) doctoral dissertation, University of California, Berkeley.

3. For the time colonization study reported here, and for the urban pathology study of Boston reported in Section 8.5, we are most grateful to Professor Murry Melbin of Boston University and to Edward Arnold Ltd., (Publishers) for allowing us to cite at some length from his work (Melbin, 1978a).

4. This section is based on an urban image study undertaken by Paul Tranter (with Don Parkes) in the Department of Geography at Newcastle University, New South Wales, as part of his honours degree in geography, (1976). It was modelled on work by Harrison and Sarre (1975) on images of Bath, England. To the extent that it was possible their experimental design was replicated, except that the thesis argued was that a time-free image study, such as their's, was imprecise because subjects might have been 'imaging' the city at different and uncontrolled times.

5. There are a number of rather complex refinements to the multidimensional analysis of repertory grids, cf Golledge and Rushton (1973), Amedeo and Golledge (1975), Levy and Dugan (1972), Lundeen (1975).

6. In the Appendix you will find what should be sufficient explanation of some of the statistical terms used here, if you are not already familiar with them.

7. Paul Tranter is undertaking research into social pathologies as part of his Ph.D. dissertation in the Department of Geography, Newcastle University, New South Wales, supervised by Don Parkes. The analyses reported here are amongst the first to be undertaken and should be treated as indicative rather than substantive (August, 1977). However, we feel that 'real' data is to be preferred to the hypothetical data which so often appear because no empirical work has been undertaken.

8. We are grateful to Dr. Annette Dobson of the Mathematics Department, Newcastle University, Australia, for passing on some of the preliminary results she computed to satisfy her own curiosity, when the data was in the first stages of coding.

9. In Schwarz's recent study of the social ecology of time barriers (1978, 1207) he puts forward a rather similar notion to the city-fast, city-slow scheme outlined here. Schwarz identifies the flow of people into high and low speed establishments (e.g. medical clinics and 'surgeries') where time is exchanged for money. The similarity lies in the ecological concept of tempo and in the mechanisms by which high status individuals are able to participate in more activities and to expend less time on those which are 'irksome' by exchanging time for money.

10. cf. Brown and Boddy (1976) and Parkes and Taylor (1976) for an exchange of letters to the Editor of *Environment and Planning* concerning some of the possible philosophical issues raised by the paper.

11. This table of generated experimental data has been included here, as it was in the original paper, in order to facilitate the interpretation of analyses which follow. If suitable computing facilities are available it is possible to alter some of the variable values, to fit other types of experimental city. (See also Taylor and Parkes, 1975).

12. Because only eight time points are used in the experiment it has been necessary to use the mid-point of the period as the anchor point and to assume that there is a continuous ebb and flow from one point to another; an empirical study would use many more time locations and space-time units.

9
Times in the built environment

9.1 Preliminaries

Rome was not built in a day. No substantial human settlement ever
has been. Whether alone on the deserted streets of a city centre in
the early morning, or pushed and jostled in the bustle of the late
afternoon 'rush hour', the massive material fabric of the city is a
symbol of both successful and unsuccessful enterprise. It can point
one backwards over the centuries, or forwards into a new and
uncertain future.

Time past and expectations about the future meet in our
experience of each moment of this material present. In this chapter
discussion centres on the influence of *big times* (Chapter 3) on the
construction of the buildings and building arrangements which
provide us with cues to temporal experience.

Section 9.2 of this chapter considers the times that are captured
in the built environment. Almost exclusively these times are of the
past. In various ways they permeate many aspects of everyday life
and they are spatially variable. However, they also persuade us of
the possibility of a future but not one which is so wide open 'as to
be able to change into anything imaginable ... the psychological
strain of such an uncertain future would be more than most people
could bear' (Lynch, 1972, 114).

Whereas section 9.2 centres on time as revealed in the
morphology of the built city (i.e. in the townscape), Section 9.3
introduces some of the underlying economic processes which have
given rise to categories of building types and their general spatial
characteristics. The focus is on the 'apparent' cyclicity of economic
systems, and the reflection of these cycles in the building sectors of
the national economy.

In the previous chapter a small time period of 24 hours was the
time boundary within which temporalities in the city were
illustrated. In this chapter the periods are much longer, in general
ranging between about three and a half years and fifty years. Once

388

again, however, time is embodied in the phenomenon of recurrence. (You will recall from discussions in Chapters 1 and 2 that it was suggested that recurrent relations among items of any sort may prove to be a basis for the identification of system specific times.)

In Section 9.4 attention turns from the national and international scale to the urban scale; to building cycles themselves rather than the underlying economic cycles as such. First, we illustrate the long waves of urban building cycles which have been identified in many cities around the world. Second, we consider the cycles of building which occur at different element-scales in the townplan; the burgage cycle which has been identified in the dynamic morphology of European cities which have a medieval heritage is one example. At a rather different scale the city of Chicago is then used to illustrate how building cycles relate to some familiar ideas of urban spatial structure.

Chapter 8 offered some alternative perspectives with which to view the city, when the time scale was restricted to 24 hours. It therefore related exclusively to timings and tempos of social and individual conduct. Chapter 9 opens up a quite different set of perspectives, focussed on the material fabric of the city in which individuals and groups spend those 24 hours. In a number of ways this chapter is rather different to those which have preceded it. What we have aimed to do here is to show how an awareness of temporal perspectives may be used to help us to gain a deeper appreciation of our environment. These perspectives are formed by the times of human experience; the sense of past which is evoked by historic landscapes. They are also formed by an awareness of the social times of our own particular society or community. Furthermore, we assist our understanding by an appreciation of the passage of waves of buoyancy and depression in the wider national and international economies. The principles outlined in Chapters 1, 2, and 3 should be kept in mind. Next time you are 'downtown' or 'uptown', or are visiting a city for the first time, reflect on the times which are sealed in the bricks and mortar.

9.2 Time records and signatures in the townscape

'Vast time, like vast space, is painful to bear, or at least it requires getting used to. It is a poor guide to current action. We look for a social image of time which enlarges, celebrates, and vivifies the present, while increasing its significant connections with the past and especially with the future' (Lynch, 1972, 134). The townscape is replete with signposts to the past and many which point to the future.[1] In both directions they initiate an *image*, the clarity of which depends on the context of the present. The future is of particular importance because our imagination of it results from our experience of the relation between now and before.[1]

390

There are many elements in the environment of the individual from which a sense of time is gained. But in urban 'places' the clues which nature can provide are generally denied us. They are replaced by Man-made artefacts which serve the same purpose but often function rather better because they reflect the social-urban world which generated them.

Idea of preservation

The built environment is a reminder to us of material environments of the past, and nowadays elaborate efforts are made to preserve individual buildings and areas of historic or cultural value. They remind us of time gone by. Yet, until a few hundred years ago, such *preservation* would have been thought quite absurd. 'In Western Europe, at least, the idea of preservation first appeared about 1500, in the form of an esoteric attraction to relict buildings, even to the point of construction of sham ruins. By the eighteenth century an affection for the structures of the past was a widespread upper class fashion, and by the nineteenth century it become part of the intellectual baggage of all middle class travellers. In the same century, first in the United States and slightly later in Europe, organized movements sprang up to preserve historic landmarks for the public' (Lynch, 1972, 29). At first, therefore, the *idea* of preservation was restricted to particular people with rather specific motives — in the United States for instance, efforts were particularly directed towards saving buildings associated with patriotic figures in an effort to foster a heightened national consciousness. It is only rather recently that the idea of preservation *for itself* has had currency among the wider population.

Use and appreciation

Reminders of the past as reflected in preserved physical structures usually come only late in a society's development, when an active distinction between *use* and *appreciation* can be made. Before this time people are too busy 'surviving' to have a surplus of free time and other resources to devote to preservation. Of course this is not to say that they do not live in a built environment which incorporates many different ages of buildings, but only that they do not consciously perceive the built environment as necessarily having properties which are worth preserving, except insofar as they relate to use.

Appreciation of the past, however, has its own rules and these rules seem to change over the years. They can be summarized under two broad categories: (a) what is worthy of preservation, and (b) how this preservation is to be carried out. What is worthy of preservation effectively means 'what is worth remembering'. Until the turn of the century the Common Man was generally considered as only being worthy of remembrance en masse. This was reflected in the buildings thought worthy of preservation and in the monuments which were erected. They were usually the preserve of the eminent and the powerful. The depth of this cultural trait should not be underestimated. (In 1978, for instance, a lot of publicity was given to the 'radical' views and actions of the widows

Figure 9.1. The Industrial Revolution captured in 1978, Beamish, County Durham, England

of men who fell in the two World Wars who were not 'entitled' to lay wreaths of remembrance at the Cenotaph memorial in London on Remembrance Day). But there has been some change of view and the status of the Common Man and the associated constellations of cultural traits and class associations are now thought to contain something of value. This changing mood of preservation has been helped by the recent upsurge of interest in social (especially local) history and a resurgence of nationalism, which has induced a more active preservation of everyday buildings of the past. The industrial archaeology movement (Figure 9.1) and the establishment of folk museums (Figure 9.9) which actively reconstruct old buildings, are indicators of the strength of this movement.

The standards against which preservation and conservation schemes are carried out do therefore change over the years, particularly as a result of shifts in wealth, values, and goals of society. Lynch (1972) has outlined three possible approaches to preservation of buildings. One is simply to save a building — perhaps moving it away from an area in which it will suffer damage. Another approach is to restore a building by doing minor repairs, perhaps including additions of what are considered 'temporally appropriate' material artefacts. The last approach is to restore the building to some 'original' state, for instance a dwelling with a tiled roof may be re-thatched, restoring it to its former appearance. In these ways 'the pattern of time may be retained, imitated or removed' (Lynch, 1972, 31).

Idea of past

What is the result of all this activity to preserve the past, or more properly to preserve the contemporary *idea* of the past? It enables life to take place in a built environment of ever increasing temporal extension into the past (and therefore into the future). 'Pastness' becomes a social resource. This trend is well illustrated by the fact that in English law, named Historic Buildings are no longer in any sense private property, to be disposed of at will. The present occupiers are regarded only as trustees for those who come after (Prince, 1978). Although most western people will still live in buildings which will be on average, younger than those that their ancestors resided in, it is necessary to realize that because of wider experience of environments beyond their own, obtained at first hand as well as through television and other visual media (Parkes and Thrift, 1978), they also receive more visual clues to the past in the present built environment. This occurs both in terms of the total number of old buildings and in terms of their absolute age-length. The resource of pastness accumulates, and conservation management operates as the curator. In Figure 9.2 the built environment of an English village appears to accumulate ever more clues to a greater span of the past. So long as its structure is not severely adjusted there need be no sense of trauma and incoherence — the past is seen as a continuous process of accumulation. In addition, contemporary 'media exposure' has heightened our awareness of this past, widening our sense of alternative environments.

The social heritage of pasteness has also been illustrated in the geographical literature, through the notion of *historical townscape* (Conzen, 1960, 1966; Carter, 1965, 1972; Carter and Davies, 1970; Johns, 1965), and its place in the study of the human geography of settlements persists in urban morphological analysis and in the recent growth of interest in phenomenological and humanistic geography.

Conzen (1966, 59) has expressed this well; 'in our preoccupation with current practical problems we are apt to overlook that a physical environment of the fullest possible historical expressiveness or historicity is an important asset to any healthy and expanding form of social life at advanced levels of civilization. It gives a sense of continuity and at the same time diversity of human effort and achievement in different periods, a 'tradition' not in any narrow but a wider sense and all the more effective for presenting itself visibly in man's artefacts on the earth's surface. It enables the individual and the social mind to take root in an area and in demonstrating the historical time dimension of human experience stimulates comparison and through it a more informed way of reasoning'.

Historical townscapes do, of course, vary in the depth of experience of the past to which they can be a key. Thus the long and varied traditions of urban life in Western Europe cannot be present

Figure 9.2. Big time
changes and con-
tinuity in the village
of Batheaston.
(a) About 1800,
from a sepia by Mrs.
Sophia Walters.
(b) About 1910.
(c) 1969. (Repro-
duced by permission
of B.M.W. Dobbie)

394

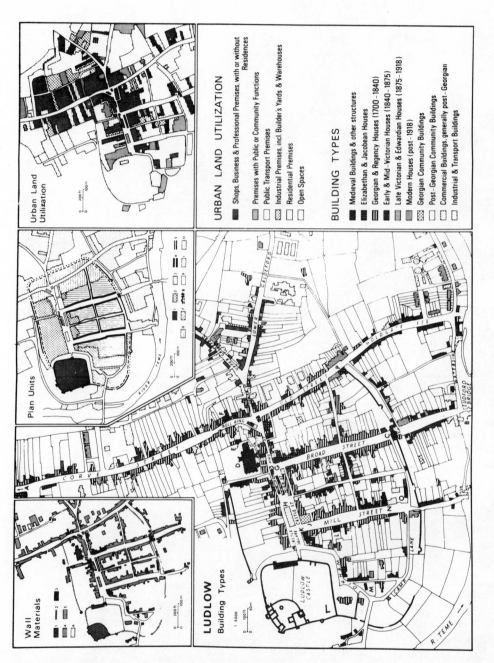

Figure 9.3. The historical townscape of Ludlow. (Conzen, 1966. Reproduced by permission of the Department of Geography, University of Newcastle Upon Tyne)

to a comparable degree in, say, the United States, Canada, or in Australia. Not only is the depth of the historical past, as reflected in the built environment, less, but the tradition of management of the built environment is also (generally) more recent. This 'pastness' of the physical environment, as captured by maps, is illustrated in Figure 9.3, where the distribution and period of building types in Ludlow is illustrated (Conzen, 1966). Figure 9.4 on the other hand, shows how the historical resource depth of the built environment of Alnwick, a town in the Border country of Northumberland, at two dates — 1774 and 1851 — varies with time. Now the maps illustrate the relative degree of historicity present in the built environment at those two times rather than the degree of historicity in the contemporary present (as in Figure 9.3).

Thus, at any time, there is a certain historical archive to draw upon of which the built environment is an important part. The preservation of these records of the *path to the present* is necessary in order to reduce the 'psychological strain' which is attendant on the anticipation of 'an uncertain future' (Lynch, 1972, 114). However, it is obviously not possible to preserve all the buildings of the past. Some buildings are simply not worthy of preservation due to their poor physical condition or they may be deemed to be poor examples of their period with better ones readily available. But in the latter point lies a further difficulty; standards, values, and knowledge about the past all change, and it is difficult to stabilise the criteria from one generation to another. A rather more awkward reason is that 'forgetting' is as important a part of life experience as remembering.

Ecological processes of adaption are features of social life, and new buildings must be constructed to serve the functions of the present as well as the anticipated future eco-systems. An unthinking servitude to the past is regressive. Finding a balance between preservation and demolition is therefore equivalent to establishing a reasoned relation between past and future. The rate at which renewal is undertaken can thereby lead to placelessness **Placelessness** (Chapter 1). If the past is erased too quickly this is more likely to happen (Ford, 1978). In such a situation, society has to grapple with a space-timing problem (Chapter 3) in order to ensure the **Space timing** persistence of reassuring relations between the elements of the built environment, over any single interval in time. City dwellers will find it unsettling to have to cope with too rapid changes in their environment.

It is possible that over recent years, in Britain at least, demolition rates have been so high that there is danger of forgetting too much rather than of suffering from a surfeit of pastness. Note, for instance, some of the changes in Newcastle-upon-Tyne illustrated in the pictures below. The first pair of photographs suggest persistence and placeness but the second pair illustrate rather a dramatic disruption in the place-continuity of townscape (Figure 9.5).

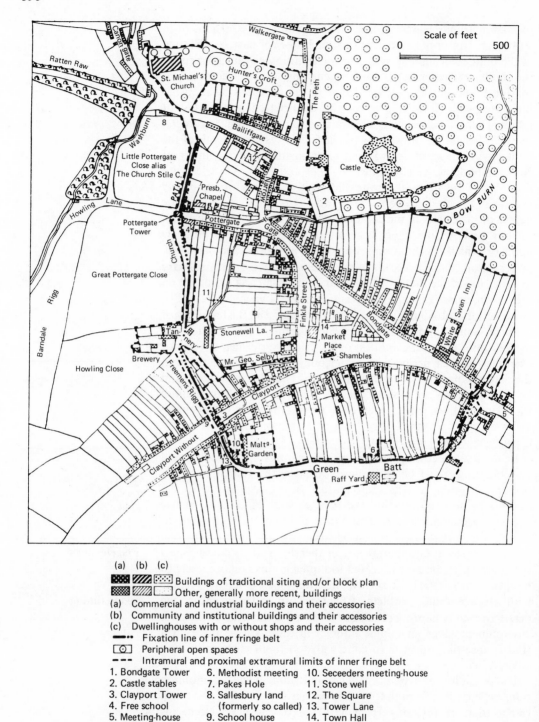

Figure 9.4. Time depth in the built environment of Alnwick in 1774 and 1851. (a) 1774. (b) 1851. (Conzen, 1960, 57, 62)

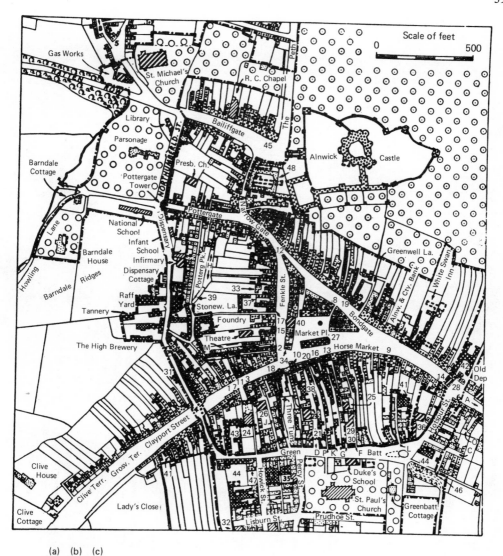

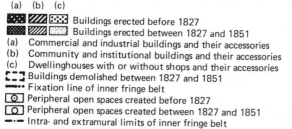

(a) (b) (c)

Buildings erected before 1827

Buildings erected between 1827 and 1851

(a) Commercial and industrial buildings and their accessories
(b) Community and institutional buildings and their accessories
(c) Dwellinghouses with or without shops and their accessories

Buildings demolished between 1827 and 1851

Fixation line of inner fringe belt

Peripheral open spaces created before 1827

Peripheral open spaces created between 1827 and 1851

Intra- and extramural limits of inner fringe belt

A Bondgate without   D St. Michael's Lane   G King's Arms Yard   K Queen's Head Yard
B Canongate   E Tower Lane   H Moore's Yard   L Roxburgh Place
C Hotspur Place   F Correction House Yard J Old Chapel Lane   M Teasdale's Yard

N Union Court
O Victoria Place
P White Hart Yard

(a)

(b)

Figure 9.5. Time-distant built environments at the same space location in Newcastle-upon-Tyne. (a) 1900. (b) 1978. (c) 1900. (d) 1978. (a) and (c) Courtesy of the Auty-Hastings Collection, Newcastle-upon-Tyne

(c)

(d)

(a)

(b)

(c)

Figure 9.6. (a) The 'New' Castle and loss of residential space 1890-1978 in Newcastle-upon-Tyne. (b) 1890. (c) 1978. The castle, of course, was a rather special residence and source of shelter. (b) Courtesy of the Auty-Hastings Collection, Newcastle-upon-Tyne

In England and Wales about 350,000 dwellings have been torn down since 1971 and those which were of cultural historic value are now, of course, lost for ever. For instance, less than one hundred years ago the residential environment of Dog Leap Stairs, in the heart of the medieval city of Newcastle-upon-Tyne, only one hundred metres from the Norman 'New Castle' looked as shown in Figure 9.6(b). By 1978 it had been reduced to the 'placelessness' of Figure 9.6(c). The lack of preservation of such niches, lost reference places for today's citizens, is a symbol of a cultural aberration which is irredeemable (see for instance Prince, 1978).

In England, at present, about 246,000 buildings are listed as of architectural or historic interest, of which 16,000 were added to the list in 1976-1977. The stock of listed buildings varies spatially and the variation, by counties, is illustrated in the map of Figure 9.7. In addition, there are also 4,355 'conservation areas' which cover many historic town centres (including those of Ludlow and Alnwick, in Figures 9.3 and 9.4). In 1976-1977 the centres of Liverpool, Newcastle-upon-Tyne, Nottingham, and Bristol were added (see Ford, 1978). These listed buildings and conservation areas are the basis of Britain's resource of 'pastness'. Persistent vigilance is required to ensure that listing continues. This in turn means marking ever more recent buildings and building complexes

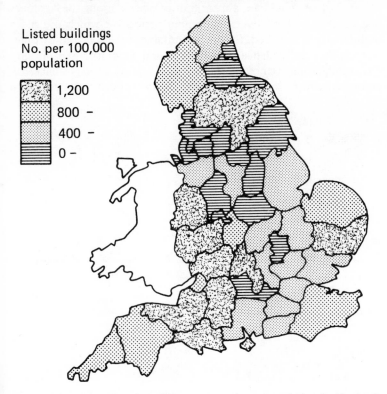

Figure 9.7. Listed historic buildings as a ratio to population in England. *(New Society,* 24.8.1978. Reproduced by permission of *New Society* and Tony Garrett)

for preservation in the years to come. As an example, in the United States a growing record of recent conservation orders is being established, with good examples being the skyscraper townscape of New York and the nineteenth century, so-called three-decker or 'Irish Battleship' tenements of Boston (Figure 9.8).

Old buildings may be a cultural resource, but they have value beyond their power of emotional evocation of times gone by. They are elements in an urban signal-system. There is a semiological element in townscapes which must be added to the syntax of built form (Hillier and Leaman, 1977). All buildings make symbolic statements about the persistence of places, and therefore about people and their ways of life (Paulsson, 1959). No more dramatic example of this can be found than in the brick-by-brick reconstruction of the centre of Warsaw following its destruction in the Second World War. The determination to show the cohesion of Polish culture was displayed through built form: the signal reads, 'Look, they were not able to change anything.'

Depending on the strength and intensity of the concentration of continuity in building form, it is possibly the case that some towns are rather more resistent to change than others. The townscape signature is writ so clear in central Bath (England) that it would be extremely difficult to insert new buildings, however inspiring in quality, which would not appear 'out of place' among the crescents and squares of the classical Georgian period for which the city's architecture is so noted. Such new buildings would be out of time, interrupting the cadence (Chapters 2 and 8) of the existing city environment and removing the essential quality of persistence. By contrast, other 'equally' historic towns appear to be more

Out of place
Out of time

Persistence

Figure 9.8. Conservation of the Boston three-deck tenement. *(The Economist,* 26.8.78)

amenable to the absorption of newly constructed buildings into their building spectrum. Ford (1978, 267) has suggested the town of Chester as an example. It has a 'temporal pot-pourri' of buildings which makes sympathetic insertion of new buildings easier. Grand unitemporal signatures of a culture, as in Warsaw or Bath, do not necessarily offer as much or more sense of place (Chapter 1) than the smaller-scale and temporally eclectic historic environments like Chester which have, in the very diversity of the times recorded in their townscape, an ability to offer just as coherent a sense of place; one which will 'stay put', so to speak, into the future.

Sense of place

The recent growth in folk museum villages seems to be further evidence of interest in the use of environmental archives where one can more easily reference the past. Such villages act as culture banks where people can 'save their past' and from time to time draw on it, in order to make up for deficiencies in the contemporary environment of everyday living. In the United States, for instance, more than 125 of these villages now exist, the best known of which is probably Williamsburg. In Australia the township of the original penal settlement of Sydney is being reconstructed using 'authentic' tools and 'original' materials. (Although the folk museum reconstruction of Sydney 'Old Town' has an inland situation, the site features of the original settlement, including the cove, are being incorporated.)

In Britain, however, it is often the existing old and occupied villages, like Lacock in Wiltshire, which perform the role of folk museum. But, in England too, there are now village museums which consist of rebuilt buildings moved from sites where they were in danger, such as the Weald and Downland Museum near Chichester (Figure 9.9). A range of these reconstituted settlements of the past is also found in Sweden, from the 'authentic' old village of Stensjo to the partially rebuilt Lund Folk Museum. Sometimes such museums are built specifically to foster national consciousness by revealing a coherent (Chapters 3 and 8) past to be drawn upon and to be displayed to international visitors. In Wales, for instance, the St. Fagan's Folk Museum buildings interact with other evironmental artefacts like old books and film and taped sound simulations, to extend the pastness experience beyond the immediate built environment. As mobility increases in modern societies, and as the rate of information circulation increases, these culture banks become havens of pastness, places of reassurance not dissimilar to churches where one can experience a *pause* in the midst of an otherwise restless and frenetic movement. These and other pauses have been put forward by Tuan as causal factors in the creation of place (Tuan, 1978; and Chapter 1.4).

Coherence

Pause as cause

The value of an awareness of the sense and record of the past seems to be beyond dispute, but the reasons why we preserve the past are not so clear. There is little coherent research about this vital element of our daily lives, and yet 'to preserve effectively, we

Past choice and change

Figure 9.9. The Weald and Downland Museum, near Chichester, Sussex

must know why the past is being retained and for whom. The management of change and the active use of remains for present and future purpose are preferable to an inflexible reverence for a sacrosanct past. The past must be chosen and changed, made in the present. Choosing a past helps us to construct a future' (Lynch, 1972, 64).

Prototypes

The sense of a personal future is an uncertain thing (and it is possibly connected with prototypes). 'We can respond to the future not merely by saving things for it and by being adaptable to it but also by creating it' (Lynch, 1972, 113). In effect what we do is to construct prototypes of future built environments. One such prototype is the English anti-urban bias expressed in the somewhat paradoxical view that the most desirable urban place is a rural one. This anti-urban bias finds expression in the phenomenon of the suburbs, in New Towns and in most British town and country planning. The fondness for rurality extends beyond the appearance of the built environment to the *pace* and spatial density of daily living. Perhaps because the past offers security in the image of the balmy and unchanging English-village-of-yore with its church, cricket ground and homely pub. No prototypes of the future have been built as yet which seem able to match this theme (Williams, 1973). Oliver Goldsmith celebrated *The Deserted Village*, in the late eighteenth century, with reflections on the past,

Urban place as rural

Pace

'A time there was, ere England's griefs began
When every rood of ground maintain'd its man'.

But this image of the sturdy, yeoman past which echoes through so much of English literature, contrasts with the fears of an uncertain unstable future,

> 'But times are alter'd; trade's unfeeling train
> Usurp the land and dispossess the swain'.

These images of idyllic 'rural timelessness' must be set beside the despair of 'industrial timelessness'. Charles Dickens' description of nineteenth century Preston (Coketown), 'a town of machinery and tall chimneys' bears this out. 'It contained several large streets all very like one another, and many small streets more like one another, inhabited by people equally like one another, who all went in and out at the same hours, with the same sound upon the same pavements, to do the same work, and to whom every day was the same as yesterday and tomorrow, and every year the counterpart of the last and the next' (Dickens, *Hard Times*). The built environment in Preston was 'placeless and undistinguished' ... 'the jail might have been the infirmary, the infirmary might have been the jail, the town hall might have been either, or both, or anything else'. There was also the greed and rapacity of capitalism to be coped with, the incessant drive for profits, 'what you couldn't state in figures, or show to be purchaseable in the cheapest market and saleable in the dearest, was not, and never should be, world without end, Amen' (*Hard Times*).

In England this past association of many cities with nineteenth century industrial capitalism has greatly influenced the form of the mental futures which 20th century planners and architects have been willing to 'image' in terms of built environments. The contemporary wisdom seems to be that the city's more deplorable associations can be cancelled out by establishing small-scale settlements in or near cities, with a rural flavour. Within such settlements solid, rural virtues would grow. The Garden Cities and New Towns in England are material evidence of the transformation of this image of the future into built form. Although such views were held more strongly in the past, there is still little evidence to a suggest that there has been any real change. Figure 9.10 is an early advertisement which seems to capture the rural image that Orwell's 'huge, peaceful wilderness of outer London' was supposed to have at the dawn of the twentieth century.

In the United States, as in most westernized countries, the idea of a rural, anti-urban future has also grown up and, as in England, has been reflected in the sprawl of suburbia. In many countries it is also associated with a further periodic retreat into the deeper countryside where a 'peaceful wilderness' is sought. Such a periodic retreat is rather similar to the initial 'migration' of commuters into the 'heart of the country' on the London District line (Figure 9.10). The main difference is in the regularity and higher frequency of

Figure 9.10. A clue to the rural image of
urban places

contemporary movement between town and country of commuters. The Scandinavian 'summer house', the Australian 'weekender', the American 'cabin' and the Russian 'dacha' still present evidence that the middle and upper classes who after all, do continue to make the decisions about the form of the built environment, are still attracted to the idea of a rural haven, to 'get away from it all'.

Reach

Today's built environment is also constructed according to ideas about the distance into the future that a society may claim as its own (i.e. in terms of its *reach*, Chapter 6). Lynch (1972) has pointed out that the builders of medieval cathedrals were so secure in their vision of the future that they often planned their building *projects* to continue for a thousand years (and of course, expected them to, as technological changes in materials and tools were not anticipated). Such long future perspectives are not entirely a 'thing of the past'. According to the stone masons involved in the construction of the world's largest 'Gothic', sand-stonecathedral, built in Liverpool, England, between 1904 and 1978, 'it will last for a thousand years'. But what about the life expectancy of the plastic

Figure 9.11. A coffee bar of the future in the present. Eldon Square shopping complex, Newcastle Upon Tyne. (Photograph courtesy of Dr. David Bennison, University of Newcastle Upon Tyne)

moulded coffee lounge in Figure 9.11? One which Dr Johnson, for all his love of coffee, would have hesitated to enter. The building of the contemporary spatial environment is, in general terms, determined by shorter, less apocalyptic timetables: the *present* lies in a much shorter future.

The historical resources that are manifest in the built environment need *conservation* in much the same way as any resource does, because there is a future in today. Conservation implies a coherent and rational programme of taking stock of a nation's built environment resources in order to enrich the choices of people in the future-present. One way to conserve these possibilities of future choice is by operating on the built fabric with an awareness of the signs or information — both spatial and temporal (as in Chapter 1) — which it carries. To practice built environment conservation, and most advanced scale societies do have official government and state subsidised departments and organizations doing just that, 'a theory of environmental disposal' is needed (Lynch, 1972, 234); one which is aware of the *times and the paratimes* that are locked in built material fabrics.

Conservation times and paratimes

Townscapes are a product of particular social values operating under particular technical constraints and opportunities. There is some evidence that these elements have greater or lesser coherence with each other in a cyclic manner. Building cycles are closely associated with economic factors and economic factors are the impelling force of greatest significance. They are, in their turn, cyclic. Because the built environment takes time to construct and requires economic inputs of capital, it seems worthwhile to draw attention to some of the characteristics of these longer, many-faceted economic cycles.

We now turn our attention to a brief discussion of some of the characteristics of economic cycles. The link with the immediate preceding discussion and that which follows in 9.3 may appear tenuous, but it is necessary to try and set the periodic variability in building activity, which in turn produces various townscapes, into a wider context. When we return more specifically to the built environment in 9.4, the link will become clearer.

9.3 The place of economic cycles

Interest in the study of economic cycles is easy enough to understand. If these cycles can be identified in terms of their periodic length and their variability in amplitude, then they provide one means of anticipating future economic behaviour. (In a sense this permits a window to be opened onto the future; a condition which we argued to be of consequence in the previous section.) Their study also provides a better understanding of the interrelations between the economy and the built environment which is our particular interest here.

It was possibly Jevons, in 1863, who gave the first powerful impetus to the statistical study of economic time series in his book *The Coal Question*. He was followed by many others. Their work resulted in the delineation of national economic *pulses*. (Pre-1914 pulses for Britain and the USA are shown in Figure 9.12, whilst Figure 9.13 shows pulses for the period 1919-1935 for the same two countries.) There has been much discussion about the nature of the so-called economic pulse and of its single or many causes. Our interest focuses on the fact that each of the monetary factors represented in these two Figures (9.12 and 9.13) is an underlying, explanatory factor in the vitality of the market place in which buildings are constructed, sold and occupied. The size, shape, style, simplicity, or sophistication of buildings as well as their spatial distribution and the temporal density of building activity, all depend on the circulation of investment capital, lending rates, and entrepreneurial willingness to take risk. In the 'city' we are surrounded by the economic environment of the past, of buildings which have always been constructed with an optimistic view of the

Circumstances

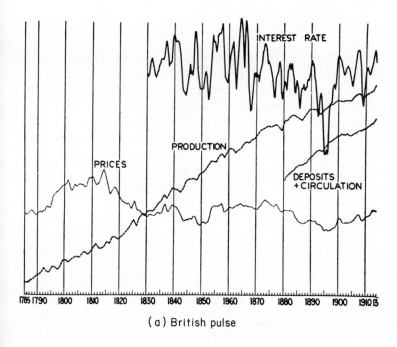

(a) British pulse

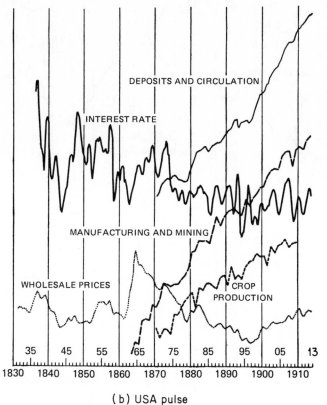

(b) USA pulse

Figure 9.12. The economic pulse of a nation (1). (From Schumpeter, 1938.
Reproduced by permission of McGraw-Hill)

410

future. (Even an air-raid shelter for nuclear war is based on optimism).

It was the large fluctuations in costs and prices following the Napoleonic Wars which gave weight to the impetus to study what have become known as economic cycles. In the late nineteenth and early twentieth century the study of these cycles was directly related to economies in a state of crisis: to economies in depression. On the whole the explanation of economic business cycles has tended not to be the province of orthodox economic theorists. They have been more concerned to develop economic principles which will 'hold in the long run' or which model economic systems in an 'equilibrium state'. The result has been that the study of economic cycles has become the province of statisticians and of political economists. Sismondi was possibly among the first of these to seriously study economic cycles although he was soon joined by Rodbertus and

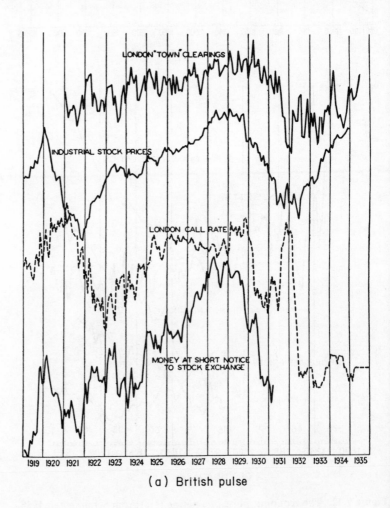

(a) British pulse

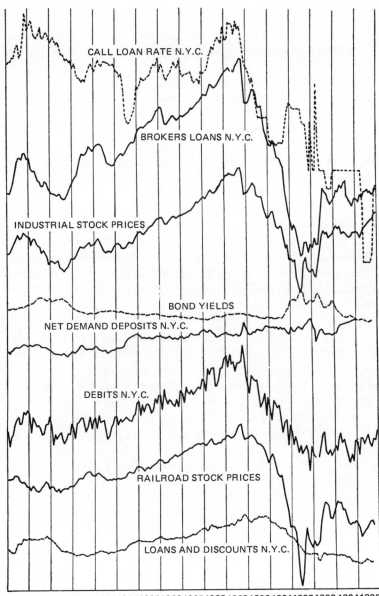

CALL LOAN RATE N.Y.C.

BROKERS LOANS N.Y.C.

INDUSTRIAL STOCK PRICES

BOND YIELDS

NET DEMAND DEPOSITS N.Y.C.

DEBITS N.Y.C.

RAILROAD STOCK PRICES

LOANS AND DISCOUNTS N.Y.C.

1919 1920 1921 1922 1923 1924 1925 1926 1927 1928 1929 1930 1931 1932 1933 1934 1935

(b) USA pulse

Figure 9.13. The economic pulse of a nation (2). (From Schumpeter, 1938. Reproduced by permission of McGraw-Hill)

Marx. Whilst they considered the cause or generator of economic cycles to lie within the very fabric of social structure itself, Jevons (1875) announced that the cause was fluctuating radiation from the sunspot cycle! (This is not quite as absurd as it may first seem).

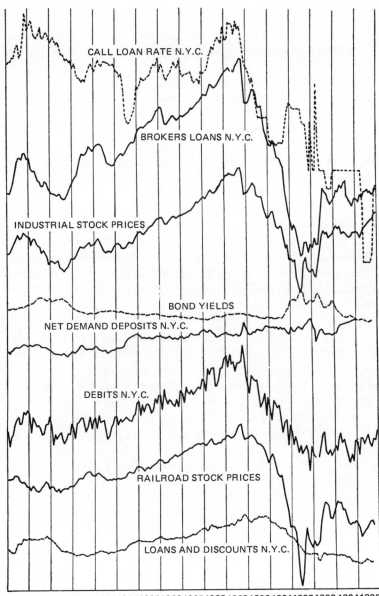

By about 1920 many explanations of economic cycles were current and Mitchell (1927) was able to list a number of competing explanatory 'theories':

(1) The weather (Jevons, Huntington)
(2) Degree of uncertainty in the market through psychological factors (Pigou)
(3) Rate of innovation (Schumpeter)
(4) Variations in savings and investment
(5) Over-production of one type of good, usually building
(6) General over-production or accumulation (Marx)
(7) Variations in banking rate of interest
(8) Variations in income flows
(9) Narrowed profits (Veblen, Marx)

This parade of explanatory factors in the movement of economic cycles created severe problems for their further study; problems which could really only be overcome by a general theory which was able to interrelate the different kinds of cycles that did seem to exist.

Wavelength and phase

Part of this badly-needed, new theoretical advance was to come as a result of an appreciation, from both statistical and economic sources, that economic cycles themselves could be classified according to their duration or wavelength and their relation to other cyclic phenomena, as phase relations. These relations were systematized by Schumpeter in *Business Cycles* in 1939. He pointed out that 'there is no reason why the cyclical process of economic evolution should give rise to just one wavelike movement. On the contrary, there are many reasons to expect that it will set in motion an indefinite number of wavelike fluctuations which will roll on simultaneously and interfere with one another in the process' (Schumpeter, 1939, 161). He outlined a three-cycle model of economic evaluation and this is interpreted in our sketch in Figure 9.14. The wave names which appear in the sketch were not necessarily used by Schumpeter, but are labels which are now used in the economic cycle literature and appear in banking and financial journals. They are convenient labels for our purposes. Three types of wave were distinguished by Schumpeter. These were:

Kondratieff wave

(i) Kondratieff waves (Kondratieff, 1978) — these are waves of about fifty years in length.

Juglar wave

(ii) Juglar waves — these are the 'standard waves' of the business cycle; they are normally considered to have a wavelength of between seven and ten years.

Kitchin waves

(iii) Kitchin waves — these are short waves of some three and a half years (forty months) period.

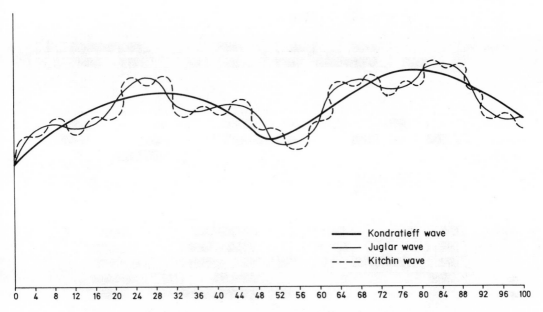

Figure 9.14. The three cycles of the Schumpeter scheme

However, Schumpeter's basic three cycle model has been disputed and is often either replaced by another type of wave, *Kuznets wave* (Thomas, 1954, 1972) or, if not rejected out of hand, has the Kuznets wave superimposed on it, making it a four-cycle system. Kuznets waves have an average span of fifteen or twenty years and appear to apply especially to the variability in activity in constructing built environments. As such they are of particular interest to us. **Kuznets wave**

It is not easy to isolate any one of these four cycle-types for a single nation, and it is an even more difficult task to isolate them for regions within a nation. Part of the reason for this difficulty is the intrusion of spatial factors which induce what is technically known as spatial autocorrelation[2] among the different time series. Another reason is that the waves spring up as the result of interaction in a system — therefore they will not always be apparent at a regional scale.

Mitchell (1927) examined the time series of various economic cycles for a number of countries between the years 1890 and 1925. By noting their conjunctures he tried to build up a picture of times of prosperity, recession, depression and revival. His results are shown in Figure 9.15. Although broad patterns do seem to emerge, it is immediately obvious that not all countries had the same periods of crisis or prosperity — there is a rather complicated pattern of leads and lags, and of phase shifts between them. These can be only partly explained by the influence of spatial factors like **Times of prosperity, recession and revival**

Distance and relative locations

414

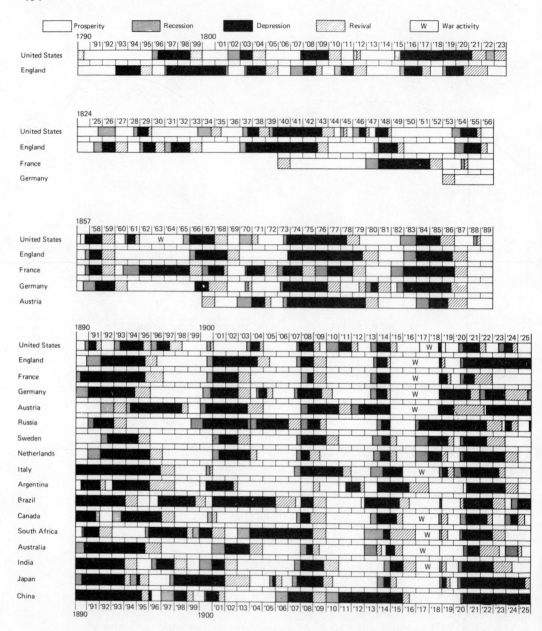

Figure 9.15. Comparative time series for 17 countries. (Mitchell, 1972, chart 26 and 27)

distance or relative locations among trading countries (based on the strength of their trading links for instance).

More recently Mandel (1975) has identified the main characteristics of four Kondratieff long waves in the world economy represented by the primary indicator of the rate of profit, but there have been critics of his claim (Day, 1976). Mandel's Kondratieff waves can be correlated with approximately fifty year periods, 1793-1847, 1848-1893, 1894-1939, and 1940-45 to beyond the present phase in the Kondratieff cycle which 'will' end about 2000. The expansive phase ended about 1966 and the world is now in the downswing. (Rostow (1978) also refers to the start of a new Kondratieff cycle around the turn of the century.) Mandel associates each of these waves with a technological innovation of consequence. Such innovations must, however, be diffused spatially (Chapter 7) in order for their impact to be of consequence. Linking cycle studies with spatial diffusion remains as a challenge to human geographer and economist alike. Cycles appear to be an endemic aspect of economic processes, as does diffusion.

Kondratieff waves and rates of profit

9.4   Cycles in the built environment

The built environment is an accumulative and a complicated feature of our experience. It is in large measure a result of the periodic variations in the 'mood' of national and international economic factors. Each of the *elements* which 'compose' it are 'produced under different conditions and according to quite different rules' (Harvey 1978, 115). Each element (a building, a road, a parking space) has a position in some *building cycle* and yet somehow all these cycles come together to produce the townscapes which we know.

The cycles of new building cannot unfold independently of the constraining forces imposed by the existing built fabric. Building cycles also occur at different spatial scales and the national cycles of one state are not necessarily independent of the national building cycles in another due, in particular, to the interrelatedness of money flows and interest rates. The economic cycles which we have briefly considered above are instrumental in the changes which occur in the townscape which we occupy.

National building cycles have been in evidence, at least in Britain, since about 1700 (Thomas, 1972). Work on the building cycle in many countries (Thomas, 1954, 1972; Gottlieb, 1976; Lewis, 1965; Richardson and Aldcroft, 1968) shows that they conform in general to movements of intermediate length, between the Kondratieff cycle and the Kitchin cycle, at the scale of the Kuznets cycle. Figure 9.16 shows cycles of residential building construction in ten countries, in the main for the years before 1939. They too have a Kuznets period.

For studies of the British building cycle Thomas (1954, 1972) has used the Kuznets cycle as a basis for the analysis of migration as a determinant of urban development before 1913; in particular for

416

Figure 9.16. Building construction waves in ten countries 1840-1955 (Gottlieb, 1976)

Atlantic economy

migration between Britain and the United States. In this 'Atlantic economy' of the nineteenth century, the long waves of changes in investment and export moved inversely to one another between Britain and the United States, as in Figure 9.17[3] The economic crises of the nineteenth century Atlantic economy were, therefore, either occuring on one side of the Atlantic or the other. Investment was therefore switched back and forth as the economic climate altered. Because this investment was so often tied up with construction of the built environment it meant that there was an uneven space-time distribution of building construction activity, closely linked to money stock and capital flows. In other words, while building activity was high in the United States, it was low in

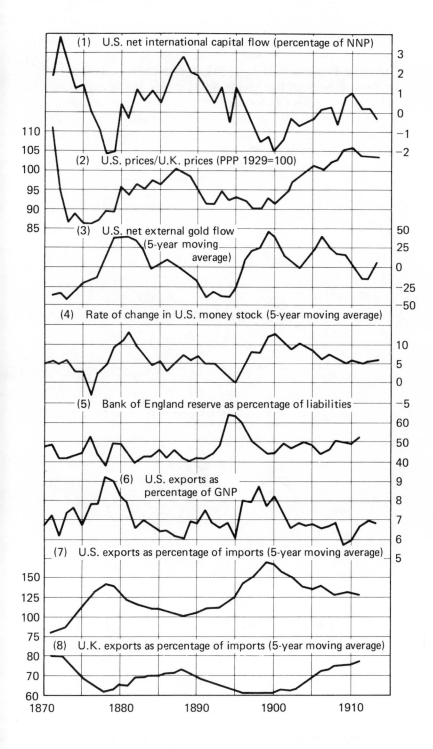

Figure 9.17. Money stock, capital, flow and other indices 1870-1913 Britain and United States. (From Thomas, 1972. Reproduced by permission of the author)

418

Britain and *vice versa*. The different rhythms of investment in the built environment, between Britain and the United States, are shown in Figure 9.18 (Harvey, 1978, and Thomas, 1972).

Some aspects of national building cycles can be exemplified by considering British experience of residential and non-residential building cycles. Residential building construction in Britain between 1856 and 1913, in sixty five British towns, had a time series profile as shown in Figure 9.19.

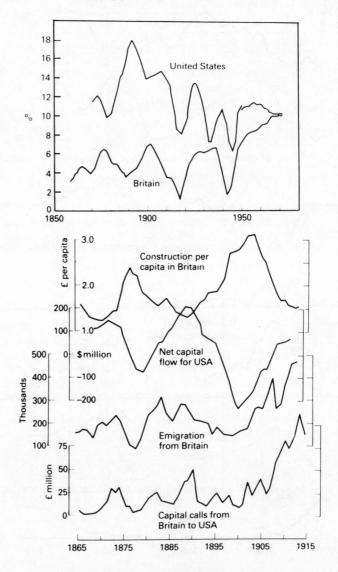

Figure 9.18. Top: Rhythms of investment in the built environment (as % GNP (USA) and % GDP (UK). (Harvey, 1978. Reproduced by permission of Edward Arnold (Publishers) Ltd.) Bottom: Uneven development of the Atlantic economy. (Thomas, 1972)

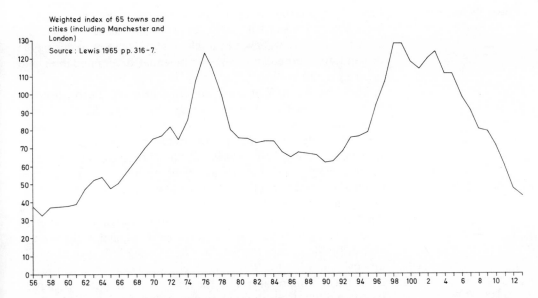

Figure 9.19. Residential building in England and Wales 1856-1913. (Data source: Lewis, 1965, 316-317)

Lewis (1965) has formulated a rather different model of the building cycle from the larger scale Atlantic model developed by Thomas. Lewis's model, although applicable to the same period and even extending to earlier times, concentrates on house building in the UK. Population and the availability of credit are the major causal factors. The argument starts with an increase in population, caused by a rise in birth rates and a change in the balance between immigration and emigration. This rise in population increases the demand for housing. This demand may be further increased if real incomes rise at a time when credit is easy to obtain. There is then a 'boom' in residential building. Because of the long gestation period in the building industry this boom will tend to continue while other economic activities experience a downturn. Indeed, redirection of investment may be to the benefit of the industry at this time. The result may be that the boom in house building gathers enough momentum to leap the gap between two business booms. However, the residential building industry will eventually slump for one of two reasons. One is the possibility of a credit withdrawal which may in fact have been induced by the building boom. This will have an impact on incomes which reduces the power of sellers in the market place and reduces effective demand. The second factor, which follows from the first, is that there will be a rise in the available housing stock and this will act as a disincentive to the investment of capital in the construction of new housing (Whitehand, 1977). The decline in the building sector in turn has de-multiplication effects on other activities; employment opportunities decline, Gross Domestic Product falls, emigration is induced, especially to countries where the phase difference is marked (e.g. USA or Australia). As emigration tends to be age and sex selective there will even be a fall in marriage rates, reducing the demand for particular types of residential accommodation and inducing a fall in birth rates.

The First World War ended a period of marked growth in the built environment of Britain, during which time there had been very little central government control over economic activity. Thereafter, however, the nature of the building cycle changed rather considerably, not only as a result of government intervention in the 'market place', but also as a result of the increasing influence of building societies. The processes and products of this post-1918 period have been extensively summarized by Richardson and Aldcroft (1968). Figure 9.20 shows gross capital formation in housing (roughly the number of buildings multiplied by the average cost of building) over the period. The growing influence of public authority housing for 'lower income' groups also became a factor of increasing importance and the series for this sector is shown, based on tables from Richardson and Aldcroft (1968, Tables 1, 3).

The strong upward trend in the value of housing, at 1930 prices, reflected the massive expansion of sub-urban, semi-detached dwellings which remain as a dominant townscape theme in all the large towns and cities in Britain. From this relatively short series it is not really possible to say much about the underlying cyclic characteristics, but peaks in 1927 and again in 1938 are of Juglar

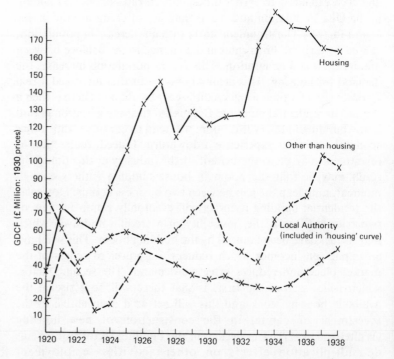

Figure 9.20. Gross capital formation in building in UK, 1920-38. (Source: Harrison and Aldcroft, Tables 1, 3. N.B. Trend not eliminated)

length in their separation. With few houses having been built in the years 1914-1918 and with the interceding years of depression, a marked shortage in housing developed, which required rapid and committed reversal. Public housing and the role of the building societies became increasingly significant factors.

In the years following the Second World War the role of building societies has become increasingly significant in the British market. Once again cyclical fluctuations are in evidence, for instance in the property boom of the late 1960's (see Matthews 1968, Vipond 1969).

If we now turn briefly to a look at the American building experience between 1860 and 1960 the imprint of the Kuznets cycle on gross new construction (at 1929 prices) can be seen rather clearly in Figure 9.21. (A detailed analysis of these waves can be found in Abramowitz (1964) for the period before the Second World War, in Albert (1962) and Campbell (1963) for the period following it, and in Gottlieb (1976) for the whole period).

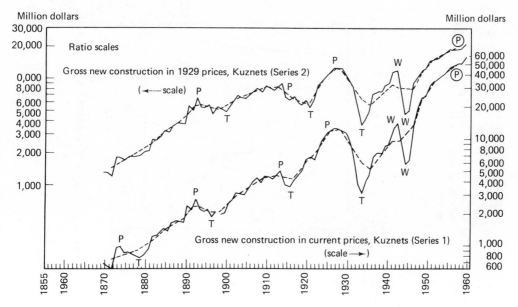

Figure 9.21. Aggregate construction: annual data and average reference cycle standings 1856-1959. (Gottlieb, 1976)

There appears to be a cyclic factor in certain characteristics of transport innovations before 1920 which of course, as a social and economic function relates closely to building purpose and arrangement. Isard (1942, 1943) has therefore proposed a transport-building cycle. He stressed the important role played by the growth of new transport innovations as generators of the waves of building activity. For the United States in the period 1820 to 1930 he identified six major transport innovations, each of which was followed by, or accompanied, a significant increase in building acitivity. Three of these relations are shown in Figure 9.22.

In Britain similar cycles in transportation and building have been identified by Richardson and Aldcroft for the period between 1860 and 1938.

422

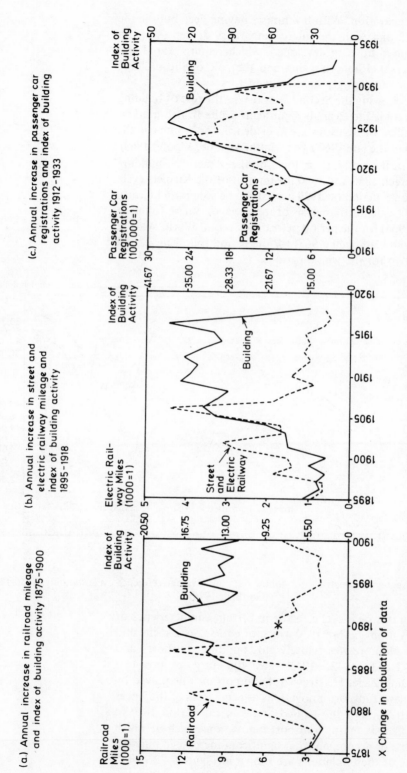

(a.) Annual increase in railroad mileage and index of building activity 1875–1900

(b.) Annual increase in street and electric railway mileage and index of building activity 1895–1918

(c.) Annual increase in passenger car registrations and index of building activity 1912–1933

X Change in tabulation of data

Figure 9.22. Transport-building cycles USA. (Isard, 1942. Reproduced by permission of John Wiley and Sons, Inc.)

9.5 Urban building cycles

Geographers have considered in some detail the differential impact of economic cycles on urban and regional systems, for instance in studies of the inter-regional business cycle (Hepple 1975) and inter-regional fluctuations in unemployment. (Haggett and Bassett, 1970; Haggett, 1971; Bassett and Haggett, 1971; Cliff *et al*, 1975; King *et al*, 1969, Sant, 1973). However, there has been less work undertaken on the building cycle for particular cities or systems of cities.

Figure 9.23 shows a summary of Gottlieb's international urban findings concerning residential building chronologies, over a period of one hundred years, before the Second World War. His conclusion is that 'local cycles were simply a local phase of a national movement, while the national movement was in turn mainly a coalescence of local cycles' (Gottlieb, 1976, 9). For the thirty-one cities he studied, Gottlieb found long cycles with an average of 19.7 years, subject to a mean deviation of five years. This figure compared with national scale long cycles averaging 19 years with a mean deviation of 4.4 years over the same period. Occasionally individual cities would 'miss' one of these cycles but this was the maximum extent of the variation in most cases. Within the period studied by Gottlieb the long swings in building activity were maintained by correspondingly long swings in migration and rates of household formation, bearing out the work of Thomas and Lewis. Slumps were due to overbuilding as suggested by Lewis. Land values in the urban areas which were studied correlated strongly with building booms and slumps, rising and falling in time with them.

Can these building cycles be related to some of the familiar spatial models or schemes of urban form? The first tentative steps to provide an answer have recently been taken. The first attempt to capture the imprint of building cycles upon urban form was made by Hoyt in 1933 (Hoyt, 1970), in his study of the variation in land values in Chicago from 1830 to 1933 (see Figure 9.24). He recognized the existence of at least five cycles of boom and collapse in residential building in Chicago, each with an average duration of fifteen to twenty years (the classic Kuznets cycle again). In the eighty years between 1852 and 1932 five sixths of all residential buildings erected were put up in the forty most active building years. Such observations enabled him to put forward a simple model of residential building activity in urban areas similar to the explanation at the national level offered by Lewis and others. The Chicago real estate cycle could be seen as "the composite effect of the cyclical movements of a series of forces which communicated impulses to each other in a time sequence.... These cycles, in the order in which they appear, are the cycles of population growth, of

424

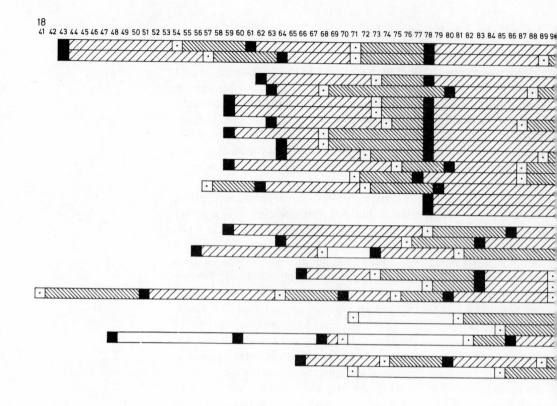

Figure 9.23. A comparison of urban cycles. (Source Gottlieb, 1976, Table 1.1)

the rent levels and operating costs of existing buildings, of new construction, of land values and of subdivision activity.'' (Hoyt, 1970; 369-72.) These cycles could be brought together in a simple model which probably applies in a general sense to nearly all urban growth, in a time when large movements of population are taking place.

An initial trigger, usually population growth and resultant migration coupled with transport innovation, causes rents to increase so that selling prices of buildings rise. This produces an upsurge in construction and a resultant land boom. Speculative building activity, however, soon pushes this construction activity above the level of effective demand, and a slump sets in, as in Figure 9.25.

The spatial consequences of this model were clear enough to Hoyt. They were the creation of a ring-like city consisting of successive bands of growth which could be broadly identified with particular periods of upsurge in the building cycle as in Figure 9.26. Hoyt was able to map this process in Chicago. That this ring-like structure is indeed the case can best be summarised by examining

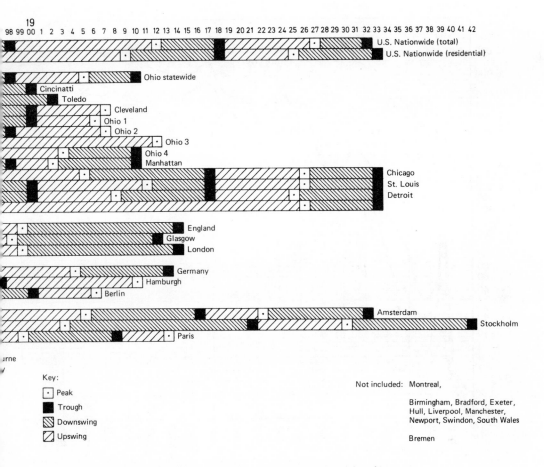

U.S. Nationwide (total)
U.S. Nationwide (residential)

Ohio statewide
Cincinatti
Toledo
Cleveland
Ohio 1
Ohio 2
Ohio 3
Ohio 4
Manhattan
Chicago
St. Louis
Detroit

England
Glasgow
London

Germany
Hamburgh
Berlin

Amsterdam
Stockholm
Paris

ırne

Key:

☐ Peak

■ Trough

▨ Downswing

▧ Upswing

Not included: Montreal,

Birmingham, Bradford, Exeter,
Hull, Liverpool, Manchester,
Newport, Swindon, South Wales

Bremen

the general contours of urban evolution in American cities in particular, recognizing that the general principles involved might also apply to other countries.

Adams (1970) has tackled these issues in relation to Midwestern cities. He asks 'What would be the residential age structure of a model city in the United States if it faithfully reflected every significant trend in residential construction activity through out a century of city building?' (Adams, 1970, 42). Figure 9.27(a) shows the number of residential units started each year in the United States from 1889 to 1960. It consists of six cycles averaging about fifteen years in length. Presumably a model city would reflect these cycles in corresponding rings of continuous urban growth, the spatial extent of each ring being dependent upon the number of houses built in each cycle together with variations in plot size from cycle to cycle. However, such a simple ring structure would be distorted by the transport system which was in operation at a particular time and by its anticipated demands. American cities (like many others) were built up during a period of significant transport related building cycles (see above). Adams considers the

426

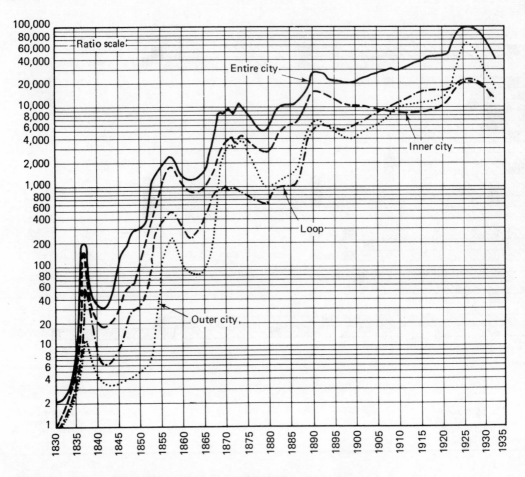

Figure 9.24. Changes in Chicago land values 1830-1933
Inner City, bounded by Belmont Ave., Lake Michigan, Pershing Rd, Kedzie Avenue
Outer City, all territory excepting inner city within city limits of 1933. (outside the loop)
Entire City, all territory within the city limits of 1933, 1 = $50,000. (Hoyt, 1970. © 1970. Reproduced by permission of the University of Chicago Press.)

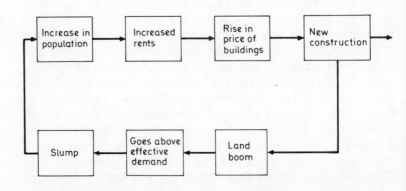

Figure 9.25. A summary of the essentials of Hoyt's model

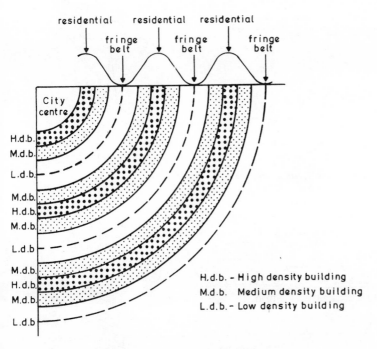

residential   residential   residential

fringe belt   fringe belt   fringe belt

City centre

H.d.b.
M.d.b.
L.d.b.
M.d.b.
H.d.b.
M.d.b.
L.d.b
M.d.b.
H.d.b.
M.d.b.
L.d.b

H.d.b. - High density building
M.d.b. Medium density building
L.d.b. - Low density building

Figure 9.26. A concentric zone scheme of building activity based on Hoyt's model of land value changes in Chicago

effect of four of these cycles on American mid-western cities. They are the era of walking and the 'horsecar' up to the 1880's (I), the electric streetcar era from the 1880's to World War I (II), the era of the recreational auto from the 1920's to 1941 (III), and the post Second World War era of extensive automobile ownership and the freeway system (IV). The margin of the city is progressively redefined by these innovations. In the case of the motor car in particular, this will be most evident in terms of time distance rather than simple Euclidean terms (see Chapter 7). Once these transport innovations were taken into account Adams found it possible to draw an idealised scheme for the model U.S. city illustrating the relative widths of the growth rings from each construction period, as well as the zones of influence of each urban transport innovation, as in Figure 9.27 (b).

The pre-1890 Old Core of the city acts as the nucleus of a successive expansion of residential areas corresponding to the length and intensity of each cycle. It is assumed that most expansion will be on the urban fringe because this reduces the risk that a new house will lose its value by being built in unattractive surroundings, because the sites on the fringe generally cost less and because the highest demand is for the 'ideal' of the rural setting: a rather persistent 'ethos' in the contemporary urban world, as suggested earlier.

428

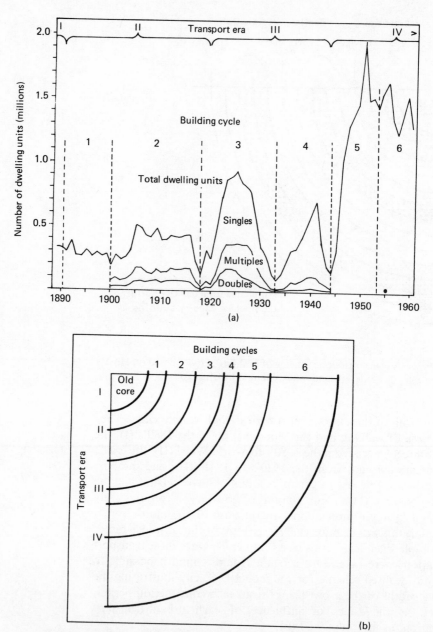

Figure 9.27 (a)-(f). Cyclic characteristics in the North American city
(a) Estimated number of dwelling units started each year, 1889-1960. Dwelling units are built as single family houses, in double houses, or in multiples. Estimated detail by type of structure is not available for units built before 1900 or after 1944. Source: United States Bureau of the Census: *Housing Construction Statistics: 1889-1964* (Washington, D.C.: Government Printing Office, 1966). Table A-1
(b) Spatial structure of a quadrant of the model city. Six post-1889 building cycles produced growth rings around the Old Core: 1890's, 1900-1919, 1920's, 1930's, 1940's, and 1950-1960 cycles. Four intraurban transport eras produced variations in internal housing arrangements and densities: the walking-horse car era; the electric streetcar period; the era of the recreational automobile; and the freeway auto era. The area of each growth ring depends on construction rates and density levels prevailing in each growth period

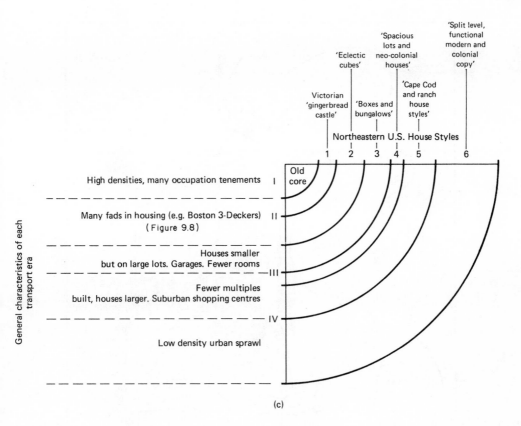

(c)

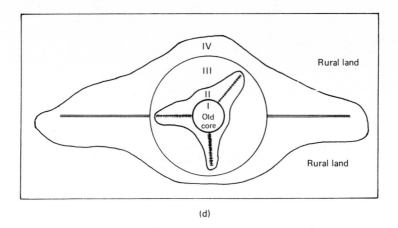

(d)

(c) Suggested period characteristics in a Northeastern American context (Rickert, 1967). For explanation of gradients, see text

(d) Expected distortions from concentric growth patterns. The highly articulated urban transport networks of transport eras two (streetcar lines) and four (freeways) promoted star shaped deviations from concentricity. Transport eras one (foot travel) and three (recreational auto) promoted transport surfaces and compact, circular urban forms. Traverses *A* through *D* indicate the variety of contrasting age gradients

430

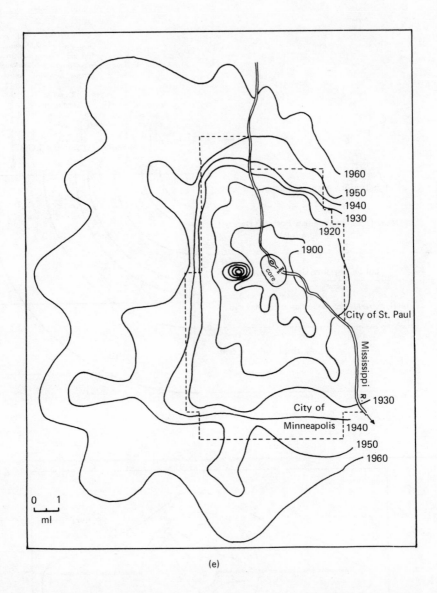

1960
1950
1940
1930
1920
1900

City of St. Paul

Mississippi R.

1930

City of
Minneapolis
1940
1950
1960

core

0   1
ml

(e)

(e) Age characteristics of residential areas along a traverse through the core of the model city, showing the percent at each point that was built during each construction cycle

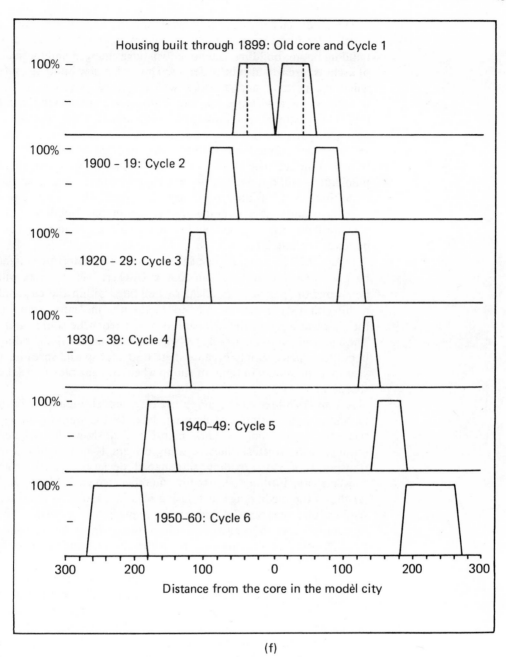

(f)

(f) Median age of housing in Minneapolis and vicinity, with isoage lines bounding growth rings from each construction cycle (a), (b), (d), (e), (f) reproduced by permission from *The Professional Geographer* (the *Annals*) of the Association of American Geographers, Volume 60, 1970, J.S. Adams

The model city will also show variation in internal appearance, which is contingent upon cyclical building activity. At each new building boom, building techniques will have changed and the style of architecture will also be different. Thus, each new building cycle will have period characteristics which will be reflected in the townscape. For instance, in the European context Whitehand (1977) has pointed out that these characteristics may include a vogue for parks (as in Victorian and Edwardian times — see Slater, 1977) or new recreational facilities and the need to adapt to new transport modes. The most severe complicating factor is that these innovations will not necessarily be adopted in distant spaces at the same time. A diffusion process will take place. Thus, large cities may have adopted a building innovation in one building cycle which will not become apparent in other smaller cities until the next boom or the one after.

Figure 9.27 (c) suggests what these period characteristics might be in a Northeastern American context (Rickert, 1967). There will be an age or 'pastness' gradient of buildings within the city and architectural styles may or may not be the best indicator owing to the characteristics of the diffusion of styles from the source area. The age gradient of an urban area will be more shallow if a building boom took place simply because there is greater spatial spread per year in a boom than in times of slump when the age gradient will be sharper.

Adams (1970) tested the concentric rings model (Figure 9.27 (b)) in Minneapolis, using Census tract data. Distortions are closely related to the transport network. Figure 9.27 (d) shows the expected outline, with streetcar lines in transport era II and freeways in transport era IV providing star-shaped deviations away from concentricity. Infilling occurred in the other two eras. Figure 9.27 (e) shows the median age of housing in Minneapolis. A distorted semi-circular concentric pattern is immediately evident. The sequential order of age of buildings found in traverses made through Minneapolis, outward from the city into the suburbs, is illustrated in Figure 9.27 (f).

At least three other factors, apart from transportation, distort the concentric model, in terms of its symmetry. The first is that at each upward swing in the building cycle demand may be greatest for particular types of land. The second is that government policy can restrict growth in particular directions or places (land use zoning in general). Thirdly, at any time, not all outward growth is contiguous, as when there is an interruption due to a 'Green Belt'. Some developments in a boom period may take the form of infilling of sites in areas already considerably 'built up', while others may leapfrog such areas of development, only to be built round at a later date. The city becomes a set of heterogeneous regions in each of which, however, distinctive architectural styles can be found at different townscape scales.

If land value changes are in phase with the building cycle, then it is likely that the proportion of institutional (i.e. educational, medical, recreational, etc.) building (as opposed to residential building) will vary with the cycle as well. Institutional buildings will be built closer to the city centre during times of slump, when land values are low, than at times of boom when housing is competing for the land and land values are correspondingly higher. In the British context Whitehand (1972, a, b) found this to be true of Glasgow. Similarly, the average plot size can be expected to vary with the building cycle, being larger in times of slump when land values are lower. Whitehand (1975), investigating middle class residential extensions of the West End of London (North Kensington), confirms this (see Figure 9.28).

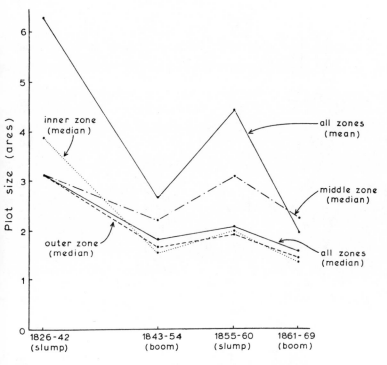

Figure 9.28. Mean and median sizes of new plots developed in relation to (universe) time. (Reproduced with permission from Whitehand, 1975. Copyright by Academic Press Inc. (London) Ltd.)

Due to the different timings of building cycles for residential and other building types, the city can be expected to conform to Hoyt's essentially annular ideal of alternating zones, of areas of greater residential building activity and greater activity by other building types. The periods which are not identified with vigorous residential development with consequent larger spatial extensions, give rise to a pattern of *fringe belts* (Conzen, 1960, 1966; Whitehand, 1967, 1972, a, b, 1974, 1975, 1977; Barke, 1974, 1976).

Fringe belts

434

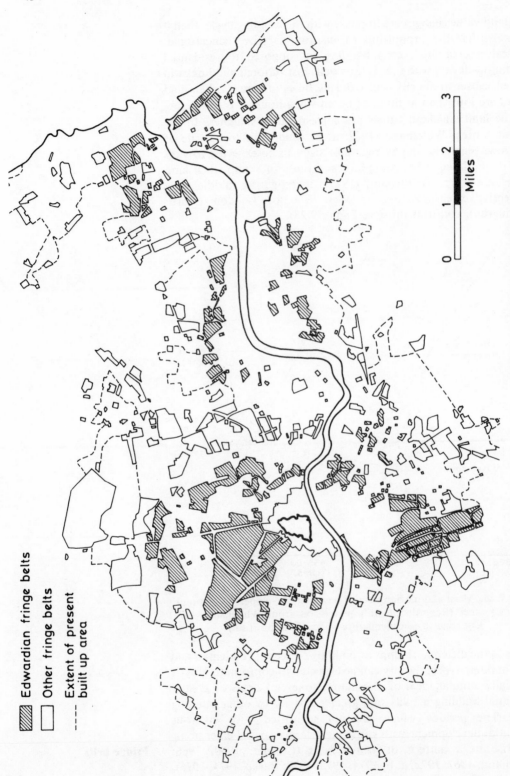

Figure 9.29. Fringe belts of Newcastle-upon-Tyne. (The heavy black line marks the approximate extent of the medieval city.) (Whitehand, 1967)

Edwardian fringe belts

Other fringe belts

Extent of present built up area

Miles
0 1 2 3

Figure 9.29 shows successive fringe belts around Newcastle-upon-Tyne (UK). Each fringe belt undergoes cycles of successive fringe belt modification (Whitehand, 1967) as each building cycle comes round: many cycles in the case of medieval areas, fewer in the case of an Edwardian or later area. These modifications will be correlated with style and predominant uses of each modifying cycle and they result in a slowly changing pattern of land use as reported, for instance, in Barke's study of Falkirk (Barke, 1974).

Phases

Conzen (1962) recognized three phases of fringe belt development with special reference to his detailed work in Newcastle-upon-Tyne. First, a fixation phase, when the town wall of mediaeval times was established, with arterial ribbon development just outside the city gates (c1300—c1700 AD). Second there is an expansion phase when both the intramural space and the extramural space grow, (c1700—c1860). Finally there is a consolidation phase (after 1860) when the extramural space is hemmed in by accretionary town growth outside. In the 'old town' functional segregation intensifies, attracting industry, warehouses and institutions, accompanied by increased site succession within its area (Conzen, 1962, 406).

Burgage cycles

The process of site intensification has been meticulously studied in relation to British towns with a mediaeval or even earlier beginning (Conzen, 1962). Conzen has called this process of site intensification the Burgage cycle because of the particular type of building plots that his empirical work concentrated upon. 'The characteristic strip-shaped plots known as burgages and representing originally the holdings of the enfranchised members of a mediaeval borough' (Conzen 1962, 387) have a cyclic pattern of intensification.

In Figure 9.30 the burgage cycle in Pilgrim Street, Newcastle-upon-Tyne, is illustrated. Initially, buildings are concentrated on the burgage heads (c1700 and before), with the burgage tails being left as yards and gardens. By careful reconstruction of the changing intensity of building phases Conzen was able to build up a series of maps and graphs which traced the burgage cycle pattern for any single burgage. They were:

1. an Institutive phase
2. a Repletive phase
3. a Climax phase
4. a Recessive phase
5. an Urban fallow

Within each city's long waves of building activity there may also be many short wave cycles which have an influence on spatial structure. Berry (1976) has examined the case of building cycles in Chicago between 1967 and 1972. He distinguished between long waves which broadly influenced the climate of urban building in general and the short waves, within these long waves, which are the

436

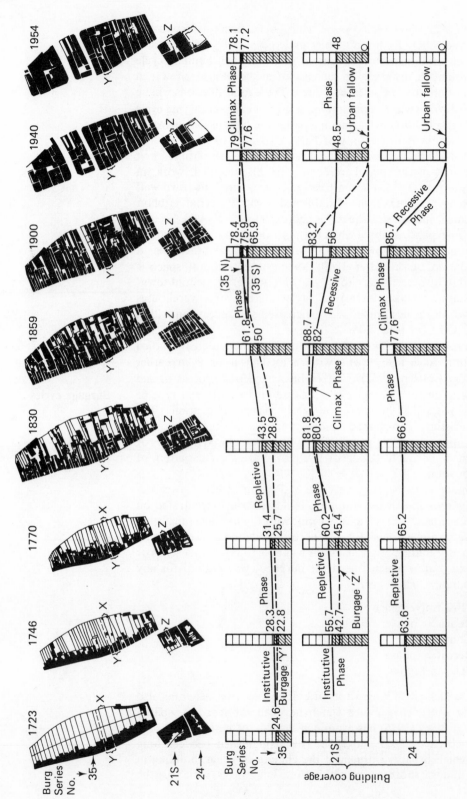

Figure 9.30. The burgage cycle in Pilgrim Street, Newcastle-upon-Tyne. (Conzen, 1962, 406). (A good example of the burgage cycle will also be found in Conzen 1960, for Fenkle's Yard, Alnwick)

unique signature of the Chicago housing market. In particular he attended to the rhythm of real estate transfers and building permits occurring month-by-month during the upswing in a general longer-term boom and at the start of the downswing as mortgages became harder to get and interest rates rose. Taken together with other urban rhythms these make up what Berry *et al.* (1976) call the city's pulse.

Figure 9.31 shows some indices of the recent property boom in the United States, whilst Figure 9.32 shows cyclical activity in the Chicago housing market during the phase of this cycle that Berry was concerned with. What effect did this boom have on the Chicago housing market and in particular on the 'dual' housing market of black and white housing? In the 1960's 481,553 new dwellings were added to Chicago's housing stock. Of these 257,590

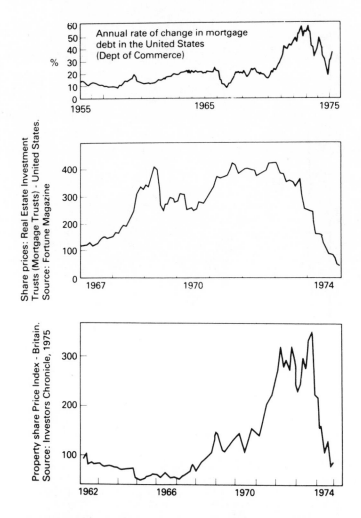

Figure 9.31. Some indices of the property boom: United States and Britain. (Harvey, 1978)

were occupied by white homeowners, 146,029 by white renters, 13,849 by black homeowners and 27,153 by black renters. 27,934 units were left vacant. Berry looked in detail at the rhythms of three different but linked housing markets in central Chicago, the white, black and white to black transitional sectors. Figure 9.32 shows these variations. Berry's work suggests that the concentric ring response to the building cycle still holds generally true, but that new mechanisms may be at work changing the reasons for the operation of building cycles and therefore the reasons for the ring-like age structure of the city.

We have considered various aspects of past, present and future as they are interrelated in the built fabric of the city. The urban present is a product of buildings and of their arrangements, remaining with us from past times but always in a temporary state

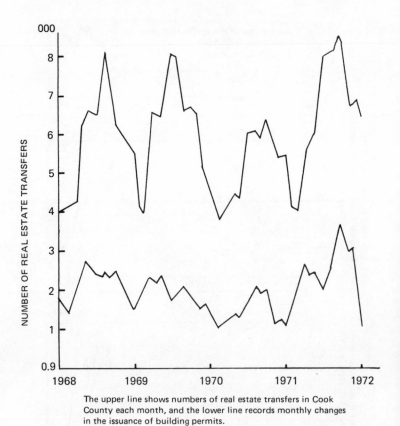

The upper line shows numbers of real estate transfers in Cook County each month, and the lower line records monthly changes in the issuance of building permits.

Figure 9.32. Cyclical activity in the Chicago housing market. (Berry, 1974. Reprinted from 'Short term housing cycles in a dualistic metropolis' by Brian J.L. Berry in *The Social Economy of Cities,* Vol.9, Urban Affairs Annual Reviews, Gary Gappert and Harold M. Rose, Editors, © 1975, pp.177, 179 by permission of the publisher, Sage Publications, Inc. (Beverly Hills/London)

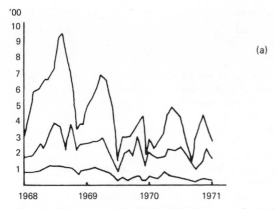

The upper line shows monthly variations in real estate transfers in the white residential area of the city; the second line charts similar trends in the ghetto expansion area, thus involving real estate transfers from white to black; and the lowest line charts real estate transfers within the ghetto.

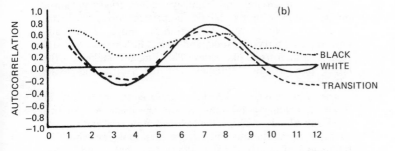

Figure 9.33. (a) Cyclical activity in the city of Chicago's housing submarkets. (b) Autocorrelations of the white, black and transitional time series plotted in A. above. (Berry, 1974. Reprinted from 'Short term housing cycles in a dualistic metropolis' by Brian J.L. Berry in *The Social Economy of Cities,* Vol.9, Urban Affairs Annual Reviews, Gary Gappert and Harold M. Rose, Editors, © 1975, pp.177, 179 by permission of the publisher, Sage Publications, Inc. (Beverly Hills/London)

so far as their future is concerned. An awareness of the contribution of built form to a sense of pastness, present order and future security seems to be a rather valuable component in our general humanistic education. Part of this awareness comes from a better understanding of the processes that have been at work, and which are likely to be at work, now and in the future. Many periodic factors have influenced the built forms of western cities. Each wave, as it breaks on the shores of a human settlement, lays down a new layer of pastness. But times and paratimes of many sorts permeate the built environment — as they do the context of nature, human life and society. Each building cycle is inseparably a part of a social structure. J.B. Priestley conjures up this image rather well in his three phase picture of England and her associated townscapes (*English Journey,* 1934, 397-401).

> 'I had seen England. I had seen lots of Englands. How many? At once, three disengaged themselves from the shifting mass. There was, first, old England, the country of the cathedrals and minsters and manor houses and inns, of Parson and Squire; guide book and quaint highways and byways England ... Then, I decided, there is the nineteenth century England, the industrial England of coal, iron, steel, cotton, wool, railways; of thousands of rows of little houses all alike, sham Gothic churches, square-faced chapels, Town Halls, Mechanics' Institutes, mills, foundries, warehouses, refined watering places. Pier Pavilions, Family and Commercial Hotels, Literary and Philosophical Societies, back-to-back houses, detached villas with monkey-trees, Grill Rooms, railway stations, slag heaps and 'tips', dock roads, Refreshment Rooms, doss-houses, Unionist or Liberal Clubs, cindery waste ground, mill chimneys, slums, fried-fish shops, public houses with red blinds, bethels in corrugated iron, good-class drapers' and confectioners' shops, a cynically devastated countryside, sooty and dismal little towns, and still sootier grim fortress-like cities. This England makes up the larger part of the Midlands and the North and exists everywhere ...
> The third England, I concluded, was the new postwar England, belonging far more to the age itself than to this particular island... This is the England of arterial and by-pass roads, of filling stations and factories that look like exhibition buildings, of giant cinemas and dance-halls and cafes, bungalows with tiny garages, cocktail bars, Woolworths, motor coaches, wireless, hiking, factory girls looking like actresses, greyhound racing and dirt tracks, swimming pools, and everything given away for cigarette coupons...'
>
> (J.B. Priestley, *English Journey*. Reproduced by kind permission of William Heinemann Ltd)

The ghost of society past observes us from every window and doorway. As Priestley, has expressed the sentiment more recently, 'we stagger beneath our inheritance' (Priestley 1977, 390). Enlightened programmes of preservation, conservation and new construction allow these oppressive loads to be lightened selectively. We may be no more successful than King Canute at turning back the waves, but we should avoid his mistake and appreciate their persistence!

Notes

1. In order to represent some of the times in the built environment, Lynch (1972) used fifty six photographs to capture what he called Boston Time. This is a useful exercise to repeat for any town or city.

2. Spatial autocorrelation refers to the association of values in areas that are near to each other. 'In general, if high values of a variable in one area are associated with high values of that variable in neighbouring areas, we say that the set of areas exhibits positive spatial autocorrelation.' (Cliff *et al.* 1975, 145).

3. Thomas finds that long cycles in Canada, Argentina, and Australia were also inverse to the British cycles.

Appendix A
A guide to some analytical methods

A.1 Preliminaries

This appendix is intended as a brief guide to some of the methods which have been referred to from time to time in the chapters of this book. However it is not intended to be a manual sufficient to itself. To satisfy such a need requires reference to one or more of the many textbooks in statistics and in quantitative geography which are available.

Section A.2 introduces some of the methods which can be used for the analysis of data based on time budgets.

Section A.3 discusses some of the basic calculations which are used in determining prism size and shape, from the time-geographic approach.

Section A.4 introduces some aspects of time series analysis. This is a complex, rapidly changing and extensive area of statistical analysis. With space-time series the statistical difficulties of time series analysis are further compounded. A short section on space-time series concludes the appendix.

We have assumed that a student reader will be undertaking, or will have undertaken, coursework in elementary statistical methods.

What follows is really an expanded glossary summarizing some of the key methods which are likely to be used if a chronogeographic approach is adopted. It is not the case, however, that quantitative methods are an essential component of the approach.

A.2 Some measures for time use studies

(i) Durations, frequencies, and sequences

Duration is the period of universe time during which an activity lasts, (unless of course, particular measures of subjective elapsed time are used to gauge psychological time duration). If a measure of the average duration of an activity is required it must be decided whether the total universe time given over to the activity is to be divided by the total number of people, or whether only participants should be used. If T_{at} is the total activity time, and N_t is all individuals, then the average duration D_{at} will be smaller than the average duration D_{ap}.

$$D_{at} = T_{at}/N_t \geq D_{ap} = T_{at}/N_p$$

If everyone participates, as for instance in an inelastic activity such as sleep, then $D_{at} = D_{ap}$.

This simple difference is easily overlooked and can give meaningless values to the computed average duration of an activity. Closely associated with the duration of an activity is the notion of *elasticity* which we will consider in a moment, (ii).

The number of occasions on which a given activity occurs over some defined period of time is its *frequency*. In conjunction with methods for the analysis of point processes (see 10.4 below) it is possible to gain an idea about the degree of regularity with which an activity recurs.

Considered in relation to other activities the frequency characteristic is part of the study of *sequence*. The difficulty with the study of sequences lies in determining which 'types of transitional behaviour are relatively independent of absolute universe time, how independent they are, and what other objective clocks might act upon transitional behaviours in the form of constraints experienced by most people and accepted as such' (Cullen and Godson, 1975, 46). The probability of occurrence of given sequences of activities in terms of changes from one type to another was considered in Chapter 5.3. A useful approach to the description of sequence structure also lies in the use of triplets or higher order runs of activities as outlined in Chapter 4 and Table 4.11. There are a number of 'runs' tests in non-parametric statistics which can be applied to the testing of sequences of numbers (Siegel, 1956).

A measure (T) of *structural difference* has been used by the Hungarian Central Statistical Office (HCSO). It allows comparison of durations of time allocated to activities by groups or between different sites. If we allow that the profile of time allocations (for a number of groups or sites), locates each group or site in a multidimensional Euclidean space with as many dimensions as there are activities, then the distance between any pair of groups or sites can be calculated by Pythagoras' theorem. The index of structural difference, T, estimates the difference amongst pairs of time allocation profiles. The value of T is a measure of similarity (or difference) and when a large number of sites (or groups) are involved it is possible, using multidimensional scaling or smallest space analysis, (Chapter 4) to allocate sites to similarity clusters based on the original matrix of T values. The formula used to compute T is

$$T = \sqrt{\left(\sum_{i=1}^{k} \frac{a_i}{24} - \frac{b_i}{24} \right)^2 \bigg/ K \cdot 100}$$

where a and b are different groups or sites, k is the number of activities and 24 is the number of hours.

The matrix of Table 4.13 is based on these T values. This formula, as the Index of Dissimilarity is used quite frequently in urban social geography to compare the spatial distributions of ethnic groups.

Multi-dimensional scaling and smallest space analysis methods cannot be reviewed here but useful guides, written by geographers, will be found in Amadeo and Golledge (1975) and Golledge and Rushton (1973). Other references have already been given in the text of Chapter 4.

The *intercorrelation of activity durations,* (or frequencies) can reveal the degree of association between activities, as shown in Table 8.10. The correlation coefficient indicates the extent to which a period of time allocated to an activity x compares with the time allocated to an activity y. Coefficients may also be factored by a principal components analysis or by any one of the many factor analytic models, to gain an indication of the underlying, but smaller number of dimensions which account for a high percentage of the variation in the correlation between activities. But attempts at factor analyses of activity durations (Cullen and Godson, 1975, Parkes, 1974; Chapin, 1973) have not always been very helpful because the interpretation of the components or factors becomes very difficult and also because of certain other properties associated with a correlation coefficient computed on a finite range.

When a group of people is sampled for their time allocations within a finite time period, such as a 24 hour day, the implication is that the more time spent on one activity the less time there is to allocate to another. This leads to the generation of negative correlations with higher values than would be found with an *open range* for the data. The problem is analogous to the more familiar problem of correlating ratio or proportional data in the closed range of percentages 0-100. Shapcott and Wilson (1976) have given special attention to computational artefacts of this kind, in the analysis of activity relations.

They propose the examination of the relationships between times allocated to different activities, taking into account what they call the 'trade-offs' induced by the fact that there are only twenty-four hours in a day; in other words the correlations which 'differ from those demanded by the twenty-four hours constraint imply that there are other factors at work determining the relationships between activities in which people engage during the day' (Shapcott and Wilson, 1976, 74).

Each individual time budget is regarded as a vector $(Y_1, Y_2, \ldots Y_i, \ldots Y_n)$ where each activity has been assigned to one of n categories and Y_i is the time allocated to activity i. The identification of the correlations which are due *only to the 24 hour constraint* and the determination of 'whether or not any residual correlations remain which could indicate other relationships between times uses' (p.75) seems to be a necessary consideration in the correlation of activity durations, *if* a strictly limited period has been used in a time budget. For the coefficients computed in the Newcastle study (Table 8.10) there was no strict 24 hour band because as you will recall, the study was based on activity durations which were considered 'normal' in the subjects' experiences and many more activities were included than could be accommodated in a 24-hour day. This is one advantage to the time 'budget' based on recall of 'normal' time allocated.

Shapcott and Wilson outline a method for testing the proposition that amounts of time spent on certain activities cause time to be spent on others, after removal of the 24 hour constraint which creates 'closed variables'. The role differences among sets of people, e.g. women and men, must also be taken into account if the sample is mixed. A set of correlations is computed known as the *estimate of the expected correlations, U_{ik}*. The observed correlations, r_{ik} are 'closed' by the 24-hour constraint and this induces negative relations at greater than the expected probability. A comparison of r_{ik} with U_{ik} is made and if there is no significant difference it is assumed that relationships, other than those caused by the 24-hour constraint, are in operation. Difficulties of this sort in the generation of correlation coefficients make the use of further analyses (like factor analysis) rather awkward.

However, this should not prohibit their adoption as so long as note is taken of the conditions which lead to spurious or otherwise misleading results.

(ii) Elasticity

The *elasticity* of an activity can be approximated by a number based on the coefficient of variation; the standard deviation about the mean duration divided by the mean and multiplied by 100 to express it as a percentage. Such a measure of activity elasticity describes the activity duration characteristics for a group or population and was illustrated in Chapter 4 (Table 4.8) and in Chapter 8 (Table 8.8). The coefficient of variation provides a measure of internal variability among groups of participants and so is based on the duration mean, D_{ap}. The Hungarian Central Statistical Office (HCSO) (Budapest, 1965) used this statistic to find out which were the 'rigid' and which were the 'elastic' activities among various combinations of men and women, workers, and employers, members of co-operatives, non-earning women, etc.

When the elasticity of an activity is calculated in relation to the percentage change in the amount of time available for an activity, a somewhat different concept of elasticity is being considered. In this case the measure is more like the classical index of elasticity, used in economics. Elasticities of activities, calculated in this way, were considered by HCSO and they posed their objective as follows. 'If the duration of activity A increases with (sic!) an hour, how much will change (increase or decrease) the duration of activity B or — expressed the same in ratios — what relative change is called forth by 1 per cent increase in duration of activity A, in the duration of activity B? (sic!) (Budapest, 1965, p.95). The Hungarians considered activity pairs in which employed time was always one of the variables. 'It was examined therefore what changes would occur in the duration of sleeping time, free time (and for women, housework) by a 1 hour (or 1 per cent) increase of working time ... The connection between the activity pairs could in every case be well described by a straight line. (Duration of activities treated as 'dependent variables' decreased proportionately to the increase in working time). Out of all the activities which they considered 'it is sleeping time which is subject to the least abbreviation, leisure time being, on the other hand, very elastic, as is shown by the high coefficient of variation' (Budapest, 1965, p.17 and p.95).

Measures of this sort seem to have potential in various aspects of urban planning, especially as they relate to the provision of recreational facilities and where quite substantial changes are being contemplated in the length and structure of the working 'week'. It clearly behoves those interested in the social structure of urban places to be aware of the different behaviours which are likely to result from changes in disposable time income.

(iii) Graph theory and activity structure

In this section some elements of graph theory will be presented and the measurement of the similarity between two or more graphs of human activity derived from a time budget will be explained. Other applications of graph theoretic methods will be found in Parkes and Wallis (1976, 1978). A useful introduction to graph theoretical methods in geography will be found in Tinkler (1977).

The use of graph theoretic methods in geography is probably best known in relation to the study of transport networks and in the study of linkages and the changes over time of linkages between settlements. In the strictest sense a graph is a particular type of network and the term graph is reserved for a network containing neither loops nor multiple edges. Various relations between any set of objects may be investigated through the theorems of 'graph theory'.

'A graph is a mathematical entity which makes explicit a relationship between objects. It consists of a collection of objects (called *vertices*) and a collection of relations between the objects (called *edges*). For example if we want to model the relationship between people in a gathering which is induced by conversations, we could take the people in the gathering as vertices and include an edge from X to Y whenever X speaks to Y: if A spoke to B twice, A spoke to C once and C spoke to A and B once each, then the graph would have two edges from A to B, one from A to C, one from C to A and one from C to B.

As a matter of terminology we say that the two vertices joined by an edge are the *endpoints* of that edge, and we say that the vertices are *adjacent*. The number of edges which touch a vertex X is called the *valency* of X; in the directed case one may also speak of the *in-valency* and *out-valency*, the number of edges directed *into* and *out* of X.

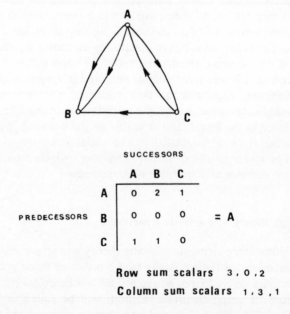

Figure A.1. A digraph and its incidence matrix

A graph or digraph can be represented by an *incidence matrix*. Given a graph on p vertices label its vertices as 1, 2, ... p in some order. Then form a matrix whose (i, j) entry is the number of edges from vertex i to vertex j. The matrix will be symmetric in the undirected case, but not usually symmetric in the digraph case' (Parkes and Wallis, 1978).

The arrows on the edges of a graph show the direction of activity relations, as predecessors or successors. Figure A.1 illustrates a simple digraph and its incidence matrix.

Various matrices are possible — different labellings of the vertices lead to different matrices — but the matrices of one graph are equivalent in an obvious way. The sum of row i and the sum of column j will be the out-valency and in-valency of vertex k, respectively. Thus in the scheme above activity A precedes two activities three times, the two activities are B and C, but B is preceded twice by A.

In some circumstances it might be useful to power the incidence matrix. Elementary matrix arithmetic related to graphs will be found in Tinkler (1977). If we square-power $[A]$ above we have:

Successors

$$
\text{Predecessors} \quad
\begin{array}{c}
 \\
A \\
B \\
C
\end{array}
\begin{array}{ccc}
A & B & C \\
\end{array}
\left[
\begin{array}{ccc}
1 & 1 & 0 \\
0 & 0 & 0 \\
0 & 2 & 1
\end{array}
\right] = A^2
$$

Row sum scalars, 2,0,3
Column sum scalars, 1,3,1.

This represents the linkages between pairs of activities which have some other activity between them, as opposed to being sequentially adjacent as in $[A]$. Higher powers may be calculated and the sums of powered matrices may also be considered. In transport networks or other commmunication linkages (say telephone traffic) such powered matrices represent accessibility or *reachability,* as it is sometimes called in operations research studies, and in studies of group authority structures (Hall, 1974). In an activity structure study, the entry a_{ij} in a square-powered matrix states that activity i may be followed by activity j, a_{ij} times, with some other activity interceding on each occasion or with the same activity occurring again, as can arise if only those activities occurring at pre-specified time locations are considered. For instance, activity B 'follows' activity A once, (row 1, column 2) through activity C, (see digraph in Figure A.1), and this is the only two step path from A to

B. In a full-scale study of activity structure such a condition might suggest the 'criticality' of *C*, (Parkes and Wallis, 1978). Participation in *C* may be critical to subsequent participation in activity *B*. Similarly, *B* might succeed *C* on two, two-step paths, as the *A* to *B* link is a multiple edge with the same directionality. Row and column sum scalars provide further descriptive information about the underlying structure of the activities under study. For instance, in Figure A.1 the row sum, which defines the predecessor status of an activity shows *B* to have the lowest status. No activities follow *B*, *B* is a *terminus* in the cycles. On the other hand, *A* and *C* have a similar successor status. Both succeed some other activity, in a two-step sequence, only once; whereas *B* does this three times.

If the incidence matrix [*A*] and its second power matrix [*A*²] are summed, the number of one and two-step sequences, which are not necessarily sequential in time as a flow, can be derived. For our example, we have:

$$A \qquad\qquad A^2 \qquad\qquad A + A^2$$

$$\begin{bmatrix} 0 & 2 & 1 \\ 0 & 0 & 0 \\ 1 & 1 & 0 \end{bmatrix} + \begin{bmatrix} 1 & 1 & 0 \\ 0 & 0 & 0 \\ 0 & 2 & 1 \end{bmatrix} = \begin{bmatrix} 1 & 3 & 1 \\ 0 & 0 & 0 \\ 1 & 3 & 1 \end{bmatrix}$$

In general, the raising of [*A*] to some power *p*, so that the sum of [*A*] + [*A*²] +, ... [*A^p*], produces an incidence matrix with no zeroes, and shows the total number of links using 1 to *p* steps, between any pair of vertices (activities). In this particular example the zeroes cannot be eliminated. The spatial location and the temporal location of all activities may be labelled and edges can be weighted to produce a weighted digraph which summarizes the duration of activities.

One reason for considering graph theory is that it is a general modelling system for relations. The vertices are 'undefined' items or objects and may be used for representation of anything whatsoever. The use of the graph underlying a situation enables us to strip off initially inessential details. With specific interest in the similarities between the activity days of individuals, the notion of *graph isomorphism* becomes useful.

Two graphs are called isomorphic if there is a map, or way of associating vertices of one with vertices of the other, which preserves adjacency. Precisely, an isomorphism between graphs G_1 and G_2 is a map ϕ from the vertex-set of G_2 which sends vertex $x\phi$ of G_2 such that (x, y) is an edge of G_1, if and only if $(x\phi, y\phi)$ is an edge of G_2: G_1 and G_2 are isomorphic if and only if there is an isomorphism between them. The relationship of isomorphism between graphs is one which needs generalization. For example,

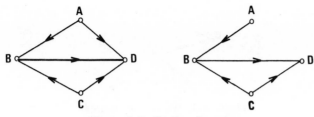

Figure A.2. Similar digraphs

the two digraphs in Figure A.2 are not isomorphic but, since they fail in only one edge, they are very similar. It is for such cases that Parkes and Wallis have tried to develop a simple measure of 'relative isomorphism'.

A digraph is strongly connected if there is a connection from every vertex to every other, and (weakly) connected if all vertices are connected in the undirected sense. There is an undirected walk from a vertex to any other vertex. A graph is often used to represent sequential events (as in critical path analysis). The existence of an edge from X to Y states that X immediately precedes Y in some sequence. The edge from X to Y represents the relation between X and Y. It is then often reasonable to say that there is a 'second stage' connection between X and Z if there are edges from X to some Y and from Y to some Z, so that X and Y are only two steps apart. More generally we define a walk from X to Y to be a sequence of vertices and edges starting with X, then an edge from X to say Y_1, then an edge out of Y_1, and so on, ending at Y. We say 'X is connected to Y if there exists a walk from X to Y. In Figure $A.3$ below, there is a walk from A to E which goes A, (A,B), B, $(B,D,)$ (either edge), D, (D,E), E (or more concisely we could say $ABDE$: this may lack precision as it is sometimes necessary to know which (B,D) edge is used, but more usually it is not).

Another such walk is $ABDCBDE$ (the repetition of vertices of edges is of no consequence). There is no walk $ABCDE$, since the edge (D,C) is directed the wrong way, although such a walk would exist in an undirected graph. To say a graph is connected means intuitively that it 'consists of one piece'. Even though two graphs are both connected, we may think that one is 'more connected' than another, in order to describe this we may define the *connectivity* of a graph to be that number n such that deletion of n vertices will leave the graph connected but deletion of $n + 1$ will result in a disconnected graph. The *edge connectivity* is a similar number defined in terms of deleting edges. In the analogous way we define strong connectivity.

If a graph or digraph is disconnected it will consist of two or more connected pieces. Each of these maximal connected subgraphs are called a component. Figure A.4 below, shows a disconnected digraph with precisely two components. The properties of connectivity are not considered here but they have been

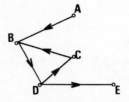

Figure A.3. Walks in a diagraph

applied to the evaluation of *critical or pivotal activities*. These activities are the ones which seem to act as an 'important' link between different types of activities (e.g. movement activity). The measure so far developed is known as a linkage coefficient Lambda of a vertex (Parkes and Wallis, 1978).

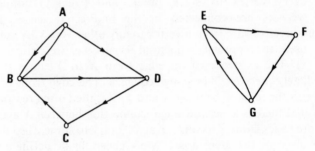

Figure A.4. A disconnected graph with two components

The similarity between graphs is measured by D_{ij}, the *relative isomorphism measure*. It is for assessing the similarity of diagraphs such as those outlined in Chapter 4. (Figures 4.3, 4.10, and 4.11) that the measure was developed.

'The isomorphism measure was computed as follows. Suppose G_1 and G_2 are two graphs with the same vertex-sets, and suppose G^k has d_{ij}^k edges from vertex i to vertex j. Then

$$D_{ij} = \frac{|d_{ij}^1 - d_{ij}^2|}{d_{ij}^1 + d_{ij}^2}$$

(where $|X|$ denotes the absolute value of X and d^k is a descriptive superscript and not an exponent) is a measure of the difference between the two graphs in the (ij) connection. If they are identical, D_{ij} will be zero for all i and j. The factor $(d_{ij}^1 + d_{ij}^2)^{-1}$ seems to be a reasonable way of weighting the relative importance of differences. The fact that one graph has two more edges (i,j) than another is relatively unimportant when the graphs have a lot of vertices joining one to another, but quite important when they do not — and this is precisely the case when $d_{ij}^1 + d_{ij}^2$ is smallest. Our isomorphism measure is just

$$\Sigma \Sigma D_{ij},$$

where the sums are taken over all pairs (ij) and where $D_{ij} = 0$ when there are no edges (ij) in either graph. This means the formula we used is

$$\triangle^{12} = \sum_{i=0}^{8} \sum_{j=0}^{8} \frac{|d_{ij}^1 - d_{ij}^2|}{d_{ij}^1 + d_{ij}^2} \quad d_{ij}^1 + d_{ij}^2 \neq 0.$$

Applied to undirected graphs, the measure would be summed over i and over j equal to or greater than i, to avoid double counting. This choice of a measure has been quite arbitrary. Two other measures were considered — one being

$$\sum \sum \frac{(d_{ij}^1 - d_{ij}^2)^2}{d_{ij}^1 + d_{ij}^2}$$

and the other being

$$\sum \sum \frac{|d_{ij}^1 - d_{ij}^2|}{\max(d_{ij}^1, d_{ij}^2)}$$

(Parkes and Wallis, 1976, 1978)

There are other problems associated with developing measures of similarity or relative isomorphism. For example, suppose G^1, G^2, G^3, and G^4 are graphs which are to be tested for isomorphism: G^1 and G^2 have n vertices each, while G^3 and G^4 have m vertices each. How should $\triangle^{12}$ and $\triangle^{34}$ be compared? In order to achieve a standard measure, $\triangle^{ij}$, the result could possibly be divided by the number of vertices G_i (and G_j) but this problem needs further investigation (Parkes and Wallis, 1978). In order to clarify the calculation of the relative isomorphism measure, consider the pair of most similar graphs G_s^1 and G_s^2 in Chapter 4, Figure 4. Three matrices have to be set up. The first is a matrix of edge differences represented by $|d_i^1 - d_j^2|$: i.e. the absolute differences among the number of edges.

Activity classes (vertices)

	0	1	2	3	4	5	6	7	Σi	
0	1	1			1				3	
1	1								1	
2										
3	1	1							2	*Edge difference*
4										*matrix*
5		1							1	(blanks are zero)
6										
7										
Σj	3	3			1				7	

The second matrix is a matrix of sums $(d_i^1 + d_j^2)$: i.e. the total number of edges between each pair of activities.

Activity classes (vertices)

	0	1	2	3	4	5	6	7	Σi	
0	3	3		4		1			11	
1	5								5	
3										
3	3	1							4	*Edge sum matrix*
4										(blanks are zero)
5		1							1	
6										
7										
Σj	11	5		4		1			21	

The third matrix is the ratio matrix of $|d_i^1 - d_j^2|$ to $d_i^1 + d_j^2$: i.e. differences divided by sums.

Activity classes (vertices)

	0	1	2	3	4	5	6	7	Σi	
0	0·33	0·33				1·0			1·66	
1	0·20								0·20	
2										
3	0·33	1·0							1·33	*Edge ratio*
4										*matrix*
5		1·0							1·00	(blanks as
6										are zero)
7										
									4·19	

The *relative isomorphism of* $\triangle^{7,10} \cong 4.2$ (see Table 4.9).

The sum of the ratio values is the measure D_{ij} and equals 4.2 approximately. This was the smallest value among the comparisons made and therefore indicated the most similar graphs, representing the individuals with the most similar activity structures. The average of a set of D_{ij} values can be used as a measure of group mean similarity and this allows a simple comparison among groups or communities. It is particularly meaningful when a measure of variance is also included and when activity classes are more numerous and the graphs more dense than those in the illustrations.

(iv) Polynomial curves and activity structure

Establishing the possible morphology, or less comprehensively the shape of an activity in terms of its variability over a period of time, is a necessary first step towards finding control programmes of one sort or another, as might be anticipated by 'planners'. In Chapter 5 a number of examples of activity time-shape were given, taken from the United Kingdom (Cullen and Godson, 1975) and from Australia (Robb, 1977). In both cases a curve-fitting exercise is undertaken in order to provide a 'best fit' curve, one which is also as simple as possible in terms of the equation which represents it. This is a necessary consideration because it is always possible to find a mathematical formula which describes a two dimensional relation precisely; such a formula may be so complicated as to shed no additional 'light' on the 'darkness' before us when the observed curve is first studied.

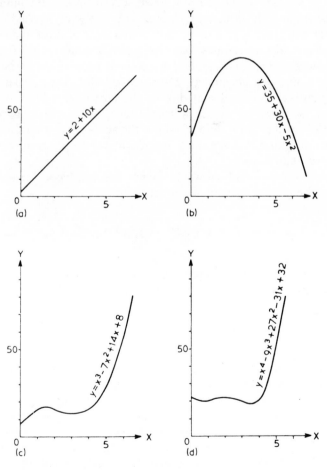

Figure A.5. Polynomials up to the fourth degree.
(a) Linear; (b) Quadratic; (c) Cubic; (d) Quartic. (Baxter; 1976)

From a time budget survey it is possible to evaluate the number of participants in various activity classes at fixed times and this value is then taken as an indicator of the changing intensity of the activity in the 'space' under consideration. Simple harmonic functions were used by Cullen and Godson with the frequency axes representing numbers of people involved. In the Australian study orthogonal polynomial functions were fitted, as illustrated in Figure A.5.

A linear equation of the form $Y = b_0 + b_1$ will be familiar to you from regression analysis of two variables. In the linear equation it is assumed that a variable Y is dependent upon particular values of X for an explanation of its own changing values and that the relationship is linear: throughout the length of the line (linear curve) the ratio of relations is constant. If we have a set of observed values x_i for the variable X, there will be a dependent value y_i for variable Y. Y is a linear function of X. In most cases, however, the relation between Y and X is not linear. Consider Y as the number of participants in an activity class and X as equally spaced time intervals. A curve with a shape more akin to one of those shown in Figure A.5 ((b), (c), or (d)) may 'fit' better. It is also possible that even more complex curves may be required because the variation in Y is not yet adequately explained by the fitted equation. The number of humps and dips in the curve are a clue to the degree. A curve of degree 4 for instance will have 3 marked 'humps' or 'dips' in it, and of course the actual shapes and slopes shown in Figure A.5 are quite specific to the values from which they were calculated.

Curves of the kind shown in Figure A.5 are based on *power* series, the general form of the equation being

$$Y = b_0 + b_1 X + b_2 X^2 + b_3 X^3 + \ldots + b_n X^n$$

where n is the order or degree of the polynomial represented by the power term. A second degree polynomial will have the form

$$Y = b_0 + b_1 X + b_2 X^2,$$

the additional term to those in a linear curve, which puts a 'bend' into the straight line first degree polynomial, is $+ b_2 X^2$. For a cubic or third degree curve the additional term is $+ b_3 X^3$, and so on. In general an Nth order polynomial is described by $N + 1$ terms. The curves are usually fitted by the method of least squares and the coefficients, $b_0, b_1, b_2, \ldots, b_n$ can be calculated by the solution of normal equations, so that in a 2nd degree polynomial we would have the following,

$$\Sigma Y = b_0 N + b_1 \Sigma X_2 + b_2 \Sigma X^2$$

$$\Sigma XY = b_0 \Sigma X + b_1 \Sigma X^2 + b_2 \Sigma X^3$$

$$\Sigma X^2 Y = b_0 \Sigma X^2 + b_1 \Sigma X^3 + b_2 \Sigma X^4.$$

The coefficients, b_0, b_1 and b_2 can be obtained as

	1	2	3	
1	N	ΣX	ΣX^2	$= b_0$
2	ΣX	ΣX^2	ΣX^3	$= b_1$
3	ΣX^2	ΣX^3	ΣX^4	$= b_2$

The largest power on the left hand side is twice the degree of the equation, e.g. ΣX^4 and b_2. Following the fitting of a polynomial curve the variance between the observed data set and the fitted curve is evaluated. It is then up to you to decide whether the percentage of variance accounted for by the polynomial transformation is acceptable at some degree n. In the Australian example of Chapter 5, based on Baxter's program (POLYI), the variance percentage for *activities fixed in space and time,* using a 4th degree polynomial, is 89.15 per cent. The full equation is

$$Y = 34 \cdot 44 - 49 \cdot 09X - 30 \cdot 46X^2 + 52 \cdot 85X^3 - 4 \cdot 14X^4$$

The 3rd degree polynomial 'explained' $89 \cdot 11$ per cent of the fit and so there is little improvement. On the other hand a quadratic (or second degree) polynomial gave only a $63 \cdot 15$ per cent fit so the improvement is sufficiently large to accept the higher order curve. The result suggests that a cubic curve might (in general) give a fairly good description of the activity shape for those activities which are fixed in their spatial and temporal locations. The technique is clearly of value in exploratory studies of activity structure.

(v) Typologies of time use

In Section (i) above we mentioned that factor analytic methods could be applied to data matrices based on correlations between activity durations. A Q-type factor analysis assumes that individuals are not arranged in terms of time use, by chance alone; some are closer to one another than others are, i.e. 'some individuals share more and some fewer common attributes' (Boh and Sakisda, 1972, 231).

One of the best known typologies in urban social geography is the Shevky-Bell social area typology which is based on the argument that there are three primary dimensions of *social space* differentiation which, in combination, produce a number of *social area types,* the number depending on the scaling of the dimensions used. The coarser the scale, the fewer the categories. A four part scale would produce 4^3 or 64 types, based on 3 dimensions. An effective typology of time use would be an advantage to urban research but few attempts have yet been made. Boh and Sakisda (1972, 229) however, have published the results of a 'largely unsuccessful attempt' at the derivation of a time use typology.

> 'If a relationship is measured only by the duration of certain activities then the sociologically least relevant measure, physical time (our universe time), has been utilized. Physical time, which is measured in units stemming from the relationship of celestial bodies, is the inner time of the Sun-Earth relationship system. The duration of an activity, of course depends on the quality of the relationships within the structure of the investigated system, that is to say, within man as a biological, social and psychological entity within a temporal framework defined by habitual social patterns of behaviour. Or in other words, besides the 'inner time' there are also the biological, the psychological and the social time each with its corresponding level of interpretation. At these levels the weights are attributed to activities with respect to their function rather than their duration. If duration is taken as the fundamental index of the significance of an activity at a certain level of interpretation, then the physical time should be transformed into the inner time of this system in specific situations. Technically this could be done by weighting the physical time at a specific level of analysis and for a specific problem.'
>
> (Boh and Sakisda, 1972, 230).'

From a time budget diary it is possible to compute the duration spent on each named activity. There is an immediate classification problem if subjects have used a free format description because a grouping of their individual activity descriptions must be undertaken straight away! An à priori classification, such as that used in the Multinational Time Budget Study and in Chapin's scheme, with as many categories as possible should be used. From such a scheme the first stage in the development of a typology of time use might be the calculation of 'difference or similarity indices' along the same lines used in the calculation of the relative isomorphism measure. Boh and Sakisda use a profile similarity index of the form,

$$r_p = \frac{2k_m - \sum\limits_{1}^{k} d_i^2}{2k_m + \sum\limits_{1}^{k} d_i^2}$$

where k_m is the number of activities expressed in minutes, d is a measure of absolute difference in the amount of time allocated to activity i,

$\sum_{1}^{k} d_i^2$ is the sum of squared differences which may be expressed

in standard scores as $z = \dfrac{x - \bar{x}}{\sigma}$, and $d = z_{a_i} - z_{a_j}$

where subscript a is the index of activity and i,j, are indexes of individuals.

A Q-type cluster analysis (Cattell and Coulter, 1966) can be applied to the matrix of r_p values which make up a matrix, of size equal to the number of individuals. The *profile similarity* indices

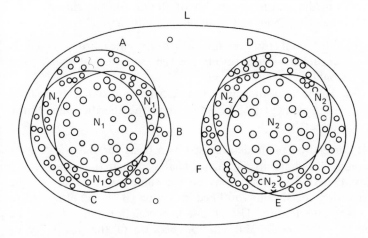

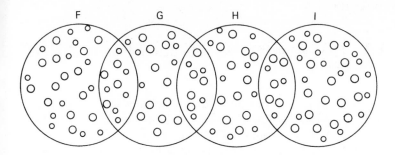

1. A–F: phenomenal clusters on higher level of r_p
2. L: Phenomenal cluster on lower level of r_p
   it includes two groups of clusters A–B–C and E–F–G
3. $N_1 - N_2$: nuclear clusters of first resp. second group of clusters
4. Segregate group of four phenomenal clusters: F–G–H–I

Figure A.6. Phenomenal, nuclear and segregate clusters. (Boh and Sakisda, 1972. Reproduced by permission of Mouton Publishers, The Hague)

following the cluster analysis, reveal clusters of *homostats,* in which individuals having a common place in an n-dimensional space (activity space) are grouped together. Within the homostat cluster there will be *phenomenal* and *nuclear* clusters. The former include all individuals determined on the basis of a selected cut-off value and have distinct characteristics. The nuclear cluster includes individuals who are located in the overlap of phenomenal clusters (*p*-clusters). Figure A.6 indicates some of these conditions.

Segregate groups may also be defined and they are composed of the clusters (and the individuals in them) which are linked by some predefined number of individuals. Segregates are clusters in which two or more phenomenal or *p*-clusters, each representing a type, are joined and they are derived by a second stage *Q*-type cluster analysis of the contiguity matrix of all phenomenal clusters.

Time use typologies can be developed only if:

(i) The total time within which activities are pursued is included — and results are representative of that period only.

(ii) The level of meaning of an activity is included. Each level of meaning must be adapted to the specific aims of the research.

If correlation methods are used as a basis for classification or typology building, as in factor analysis, then the effects of a 'closed' period such as a day on the value and direction of the correlation coefficients (outlined in section (i) above) also need to be taken into account. Use of data in correlation methods must be treated with considerable caution because of the possibility of introducing uncontrolled correlations due to spatial auto-correlation.

A.3 Time-geography and the determination of individual prisms

A central aspect of the time-geographic approach is the notion of a space-time prism and the simulation of possible activity patterns. The *prism* may be thought of as an '*angular bubble*' within which human activity is prescribed. In this section the derivation of a prism is outlined.

The prism defines the space-time 'volume' within which lie the set of possible individual paths. The summary given here is based directly on Lenntorp (1976, Chapter 4).[1] The determination of an individual's prism is a critical component in the simulation model — PESASP. The length of an individual path is the sum of a set of sub-paths, measured with the aid of Pythagoras' theorem as the space is co-ordinated on a regular grid network. The length of the ith sub-path is

$$(T_i^2 + S_i^2)^{1/2} \qquad\qquad (1)$$

where T_i = the difference in time between the origin and destination in space-time for the ith sub-path. S_i = the difference in space between the origin and destination in space-time for the ith sub-path

The sum of the n sub-paths is the complete path, (LSP) known as the *length of the sample path,*

$$\text{LSP} = \sum_{i=1}^{n} (T_i^2 + S_i^2)^{1/2} \qquad\qquad (2)$$

The units of measure used for time and space dimensions express the LSP units but LSP may be expressed either in time or space units alone, in this case the other unit is transformed by the *velocity* of movement,

$$V = S/T \text{ where } V \text{ is the velocity} \qquad\qquad (3)$$

and is defined for all values of T that are greater than zero. The individual's reach can be determined given the maximal length of an individual path based on the available time interval (T) and maximal velocity (V). $T^2 + S^2$ from (2) above can also be written $T^2 + V^2T^2$ because the difference between origin and destination is a function of the velocity and of the time which is available. All positions in space-time which satisfy the parameters of the longest path lie within the prism. (PPA) is the potential path area that defines the activity area for the individual and is formed by the prisms projection in space. Its calculation and time-geographic geometry will now be outlined according to three different conditions.

(i) Destination undefined

An individual travelling at maximal velocity will have covered VT distance units after T time units, following (3). With movement possible in any direction, a circle of radius VT is defined and the *activity area* or *potential path area*, PPA, lies within it. Its *area* is given by

$$\text{PPA} = \pi (V\,T)^2 \qquad\qquad (4)$$

T = available time interval
V = maximal velocity

In three dimensional space-time (i.e. 2 space dimensions and time), the prism takes on the shape of a cone with its apex at the origin of the individual's movements, say, home. The volume of the cone, or prism as it will be called because of the two dimensional representation which is usually used in diagrams, is given by:-

$$PPS = \frac{\pi}{3} V^2 T^3 = \frac{T}{3} PPA \qquad (5)$$

Thus formula (5) allows the calculation of an individual's prism, given maximal velocity and time available, and with destinations undefined.

(ii) Destination defined

An individual may either return to his origin when completing a sample path or terminate the path at another destination. In the former case, after T time units he/she must have the same space location. With movement away from the origin at maximal velocity he must commence the return journey at $T/2$, also at maximal velocity. The PPA is therefore determined in terms of half the available time interval,

$$PPA = \pi \left(\frac{VT}{2}\right)^2 \qquad (6)$$

The prism (or cone) in fact takes the form of two equal-sized cones having a common base (Figure A.7).[2]

$$PPS = \frac{T}{3} PPA \qquad (7)$$

where PPS is prism volume.

The joint variation of prism volume (PPS) and activity area (PPA) is shown in the graph of Figure A.8 and the associated hypothetical data. The other form of defined destination is one in which the destination differs from the origin, in other words dS, the distance in space between origin and destination is some positive number greater than 0, $dS \geq 0$ and the direction of movement has to be considered. An initial movement away from the destination will require an earlier change of direction than if the original direction was towards the destination. When the origin and destination are not at the same location the prism is distorted in the direction of the destination, Figure A.9.

The two dimensional prism image, of the three dimensional space-time cone is shown in the plane, blocked in. The activity area is now calculated as

$$PPA = \frac{\pi}{4} V^2 T \ \left(T^2 - \left(\frac{dS}{V}\right)^{1/2}\right. \qquad (8)$$

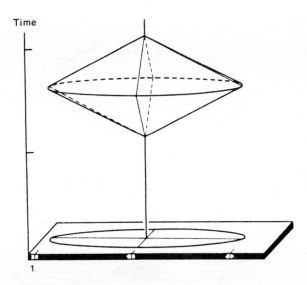

An individual's prism and activity area, assuming a given time interval and maximal velocity and an origin and destination which coincide in space. The time interval between origin and destination amounts to 100 units (the distance between adjacent markers on either the time or the x-axis) and the maximal velocity is two space units per unit.

Figure A.7. Destination and origin the same: prism and PPA. (Lenntorp, 1976. Reproduced by permission of Bo Lenntorp)

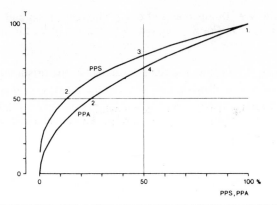

Point no.	Time interval (T)	Prism volume (PPS)	Activity area (PPA)
1	100.0	100.0	100.0
2	50.0	12.5	25.0
3	79.5	50.0	63.0
4	70.5	35.5	50.0

Assuming a constant velocity which can be arbitrarily chosen. Values at the numbered points are tabulated below the figure. ($dS = A = 0$)

Figure A.8. Relative change in prism volume and actitity area as a function of T. (Lenntorp, 1976. Reproduced by permission of Bo Lenntorp)

Time

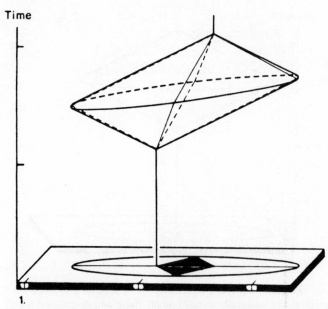

1.

An individual's prism and activity area, assuming a given time interval and maximal velocity and an origin and desination which do not coincide in space. The time interval between origin and destination amounts to 100 units (the distance between adjacent markers on either the time of the x-axis) and the maximal velocity is two space units per time unit. The distance in space between origin and destination is 50 units with (parallel to the x-axis)

Figure A.9. Destination and origin different: prism and PPA. (Lenntorp, 1976. Reproduced by permission of Bo Lenntorp)

and Figure A.10 shows the relative change in the prism volume and activity area as a function of the distance in space, dS, between origin and destination. The cones which form the prism have their apexes at the origin and destination respectively and the prism volume (PPS) is

$$ \text{PPS} = \frac{T}{3} \left(1 - \left(\frac{dS}{VT}\right)^2\right) \text{PPA} \qquad (9) $$

(iii) Individual prism with stationary activity time included

In this situation an individual is required to remain at a station during part of the available time interval T. The first of the two alternatives above will now be considered. Moving from an origin at the maximal velocity V the individual must be at the same location after T time-units have elapsed but for A time-units he is constrained, by any one of the three major time-geographic

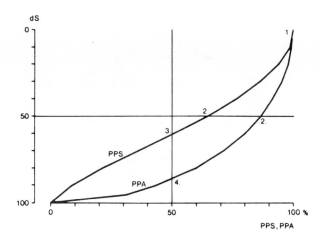

Point no.	Distance in space between origin and destination (dS)	Prism volume (PPS)	Activity area (PPA)
1	0	100.0	100.0
2	50.0	65.0	86.5
3	61.0	50.0	79.0
4	86.5	12.5	50.0

A constant velocity is assumed which can be arbitrarily chosen, though a relation exists between velocity, length of the time interval, and the largest spatial distance between origin and desination. In the figure dS is expressed in percent of its maximal value ($= V \cdot T$). Values at the numbered points are tabulated below the figure. ($A = 0$)

Figure A.10. Relative change in prism volume and activity area as a function of distance in space. dS between origin and destination. (Lenntorp, 1976. Reproduced by permission of Bo Lenntorp)

constraint types outlined in Chapter 6, from leaving a particular station; for instance, a place of work. Movement time is now (T-A) units. For half of the time T-A he can move away from the origin; the other half he requires for his return journey: at maximal velocity (say the average speed a bus travels at). The area of the circle which prescribes his activity area is defined by a radius of T-A units and the velocity (or speed) of movement by,

$$PPA = \pi \left(\frac{V(T\text{-}A)}{2} \right)^2 \tag{10}$$

Now, however, the prism is made up of three components — two cones with their apexes at the origin and destination respectively and their bases separated by a cylinder with A (the waiting or constrained time) as its height, (Figure A.11).

466

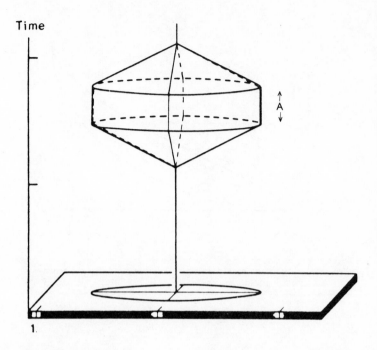

Time

The time interval between origin and destination is 100 units (the distance betweeen adjacent markers on either the time or the *x*-axis) and the maximal velocity is two space units per time unit. The stationary activity time amounts to 30 units.

Figure A.11. Prism and activity area with stationary time (*A*): coincident origin and destination. (Lenntorp, 1976. Reproduced by permission of Bo Lenntorp)

Once again the volume of the prism (PPS) may be calculated,

$$\text{PPS} = \frac{T + 2A}{3} \text{ PPA} \tag{11}$$

The relation between the size of **A**, the waiting time cylinder, the prism volume (PPS) and the activity area or potential path area (PPA) is shown in Figure A.12. From these extracts from Lenntorp's study, the principles of prism determination should be fairly clear. However, we have only referred to movement at constant and maximal velocity. The everyday world is not quite like this and it is also necessary to compute prisms in terms of variable velocities. In Figure 6.8 the prisms for an activity programme (Lenntorp, 1976) involving variable velocities were illustrated and the result is simply a more complex system of prisms.

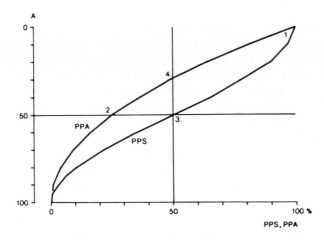

The duration is expressed in percent of the total time interval T. A constant velocity is assumed which can be arbitrarily chosen. Values at the numbered points are tabulated below the figure. ($dS = 0$)

Point no.	Duration of stationary activity (A)	Prism volume (PPS)	Activity area (PPA)
1	0	100.0	100.0
2	50.0	50.0	25.0
3	50.0	50.0	25.0
4	29.5	79.0	50.0

Figure A.12. Relative change in prism volume and activity area as a function of duration of stationary activity, A. (Lenntorp, 1976. Reproduced by permission of Bo Lenntorp)

A.4 Time series analysis

This is a complex and relatively recent field of statistical analysis. Many of the methods which are now used were only developed in the last 30 years or so (Chatfield, 1975) and in fact 'prior to 1970 there were few practical books on time series and there seems little doubt that this hindered the development of the subject' (Chatfield, 1977, 492). Chatfield's book should be referred to for a succinct and comprehensive introduction covering a difficult area of statistical analysis in a detail which is far beyond the scope of this short summary. The development of high speed computing systems has made it possible to consider many alternative approaches to the analysis of a time series. In this section we will only consider time series analysis in terms of its descriptive properties. Strictly, time series analysis means the estimation of the properties of the generating process from the given sample (Hughes, 1970, 3). Issues of statistical inference and the underlying probability models related to estimation procedures are beyond the scope of this chapter.

What is *a time series?* It is a 'collection of observations made sequentially in time' (Chatfield, 1975, 1). There are a great many models and techniques for identifying patterns in time series behaviour and for estimating generating processes. They range from basically descriptive techniques which make some fairly low level assumptions and which can be used for exploratory analysis to more precise but also more demanding models.

First of all we distinguish between *continuous* and *discrete* 'observations'. The former are made continuously in time and the latter are only taken at specified (usually regular) times; for instance, at the same clock time each day. We will be concerned only with discrete time series but there is also a type of time series for 'events' which occur 'randomly' in time as 'points' and such events as natural disasters or accidents or crimes may be treated as *a point process* in circumstances where only one such event can occur at a time. Figure A.13 shows the time point of occurrence of distress calls from an urban area to a single reception centre. As only one call could be handled at a time this might be treated as a *point process* or *'event series'*. Here the distribution of the individual events in a period of time, say a day, is the subject of interest and in particular the *interval* between occurrences, or their spacing in time. Point processes and analyses of series of events are not considered further here, but are covered by Cox and Lewis (1966) and briefly by Davis (1973).

A *deterministic* time series is one in which observations of events are such that future values may be predicted exactly from past events 'but most time series are *stochastic* in that the future is only partly determined by past values' (Chatfield, 1975, 6).

Why might one be interested in analysing a time series? A number of objectives may be identified; description, explanation, prediction and control. Whatever the objective, the first and most important step is to *plot* the data but care must be taken to consider the distortion that can be induced simply by the *scale* which is chosen. Figure A.14 shows an example from Chatfield (1977).

A plot will display many of the features which become critical in the selection of a particular technique for the analysis. Extreme values which are known as *outliers* can be identified and *smoothed* or *filtered* (see below) if there is a known reason for their existence such as measurement error. Sometimes filtering is required in order to satisfy certain demands of the 'model' to be used. Control objectives will not generally be involved in urban social geography,

469

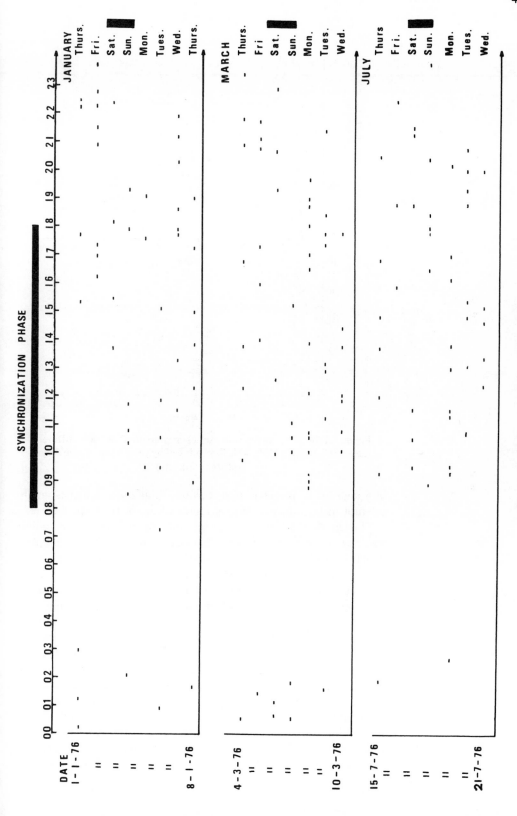

Figure A.13. Some point process records

470

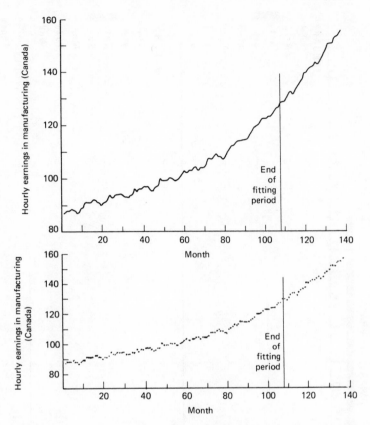

Figure A.14. The same time series represented at two different scales. (Chatfield, 1977, 493. Reproduced by permission of the Royal Statistical Society)

although it is possible that certain simulation schemes might attempt to introduce a control factor of one sort or another.

The length of a time series (essentially the number of time points upon which it is based) will also determine the sort of analysis which is available, as will a number of other characteristics which will often be apparent in the plot; we will consider these characteristics as types of variation in the time series. Four components of variation which appear in many economic and social time series are *trend, seasonal variation, other cyclical variation* and *irregular fluctuation,* or *noise.* The *decomposition* of a time series into these component parts is one significant objective of time series analysis. Each of these components is represented in the plot of the hypothetical series shown in Figure A.15.

The existence of these components, especially the trend or a dominant 'seasonal' variation, say between summer and winter, means that the series is *non-stationary* and a *stationary* time series is a requirement for some methods of analysis. Various methods exist for removing the trend, which is simply a tendency for the

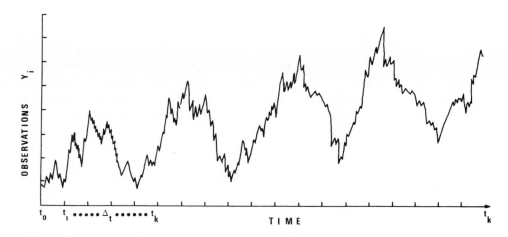

Figure A.15. Hypothetical time series showing trend, seasonal variation, other cyclic variation and irregular fluctuation

successive values of an observation to increase (upward trend) or decrease (downward trend). A stationary series will show no systematic change in the mean, no systematic change in variance, (i.e. the size of the mean variability of the observed values around the mean) and will have had all strictly periodic variations removed. In other words, the first two moments of the data, the mean ($\bar{x}$) and the variance σ^2 are not functions of time, they do not change during the passage of time in the period being considered, and they are not *expected* to change in any period of similar length in the future.

Transformations of various sorts are available to stabilise the variance so that it does not change systematically as the mean changes for data with a trend, and if seasonal effects are varying systematically with the mean they too can be transformed. The trend can be identified by curve fitting methods (A.2 (iv) above) and linear regression of the Y_i observations against time (t) is a useful first approximation. It would be wrong however, to think that only linear trends existed. If a straight line does not 'fit' the data adequately by accounting for a high proportion of variances and leaving only random variation or white noise, then higher order polynomial curves might be tried. Having found a best fitting line the trend is removed and the chosen analysis is performed on the residuals. The residuals are the observations which are not accounted for by the correlation between observation value and the elapsed time. A linear regression would have the familiar form $Y_t = A + kT$, where Y_t is an observation at time t, A is the intercept giving the value of the observation at the start of the period, k is the slope of the line or trend (horizontal with no trend) and T is the elapsed time.

Other methods of trend removal involve smoothing or filtering by fitting a moving average to the data but simple moving averages should be treated with caution as they can create problems in the calculation of the *autocovariation* and *autocorrelation* coefficients which are a cornerstone of time series analysis. Differencing also removes trend rather effectively. This method of filtering involves the generation of a new series of values, $y_1 \ldots, y_{n-1}$ from the original series which contains trend $x_1, \ldots x_N$ by subtracting the value of x_t from its next value x_{t+1},

$$y_t = x_{t+1} - x_t = \nabla x_{t+1}$$

If this *first order difference* is insufficient to remove the trend then a second-order difference may be used. (See Chatfield, 1975, 21).

There are no hard and fast rules to guide the selection of an appropriate filter, most texts on time series analysis point to the importance of 'experience' in the interpretation of the plot as a critical guide to the selection of filters. For our purposes the important point to have grasped is that unless the objective of the analysis of a time series is the study of trend itself, as for instance in forecasting, or unless interest is to be focussed on regular 'seasonal' or other low frequency cycles, it will be necessary to remove their effects in order to achieve a stationary series so that further analysis may be undertaken. It should be added, however, that methods are being developed for the analysis of non-stationary time series in order to achieve similar goals to the stationary models, for instance when spectral analyses are used.

(i) Descriptive analysis in the time domain

One of the most important techniques which is applied to the analysis of a time series is the measurement of autocorrelation and its component, autocovariance. The autocorrelation coefficient is a guide to the cyclic or recurrent properties of a time series and is applicable to stationary series only. The coefficients measure the correlation between repeated observations of the same variable at different time distances apart. You will be familiar with the correlation coefficient r which measures the relative agreement among two variables x_i and y_i (or more of course),

$$r = \frac{\sum (x_i - x)(y_i - y)}{\sqrt{[\sum (x_i - x)^2 . \sum (y_i - y)^2]}} = \frac{\text{Covariance}}{\text{Product of standard deviations}}$$

For the autocorrelation of a discrete time series, instead of having two variables we have a single variable compared with itself. Figure A.16 will help you to understand this idea.

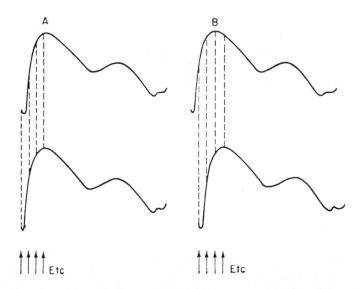

Figure A.16. Illustration of autocorrelation process. (a) is the auto-correlation with no lag, (b) the autocorrelation with a lag of one. (Reproduced with permission from Oatley and Goodwin, 1971, 26. © Academic Press Inc. (London) Ltd.)

Autocorrelation, which means 'comparison with itself' is one way of finding out if a rhythm exists in a time series. A discrete series of N observations $x_1, \ldots x_N$, can be formed into $(N-1)$ pairs of observations, namely (x_1, x_2), (x_2, x_3), $\ldots$ (x_{N-1}, x_N). The self-comparisons may be made at various lags. The lag is simply the amount of time by which a series is shifted so that comparisons can be made. For instance, at lag one, the value at time t_0 will be compared with the value at t_1, the value at t_1, with that at t_2 the value at t_{N-1} with that at t_N, for a lag of 2 the gap will be two time units. The lag may be in either direction, as shown below, when two variables are involved (Veldman, 1969).

	No lag	Lag = 1			No Lag		Lag = 1		Lag = 1 (reversed)	
	Autocorrelation				Normal		Cross-lag correlation			
	X	X	X		X	Y	X	Y	X	Y
t_0	25	25		t_0	25	20	(25)			(20)
t_1	24	24	24	t_1	24	19	24	20	25	19
t_2	22	22	22	t_2	22	25	22	19	24	25
t_3	26	26	26	t_3	26	20	26	25	22	20
t_4	23	23	23	t_4	23	18	23	20	26	18
								(18)	(23)	
	$N=5$	$N=4$			$N=5$		$N=4$		$N=4$	

The formula for calculation of this autocorrelation coefficient is directly comparable with the zero order correlation for two variables.

$$r_k = \frac{\sum\limits_{t=1}^{N-k} (x_t - \bar{x})(x_{t+k} - \bar{x})}{\sum\limits_{t=1}^{N} (x_t - \bar{x})^2}$$

where k is the number of lags or distance apart, in time, of the observations. The autocovariance is simply a non-standardized autocorrelation coefficient; there is no division by the variance.

Figure A.17 shows a plot of a set of autocorrelation coefficients against their lag and is known as a correlogram. The raw data time series from which the correlogram was plotted is shown in Figure A.17(a). The observed data are telephone calls to a welfare agency at half-hour intervals over 7 days. The shape of the correlogram is an indicator of the nature of the time series, but care must be exercised in its interpretation. A summary of some idealized time series and their associated correlograms will be found in Figure A.18. Any cyclic fluctuation which appears in the time series will be reflected in the correlogram, if the time series is stationary and at the same frequency. The confidence bands for determination of which lagged coefficients are significant can be calculated as a band $\pm 2/\sqrt{N}$ either side of $r_k = 0$. As a rule the lag should not exceed $N/4$, e.g. a series of length 48 would not be lagged above 12. Clearly it therefore becomes important that the lack of existence of cycles in the time series should not be assumed from the study of the correlogram if the lag is small, and the series is short.

(ii) Descriptive analysis in the frequency domain

When we consider the autocorrelation of discrete events, equally spaced in time, the data are in the time domain. In this section we consider the frequency domain: the identification of the component frequencies or harmonics in a time series rather than its overall shape. Davis (1973) uses the diagram in Figure A.19 to illustrate the transformation of a signal (or time series) in the time domain to one in the frequency domain.

First we consider 'harmonic' or *Fourier analysis* and then the related method of *spectral analysis*. Whereas Fourier analysis is applied directly to the raw data or, if necessary, to a filtered version of it, spectral analysis involves use of the autocovariance or autocorrelation function *and* 'a Fourier analysis'. The Fourier equation is used to transform the autocorrelation or covariance values at successive lags into their various frequency components as

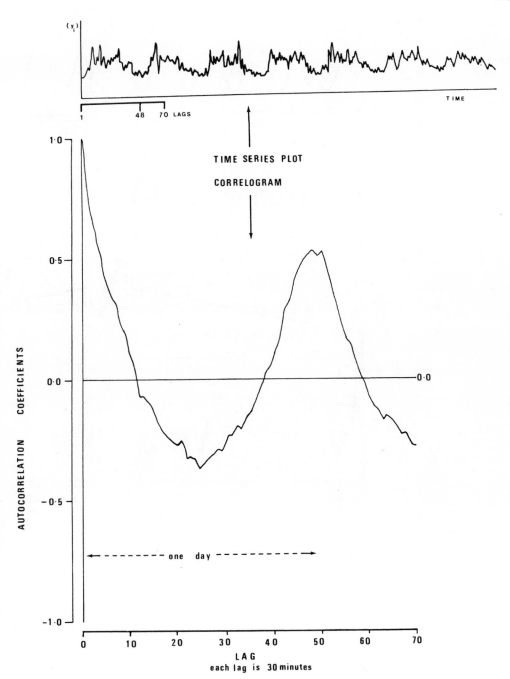

Figure A.17. A time series and its correlogram ($N = 336$ and lag $= 70$)

a *power* or variance spectrum. Frequency domain analyses aim at finding the hidden periodicities (the information in the data) and in eliminating the noise or randomly fluctuating variations from a time series.

476

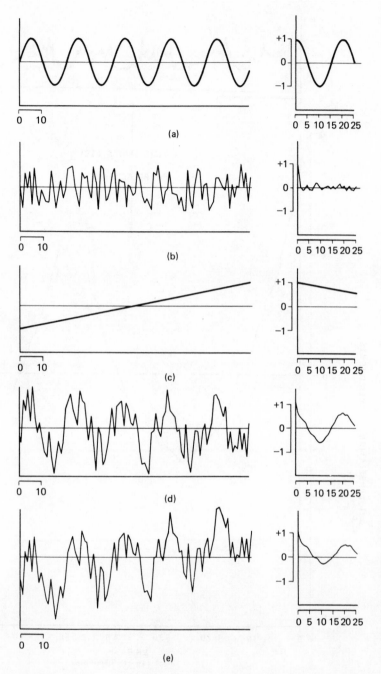

Figure A.18. Some idealized *time series and their correlograms*. Graph (d) should be compared with Figure A.17(a) and (b). Some idealized time series and their autocorrelation functions (a) Sine wave with wavelength 20 units. (b) Sequence of random numbers or 'noise'. (c) Sequence of linearly increasing numbers or 'trend'. (d) Sine wave plus random noise (sequence (a) plus sequence (b)). (e) Sine wave plus random noise plus linear trend (sequence (a) plus sequence (b) plus sequence (c)). (Davis, 1973, 235. Reproduced by permission of John Wiley and Sons Inc.

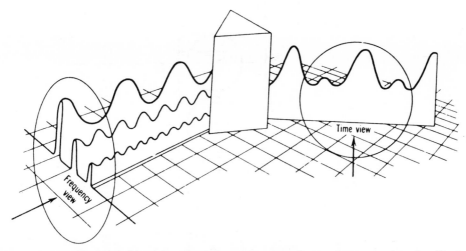

Figure A.19. Transformation of a signal from time domain to a frequency domain. (Davis, 1973, 272. Reproduced by permission of Hewlett-Packard Company.)

The method of harmonic or Fourier analysis describes regularly periodic phenomena in terms of sine and cosine waves. Figure A.20 illustrates the generation of a sinusoidal wave form from the constant velocity rotation of a circular disc. The sinusoidal wave may be represented by a Fourier series and this makes it possible to use the properties of the Fourier series to transform any signal from a function of time to a function of frequency. For the example given in Figure 8.10 the Fourier curve, z, which was fitted to the raw data shown in the lower part of the diagram was computed from

$$\zeta = \sum_{k=0}^{M} (a_k \cos_k \Theta + b_k \sin k\Theta)$$

(Baxter, 1970); a_k and b_k are coefficients related to the phase of each point, the angle Θ by which it is separated from the origin or starting time of the period.

The Fourier transform for a *discrete series* is a *sum* (or addition) of the products of the cosine (and sine) terms summarizing the pure sinusoids; multiplied at each of the discrete observation points by the values of the time series being analysed. This can be illustrated by Figure A.21. If the series being analysed is in fact a single pure sinusoid, then only a single cosine or sine term will be required but usually the series will be sufficiently complex to require both. The 'goodness of fit' for a Fourier curve of M harmonics can be calculated as a measure of the variance σ^2 between the original data and the Fourier transformed data. (Chatfield, 1975; Davis, 1973; Baxter, 1970). The Fourier analysis reveals the hidden harmonics in the original signal as integer (whole number) multiples of the fundamental interval, which is the total length (2π) of the time

478

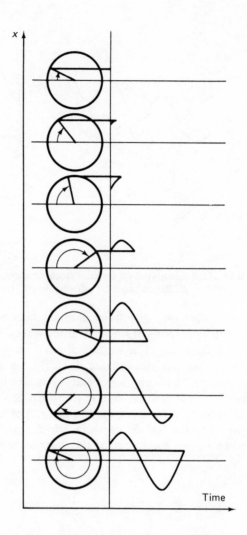

Figure A.20. The generation of a sine function from a vector rotating at a constant velocity. (Sollberger, 1965, 15. Reproduced by permission of Elsevier/North-Holland Biomedical Press.)

series being studied. Any harmonics will also have their amplitudes (heights above the mean line) identified but only those cycles which are multiples of the total period. It is always possible to fit a Fourier transformed time series to the original series with a perfect fit (as with the polynomial curves mentioned in A.2(iii)) but this is not the objective; rather it is to use the properties of the Fourier series to identify the major harmonics which account for significant proportions of the total variance in terms of an additive assembly of 'average curves', which are sine and cosine curves in this case.

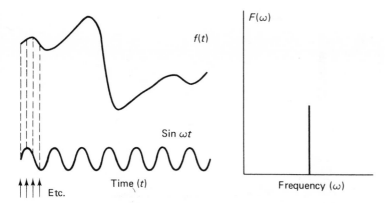

Figure A.21. Scheme for obtaining the Fourier sum. On the left, a signal *f(t)* for which the spectrum is to be found, and a sinusoidal signal sin *ωt*, are both sampled regularly at sample points shown by the arrows and dotted lines. The signal *f(t)* and the sinewave are now multiplied and summed.

The result, plotted as the height of a line on the spectrum at the frequency *ω* of the sinusoid used, is shown on the right of the figure. (Reproduced with permission from Oatley and Goodwin, 1971, 24. Copyright by Academic Press Inc. (London) Ltd.)

Spectral analysis combines the time domain autocorrelation analysis and the frequency domain Fourier analysis because it is a Fourier transformation of the autocovariance function or the autocorrelation function, to produce what is known as a *power spectrum* or *variance spectrum,* i.e. the relationship between signal *frequency* and *variance* produce a 'continuous' series of cycles which are not merely harmonic components of the fundamental interval. The spectral analysis of the correlogram shown in Figure 10.17 has its power spectrum shown in Figure 10.22. We shall interpret it in a moment.

Spectral analysis is 'concerned with estimating the spectrum (the frequency-variance components) of a stationary stochastic (chance related) process. The standard approach, based on Fourier or harmonic analysis, is essentially non-parametric in that no model is assumed *à priori*, but rather the estimated spectrum may help in suggesting an appropriate parametric model'. (Chatfield, 1977, 499). There are many algorithms for computing the spectral density function (the variance carried by a frequency) but they vary essentially in terms of the various filtering procedures which are used to adjust the time series to stationarity and in the *window* size and type (essentially the shape and size of the lag) which is used to focus the analysis upon; Chatfield (1975) provides a comprehensive discussion which is beyond our facilities here.

The power spectrum shown in Figure A.22 may be loosely interpreted as follows: 336 half-hourly observations (*t*) for 48 × 7 days over a year of the number of distress calls (Y_i) to an urban welfare agency indicates a clear circadian (approximately 24 hour,

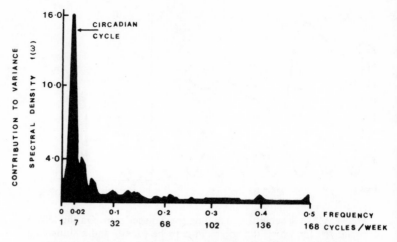

Figure A.22. A power (or variance) spectrum for urban distress signals over a one week period based on one year of signal records. $n = 336$ equally spaced time points with 30 minute intervals. Parzen 2 window and lag 100

48 half-hour) cycle with higher frequencies, i.e. shorter length cycles of period 12 hours, 8 hours, 5 hours, 3½ hours, 2½ hours and 1 hour 12 minutes approximately. The dominating circadian rhythm is acting as a 'seasonal' factor and is so dominant that the higher frequency cycles are essentially harmonics of its 24 hour period. Removal of this 24 hour cycle should be undertaken and the spectral analysis undertaken on the residual values to find the independent cycles, if any. Such an exercise is quite straightforward and will not be considered further. The peak value shown at frequency $(f) = 0 \cdot 02020$, has a cycle every $49 \cdot 50$ (half hours) within a period of 336 units; which is $6 \cdot 79$ or approximately 7; the number of days in the week. Recall that the length of the observation period is made up of 7 days, at half hourly intervals. The frequency (f) is obtained from $\omega/(2\pi)$, and so it is the number of cycles per unit time. The reciprocal of the frequency, i.e. $1/f$ will give the length or period of the cycle, thus $1/0 \cdot 0202$ is $49 \cdot 5$. In this example 'unit time is' $1/336$ of a week and the values of ω in radians per week are $\omega = 336\pi j/Q$, where Q is the size of the lag window which was used, in this case 100, and j is the lag.

In our example we see that the low frequency (i.e. relatively long period length for a cycle) accounts for most of the variation, and is given by the spectral density expressed on the ordinate. The highest frequency (on the abscissa) which we can identify here is known as the Nyquist Frequency and it is $2\triangle t$, i.e. $2 \times$ ½ hour or one hour. If there are in fact higher frequencies than this then a smaller time interval $(\triangle t)$ must be used. The value of a properly executed spectral analysis, in the absence of a well developed theoretical basis for the explanation of rhythms lies in its ability to indicate the possible shape of the timing of events and this allows hypotheses and theories as to why this shape exists to be developed.

If cycles do in fact exist something, some zeitgeber, is generating them. Various statistical tests can be applied to the resulting spectral density values to estimate the probability that computed amplitudes and the variance explained by them have not occurred by chance.

(iii) Time series of more than one variable

The methods of analysis which can be applied to a single time series can be adapted to allow two or more time series to be compared. The autocorrelation function in which observations are compared with themselves at various lags can be adapted to the correlation between two (or more) series and this was illustrated in Table 8.15. The method is then known as *cross-lag correlation*. The spectral analysis of a cross-lag correlation produces a measure of the *coherence* among the frequencies and amplitudes of a rhythm. To what degree are they in phase? Which variable leads and which one follows? Is there a causal relationship between them and if so does this exist inherently in the variables being analysed or is the relationship between them a partial function of some other variable which may or may not, in itself, be regularly cyclic; but depending on its strength, induces a signal of differing length? The methods of cross-lag correlation and associated methods of cross-spectrum analysis can be found in Chatfield (1975) and other statistical texts which deal with bivariate processes in the time and frequency domains.

(iv) Cosinor analysis

From time to time in the text, especially in Chapter 2, we drew attention to the importance of circadially variant processes in *lifetime* especially the somatic and psychomatic functions which have been found, by biologists (e.g. Smolensky, Halberg, and Sargent, 1972) and psychologists (Orme, 1969, 1978) to vary with an approximately 24-hour cycle. A technique which has been developed by chronobiologists to aid in the identification of circadian and circannual rhythms is *cosinor analysis*. A brief introduction to elementary aspects of the method is given below. There is an extensive statistical literature about the method but for our purposes Halberg, Tong, and Johnson (1967), Reinberg (1974), Halberg (1974), and Luce (1970) provide sufficient further reading.

If the rhythmic character of a phenomenon is known or can be assumed to have a single cycle in a period, such as a 'day', then it is possible to 'fit' a sine or cosine curve to it as a best fit approximation which allows characteristics of the shape of the curve such as the *amplitude, phase* and *slope* to be determined. Cosinor analysis involves a *cosin*(e) vec*tor* representation of the time series which is essentially the illustration of cycle characteristics in a clock-like diagram as shown in Figure 8.13.

However certain statistical aspects of the cosinor method relating to the representation of *confidence arcs* and *ellipses* are not discussed here (see Halberg, Tong, and Johnson 1967). An illustration of their diagrammatic representation will be given after the first steps in a cosinor analysis, the fitting of a cosine curve and the generation of some of the parameters for rhythm detection have been explained.

The fully worked example, relating to the first stage in a cosinor analysis is taken from a paper prepared by Koukkari, Duke, Halberg, and Lee (1974) as a student exercise in botany. For our purposes you might consider the data to relate to equally spaced observations on the number of unoccupied dwellings in a single urban 'neighbourhood', over the course of a 24 hour period. Only eight observations are used in order to keep the number of arithmetic steps to a minimum. In the example given in Chapter 8 and applied to the record of distress calls, made by telephone to an urban welfare agency every call made over the period of a year was used for distress calls concerned with two of the major social pathology categories. Their detail does not matter at the moment.

The formula is

$$Y_i = M + A \cos (\omega t_i + \Phi) + e_i, i = 1, \ldots n.$$

where y_i is a single observation at time t_i

$\quad n$ is the number of observations

$\quad M$ is the *mesor* or rhythm-determined average value (midway between the highest and lowest point of the mathematical function used to approximate the rhythm). This may or may not be equal to the calculated mean.

$\quad A$ is the amplitude; the difference between the maximal and the mesor (mean value of a sinusoidal function used to describe the rhythm).

$\quad \phi$ is the *acrophase* or lag from zero time to the peak of the best fitting cosine function used to approximate the rhythm in the original data and is expressed in degrees, radians or other designated units such as the actual time of the peak.

One cycle of the rhythm equals one complete circle, for example $360° = 24$ hours.

$\quad \omega$ is the angular velocity, or $360°/\tau$, where τ is a fixed period assumed *á prior* to characterize the data. In our example $\tau = 24$ hours, $\omega = 360°/24 = 15°$/hr or $0·26$ radians. 1 hour is $15°$ of a complete cycle.

$\quad e_i$ are error terms which we will assume to be zero. Where they do not exist they are assumed to be independently distributed with mean zero and an unknown variance, similar for all points.

483

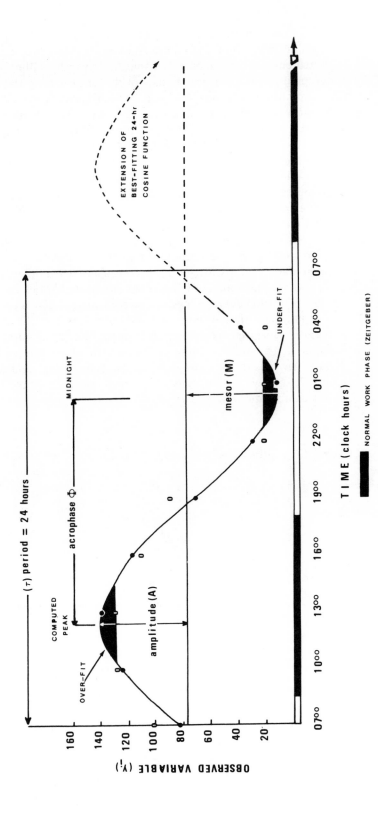

Figure A.23. The output from the first stage in a cosinor analysis

The computational steps for the first stage of a cosinor analysis are illustrated following Figure A.23 which shows the shape of the fitted curve from the data in Table A.1, as well as illustrating the graphical characteristics of some of the terms used in the method.

Amplitude A (Figure A.23 and Table A.1) can be 'tested' with an F-ratio by finding the value $F_{(2, n-3)}$ at a predetermined level of significance (e.g. $0 \cdot 05$, $0 \cdot 01$) using the F-tables found in most statistics books; if the calculated F-value exceeds $F_{(2, n-3)}$ the amplitude is statistically different from zero. In the worked example $F > F_{(2,5)} (0 \cdot 05) = 5 \cdot 79$, reject the hypothesis that $A = 0$. In other words there is a statistically significant difference between the amplitude of the data and *zero* and the detection of a 24-hour rhythm can be treated with a rather high degree of confidence. (Of course there may be other 'hidden' periodicities in the data such as would be found by a harmonic or spectral analysis and in addition there may be an oscillation in the acrophase when other 'samples' are taken. The number of degrees of arc through which the peak of the best fitting curve swings can be computed to give a *confidence arc* and the variation to be expected in the length of the amplitude vector, which represents the height of the best fitting curve, can be computed as an error ellipse the long axis of which is the confidence arc. An illustrative example from Halberg, Tong, and Johnson (1967) is shown in Figure A.25 to which reference should be made for computation methods.)

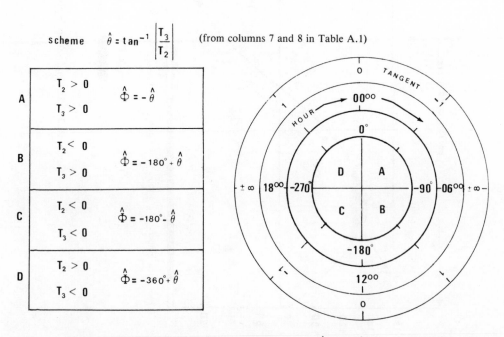

Figure A.24. Computation of the ACROPHASE $\hat{\Phi}$ from $\hat{\theta}$ in Table A.1. (After Koukkari, Duke, Halberg, and Lee, 1974, 293)

Table A.1. Computational steps in the first stage of a cosinor analysis (After Koukkari *et al.*, 1974, 292, 295)

1	2	3	4	5	6	7	8
Time t_i	Data y_i	$x_i = 15 \cdot t_i$ $(\omega \cdot t_i)$	$a_i = \cos x_i$	$b_i = \sin x_i$	y_i^2	$c_i = a_i \cdot y_i$	$d_i = b_i \cdot y_i$
16^{00}	110	240	$-0 \cdot 5$	$-0 \cdot 8660$	12100	$-55 \cdot 00$	$-95 \cdot 26$
19^{00}	90	285	$0 \cdot 2588$	$-0 \cdot 9659$	8100	$23 \cdot 292$	$-86 \cdot 931$
22^{00}	20	330	$0 \cdot 8660$	$-0 \cdot 5$	400	$17 \cdot 32$	$-10 \cdot 0$
01^{00}	20	15	$-0 \cdot 9659$	$0 \cdot 2588$	400	$19 \cdot 318$	$5 \cdot 176$
04^{00}	20	60	$0 \cdot 5$	$0 \cdot 8660$	400	$10 \cdot 00$	$17 \cdot 32$
07^{00}	100	105	$-0 \cdot 2588$	$0 \cdot 9659$	10000	$-25 \cdot 88$	$96 \cdot 59$
10^{00}	130	150	$-0 \cdot 8660$	$0 \cdot 5$	16900	$-112 \cdot 58$	$65 \cdot 00$
13^{00}	130	195	$-0 \cdot 9659$	$-0 \cdot 2588$	16900	$-125 \cdot 567$	$-33 \cdot 644$

$\sum_{i=1}^{n} y_i = 620$

total:

$T_1 = 65200 = \sum_{i=1}^{n} y_i^2$

$T_2 = -249 \cdot 097 = \sum_{i=1}^{n} c_i$

$T_3 = -41 \cdot 749 = \sum_{i=1}^{n} d_i$

(1) $\bar{y} = \dfrac{1}{n} \sum_{i=1}^{n} y_i$ (overall mean)

(2) $\hat{M} = \bar{y}$ (under assumptions)

(3) $\hat{A} = \left\{ \dfrac{4}{n^2} (T_2^2 + T_3^2) \right\}^{1/2}$

(4) $\hat{\theta} = \tan^{-1} \left| \dfrac{T_3}{T_2} \right|$

(5) $\hat{\phi} = K\pi + \hat{\theta}$ (See Figure A.24)

(6) $Q_1 = \dfrac{2}{n} (T_2^2 + T_3^2)$

(7) $Q_2 = (T_1 - n\bar{y}^2 - Q_1)$

(8) $F = \left(\dfrac{Q_1}{Q_2} \right) \dfrac{n-3}{2} = \dfrac{\left(\dfrac{Q_1}{2} \right)}{\left(\dfrac{Q_2}{n-3} \right)}$

(1) $\bar{y} = 77 \cdot 5$

(2) $\hat{M} = 77 \cdot 5$

(3) $\hat{A} = \dfrac{4}{64} \{(63792 \cdot 294)\}^{1/2} = (3987 \cdot 02)^{1/2} = 63 \cdot 15$

(4) $\hat{\theta} = \tan^{-1} 0 \cdot 1676, \hat{\theta} = 9 \cdot 5°$

(5) $\hat{\phi} = -189 \cdot 5°$ or $12 \cdot 38$ clock time

(6) $Q_1 \ 15948 \cdot 076$

(7) $Q_2 = 65200 - 48050 \cdot 00 - 15948 \cdot 076$
$= 1201 \cdot 924$

(8) $F = \dfrac{7974 \cdot 038}{240 \cdot 385} = 33 \cdot 17$

$F > F_{(2,5)} (0 \cdot 5) = 5 \cdot 79$; reject hypothesis: $A = 0$

a) hat, ^ used on top of letters (e.g., $\hat{M}$ reads *M hat*) denotes estimate of corresponding parameter;

b) $|A|$ denotes absolute value of A. (e.g. $|-3| = 3$, $|3| = 3$);

c) $\theta = \tan^{-1}x =$ arc tan x stands for the value θ in degrees of which tangent gives x in radians (from trigonometric table), (e.g., $\theta = \tan^{-1}1$; then $\theta = 45°$);

d) in $\hat{\phi} = K\pi \pm \hat{\theta}$, K is a constant equal to 0, -1, or -2 according to the signs of T_2 and T_3 at the bottom of columns 7 and 8 and also $\hat{\theta}$ takes $\ll - \gg$ or $\ll + \gg$.
Computation of $\hat{\phi}$ from $\hat{\Theta}$ is shown in Figure A.24 following.
(e.g., if $T_2 < 0$ and $T_3 < 0$, $K = -1$ and θ takes $\ll - \gg$ sign and $\hat{\Phi} = -\pi - \hat{\Theta} = -180° - \hat{\Theta}$);

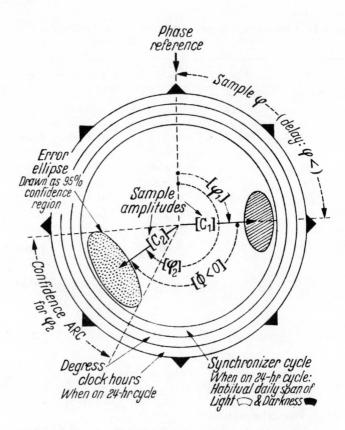

Figure A.25. An illustration of confidence arcs and error ellipses. C_1 and C_2 are the vectors equivalent to the amplitude (A) and in Figure 8.13 these are the 'small hands' of the clocks. C_1 and C_2 are two different samples and the advantage of the cosinor scheme for comparative purposes should be clear. (Halberg, Tong, and Johnson, 1967)

The cosine curve having been fitted the next stage is to identify the location of the acrophase and plot it using a vector of length proportional to the amplitude at its peak. The cosinor analysis makes its major contribution in the polar plot of the acrophase and its relation to any other time *marker* which may have empirical or theoretical significance. A number of different samples or individual phenomena may be plotted on the same chart. It is then possible to compute the confidence arcs and error elipses around the computed acrophase because there will always be some variability in a sample of observations. The elliptical confidence region which is calculated (and shown as an example in Figure A.25) is based on the length of the vector (or directed line of specified length) representing amplitude. The confidence arc represents the average range of the distribution of acrophase times and of the amplitudes. The smaller the area of the ellipse the smaller the variability in the range of observations and, furthermore, the nearer the ellipse is to the pole (or centre of the circle) the greater the variability in the likely location of the peak. A flat line or 'white noise' signal would have an acrophase which could point in any direction and the error ellipse would be right at the centre of the circle.

The cosinor method can be applied to the testing of any one of the frequencies determined by a spectral analysis. The confidence arcs and error ellipse as shown in Figure A.25 provide an estimate of the significance of the three critical parameters, the *period* (τ), the *amplitude* (A) and the acrophase (Φ).

(v) Space-time series

Geographers have only recently tried to adapt time-series analysis to spatial problems. Most of the work that has been carried out in geography has been involved with regional and spatial variations in the short and medium terms (0-10 years) and with the regional impact of economic cycles (King and Clark, 1978). Since the main regional or sub-regional time series data in the U.K., North America, and elsewhere are related to labour market variables (employment, unemployment, wage-rates) or notification of various infectious diseases, these have been the areas that geographers have mainly turned to (Cliff, *et al.*, 1979), although work on population forecasting has also proved popular (Bennett, 1976; Curry, 1971; Curry and Bannister, 1974).

The two main approaches to the statistical modelling of time series are both used in the geographical literature. These are:

(1) time-domain methods
(2) frequency-domain or spectral methods.

(1) Time-domain methods

These are essentially extensions of correlation techniques and the general linear regression model to incorporate lead-lag relations and serial correlation. One of the best known of such models is the Box-Jenkins model (Box and Jenkins, 1970). Following Cliff (1977) we suppose that our system of interest is composed of $i = 1, 2, \ldots,$ n regions and $t = 1, 2, \ldots, T$ periods. Let x_{it} be the value of the variate (e.g. a particular level of unemployment) in the region i at time t and let ϵ_{it} be a set of independently distributed random shocks with a mean of zero and variance σ^2 representing *white noise*. It is then possible to formulate a space-time forecasting model which has autoregressive (AR) and random shock or moving average (MA) components as:

$$x_{it} = \alpha + \text{\ss}x_{i,t-1} + \gamma \underset{j}{\Sigma} w_{ij}x_{j,t-1} + e_{it} \qquad (\text{V}.1)$$

$$\underbrace{}_{a} \quad \underbrace{}_{b}$$

a = temporal autoregressive component
b = lagged in time spatial covariances

where
$$e_{it} = \epsilon_{it} + b\epsilon_{i,t-1} + e\underset{j}{\Sigma} w_{ij}\epsilon_{j,t-1},$$
$$(\text{V}.2)$$

is the random component.

In this model the quantities α, $\text{\ss}$, γ, b and c are parameters which have to be estimated. If we were trying to forecast unemployment then the variation in $x_{i,t}$ (unemployment) in the present would be a function of the past values of unemployment ($x_{i,t-1}$), the weighted sum of past values in the j other regions believed to influence unemployment in i at t (the space-time covariances in the equation) weighted according to what is believed to be their importance, and a random component to take into account essentially non-predictable happenings, for instance a sudden factory closure or opening (Cliff, 1977).

Models of this type can become quite complex as the effects of local economic cycles and serial correlation are taken into account. Recently a number of extensions to the model have been taken into account and these include the Kalman filter and distributed lag models.

Regional responses and relationships are assumed to be stable over time in these models, but in many cases (for example, in the British economy after 1967) there is strong evidence of shifts in regional relationships. Such changes violate these models' assumptions of *stationarity* (that is the process is assumed to be spatially and temporally invariant in terms of its mean and covariance structure).

Whilst it is possible to satisfy this stationarity assumption by differencing the more usual procedure now is to borrow a procedure from control engineering and allow the model structure to vary through time and/or over space by using the *Kalman filter* (see Martin, 1978; Bennett, 1979; Bennett and Chorley, 1978). In this approach the general linear model equation is shadowed by a second equation which defines the way in which the parameters can vary over time. The Kalman filter therefore provides recursive procedures for updating parameters. Apart from varying the parameters it is also possible to improve the basic model in (V.1) by placing greater emphasis on regions and time periods which are 'near' (in some sense) to a given region we want to forecast. To do this we might want to discount older observations. This can be done by using the *distributed lag* model of Dhrymes (Cliff, 1977; Hepple, 1978).

A particular research theme has been to trace short-run reaction paths in urban and regional systems — for instance, the spread of an economic cycle or an infectious disease from city A to city B to city C and the possible combinations of feedbacks, for instance from city C to city B, from city B to city A or from city C to city A. However causal modelling of these flows and feedbacks poses formidable identification or specification problems unless there is already a lot of information to draw on, for instance about economic linkages in an area. If there is not then the resulting *identification* problems are twofold (Cliff, 1977):

(a) What elements (autoregressive and moving average) should the model equation contain?
(b) What order of temporal and spatial lags should be included in the model?

Obviously theory should assist such questions but in its absence it is possible to examine the structure of the autocorrelation and partial autocorrelation functions formed by the variable(s) x_{it} in time and space (Bennett, 1978; Martin and Oeppen, 1975) to see at what point the autocorrelations go to (near) zero.

(2) Frequency-domain methods

An alternative approach to the problems of tracing short-run reaction paths in urban and regional systems has been to use a correlation-based analysis to identify significant urban and regional time series linkages rather than explicit paths. Time-domain lagged correlations can be used for this task but more commonly frequency-domain or spectral methods are used. These involve studying time series in terms of different wavelengths or periodicities, that is their spectrum. In geography some work has been carried out on the spectral analysis of a single univariate time series, or of spatial transects which can be treated in such a manner, for instance Tobler's (1969) study of U.S. Highway 40. Then again a limited amount of work has been produced on the spectral analysis of two dimensional spatial data (Rayner, 1971). However most work has been carried out on the patterns in space-time data sets, in particular on comparisons between *pairs* of regional time series. The method to deal with this latter problem is called *cross-spectral analysis* (Granger, 1969). Since cross-spectral analysis assumes only temporal stationarity and not spatial stationarity it is relatively easy to apply.

Cross-spectral analysis usually begins by computing the cross-covariance between two time series. This information is needed in order to find out whether two series are moving together (*in phase*) or whether one series *leads* or *lags* the other (Haggett *et.al.*, 1977). This obviously gives important clues as to the direction of particular economic, epidemiological or other impulses. However other measures can also be computed from the *sample cross-spectrum* (the Fourier transform of the sample cross covariance function). Following Haggett *et.al.*, 1977, these are:

(i) *coherence* — this is the spectral analysis equivalent of the coefficient of determination. It is the square of the correlation coefficient between the two series being studied at the chosen frequency, *w*. A plot of coherence against frequency gives a coherence diagram which shows the strength of the relationship (0 = no relation, 1 = perfect correlation) between the two series at corresponding frequencies.

(ii) *phase angle* — the difference between the phase shifts of the two series of frequency, *w*. A phase diagram can be constructed in which the phase angle is plotted against chosen frequency.

(iii) *gain* — the gain can be interpreted as the regression co-efficient of one series on the other at the chosen frequency, *w*.

Much work has been carried out in geography using cross-spectral analysis, for instance in analysis of regional variations in unemployment patterns to find out the linkages and correlations between unemployment time series (e.g. Bassett and Haggett, 1971; Cliff *et al.*, 1975; Hepple, 1975). The method can also illuminate data in historical geography, for instance it has been used to look at bankruptcies by county in the Northeast in nineteenth century England as a reflection of interregional trade cycles (Hepple, 1975) and also to look at fluctuations in wheat prices in eighteenth century England (Granger and Elliott, 1967).

Cross-spectral analysis is not, however, without problems. It is dangerous to use the method with short time series, and certainly should not be attempted with very short series of less than 100 observations. There are also problems with the Nyquist frequency (see above). Haggett *et al.*, (1977, 413) list other estimation problems which should be noted *before* attempting to use this method. Although there are ways around some of these estimation problems (for instance by first differencing) there are some situations in which cross-spectral analysis cannot be safely used.

For a general introduction to space-time series as a whole the most comprehensive work is that of Bennett (1979).

Notes

1. We extend our thanks to Dr. Bo Lenntorp for allowing us to cite his monograph, *Paths in Space-Time Environments, A Time-Geographic Study of Movement Possibilities of Individuals,* so extensively.
2. The rotated projections which Lenntorp also uses have not been included.

Bibliography

Abbé, M. (1936). The spatial effect upon the perception of time, *Japanese Journal of Experimental Psychology,* **3,** 1-52.

Abbé, M. (1937). The temporal effect upon the perception of space, *Japanese Journal of Experimental Psychology,* **4,** 83-93.

Abe, S. (1935). Experimental study on the correlation between time and space, *Tohakn Psychologia Folia,* **3,** 53-68.

Abler, R. (1976). Effects of space-adjusting technologies on the human geography of the future, see also: Monoculture or miniculture? The impact of communication media on space, 122-131, in, Abler, R., Jannelle, D., Philbrick, A., and Sommer, J. (Eds.) *Human Geography in a Shrinking World,* Belmont, California, Duxbury, 35-56.

Abler, R., Adams, J., and Gould, P. (1971). *Spatial Organization: the Geographer's View of the World,* Englewood Cliffs, Prentice-Hall.

Abler, R.F., Adams, J.S., and Borchert, J.R. (1976). *The Twin Cities of St. Paul and Minneapolis,* Cambridge, Mass., Ballinger.

Abramowitz, M. (1965). *Evidence of Long Swings in Aggregate Construction since the Civil War,* New York, National Bureau of Economic Research.

Abramowitz, M. (1968). The passing of the Kuznets cycle, *Economica,* **35,** 349-67.

Abrams, M.(1978). Time and the elderly. *New Society,* **46,** 846/7, 685-6.

Abu-Lughod, J. (1969). Testing the theory of social area analysis: the ecology of Cairo, Egypt, *American Sociological Review,* **35,** 198-212.

Adams, J.S. (1970). Residential structure of midwestern cities, *Annals of the Association of American Geographers,* **60,** 36-62.

Alberts, W.W. (1962). Business cycles, residential construction and the mortgage market, *Journal of Political Economy,* **70,** 263-81.

Alexander, S. (1920). *Space, Time, and Deity,* Bk.1, Space-Time (1920) from Macmillan (1966 edition).

Alonso, W. (1964). The historic and structural theories of urban form: their implications for urban renewal, *Land Economics,* **40,** 229-30.

Althusser, L. (1970). The object of capital, in, Althusser, L., and Balibar, E., *Reading Capital,* London, New Left Books, 71-198.

Amedeo, D. and Golledge, R.G. (1975). *An Introduction to Scientific Reasoning in Geography,* New York, John Wiley.

Amin, S. (1977). *Unequal Development,* Hassocks, Sussex, Harvester.

Anderson, J. (1971). Living in urban space-time, *Architectural Design,* **41,** 41-44.

Anderson, J. (1971). Space-time budgets and activity studies in urban geography and planning, *Environment and Planning,* **3,** 353-368.

Angel, S. and Hyman, G. (1970). Urban travel times, *CES WP 65,* Centre for Environmental Studies, London.

Angel, S. and Hyman, G. (1976) *Urban fields: a geometry of movement,* London, Pion.

Augustine, St. *Confessions,* Bk.11, Ch.14, p.17.

Bannister, D. (1962). Personal construct theory: a summary and experimental paradigm, *Acta Psychologica,* 20.

Bannister, D. and Mair, J.M.M. (1968). *The Evaluation of Personal Constructs,* London, Academic Press.

Barke, M. (1974). The changing urban fringe of Falkirk: some morphological implications of urban growth, *Scottish Geographical Magazine,* **90,** 85-97.

Barke, M. (1976). Land use succession: a factor in fringe-belt modification, *Area,* **8,** 303-6.

Barker, D. (1978). The paracme of innovations: the neglected aftermath of diffusion or a wave goodbye to an idea? *Area,* **9,** 259-64.

Barker, R.G. (1968). *Ecological Psychology,* Stanford, Stanford University Press.

Barnes, J.A. (1971). Time flies like an arrow, *Man,* **6,** 537-52.

Barth, F. (1959). The land use pattern of migrating tribes of South Persia, *Norsk Geografisk Tidsskrift,* **17,** 1-11.

Bassett, K. and Haggett, P. (1971). Towards short-term forecasting for cyclical behaviour in a regional system of cities, in Chisholm, M., Frey, A.E., and Haggett, P. (Eds.), *Regional Forecasting,* London, Butterworth, 389-413.

Baxter, R.S. (1976). *Computer and Statistical Techniques for Planners,* London, Methuen.

Beavon, K.S. (1977). *Central Place Theory: A Re-interpretation,* London, Longman.

B.B.C. (1965). *What the People are Doing, The People's Activities,* BBC Audience Research Department, London, British Broadcasting Corporation.

B.B.C. (1978). *The People's Activities and Use of Time.* BBC Audience Research Department, London, British Broadcasting Corporation.

Becker, G.S. (1965). A theory of the allocation of time, *Economic Journal,* **75,** 493-517.

Bell, W. (1952). *A comparative study in the methodology of urban analysis,* unpublished Ph.D. dissertation, University of California, Berkeley.

Benjamin, A.C. (1968). Ideas of time in the history of philosophy, in Fraser, J.T. (Ed.) (1968), 3-30.

Bennett, R.J. (1976). The representation and identification of spatio-temporal systems: an example of population diffusion in northwest England, *Transactions, Institute of British Geographers,* **66,** 73-94.

Bennett, R.J. (1979). *Spatial Time Series: Analysis, Forecasting and Control,* London, Pion.

Bennett, R.J. and Chorley, R. (1978). *Environmental Systems, Philosophy, Analysis and Control,* London, Methuen.

Berry, B.J.L. (1974). Short-term housing cycles in a dualistic metropolis, in Gappert, G., and Rose, R. (Eds.), *Contemporary Urbanization,* California, Sage, 165-82.

Berry, B.J.L., Cutler, I., Draine, E.H., Kiang, Y., Tocalis, T.R. and Devise, P. (1976). *Chicago: transformations of an urban system.* Cambridge, Mass; Ballinger.

Blake, M.F.J. (1971). Temperament and time of day, in W.P. Colquhoun (Ed.) (1971), 109-48.

Blatt, S.J. and Quinlan, D.M. (1972). The psychological effects of rapid shifts in temporal referrents, in Fraser, J.T., Haber, F.C., and Müller, G.H. (Eds.) 506-522.

493

Blum, A.F. (1964). Social structure, social class, and participation in primary relationship, in Shostak, A., and Gomberg, N., *Blue-collar World: Studies of the American Worker,* Englewood Cliffs, N.J., Prentice-Hall, 195-206.

Blumer, H. (1969). *Symbolic Interactionism: Perspective and Method,* Englewood Cliffs, N.J., Prentice-Hall (Cited by Hugil, P.J. (1975)).

Boal, F.W. (1976). Ethnic residential segregation, in, Herbert, D.T., and Johnston, R.J. (Eds.) 41-80.

Boh, K., and Sakisda, S. (1972). An attempt at a typology of time, in, Szalai, A. (Ed.), *The Use of Time,* The Hague, Mouton, 229-48.

Bourdieu, P. (1977). *Outline of a Theory of Practice,* Cambridge, Cambridge University Press.

Box, G.E.P. and Jenkins, G.M. (1970). *Time Series Analysis, Forecasting, and Control,* San Francisco, Holden Day.

Bracey, H.E. (1962). English central villages: identification, distribution and functions, *Lund Studies in Geography, Series B, Human Geography,* **24,** 169-90.

Bradley, P.N., Raynault, C. and Torrealba, D. (1977). *The Mauritanian Guidimaka: a Critical Analysis Leading to a Development Project,* London, War on Want Publications.

Bridgeman, P.W. (1932). The time scale: the concept of time, *The Scientific Monthly,* **35,** 97-107.

Brög, W., Heuwinkel, D. and Neumann, K. (1977). *Psychological determinants of user behaviour,* Paris, O.E.C.D.

Brown, B.A. and Boddy, M.J. (1976). A Cartesian view of the city: comments on the recent article by Taylor and Parkes, *Environment and Planning A,* **8,** 599-601.

Brown, F.A. Jr. (1969). A hypothesis for extrinsic timing of circadian rhythms, *Canadian Journal of Botany,* **47,** 287-98.

Browning, C.E. (1964). Selected aspects of land use and distance, *Southeastern Geographer,* **4,** 29-40.

Budapest (1965). The 24-hours of the day, *Hungarian Central Statistical Office,* (English version).

Bullock, N., Dickens, P., Shapcott, M. and Steadman, P. (1974). Time budgets and models of urban activity patterns, *Social Trends,* **5,** 45-63.

Bunge, W. (1974). The human geography of Detroit, in Roberge, R.A. (Ed.) *La Crise Urbaine: A Challenge to Geographers,* Ottawa, University of Ottawa Press, 49-69, (see Smith, D.M. (1977)).

Bunge, W. (1975). Detroit humanly viewed: the American urban present, in R. Abler, D. Jannelle *et al., Human Geography in a Shrinking World,* North Scituate, Mass., Duxbury Press, 147-81.

Burnett, K.P., and Thrift, N.J. (1979). New approaches to travel behavior in Hensher, D., and Stopher, P. (Eds.). *Behavioral Travel Demand Modelling,* London, Croom Helm.

Buttimer, A. (1976). Grasping the dynamism of lifeworld, *Annals of the Association of American Geographers,* **62,** 277-92.

Calkins, K. (1970). Time: perspectives, marking and styles of useage, *Social Problems,* **17,** 6. Accordian effect of time stretching quoted by Calkins from work of Davis, F.

Campbell, B. (1963). Long swings in residential construction: the postwar experience. *American Economic Review,* **53,** 508-18.

Capek, M. (Ed.) (1976). *The Concepts of Space and Time: their Structure and their Development,* Boston Studies in the Philosophy of Science, Vol.22, Dordrecht, D. Reidel Publishing.

Caplowitz, D. (1963). *The Poor Pay More,* Glencoe, Free Press.

Carlstein, T. (1974). *Time Allocation,* University of Lund, Department of Geography, (mimeo).

Carlstein, T. (1975). Time allocation: on the capacity for human interaction in space and time, Department of Geography, Lund (mimeo).

Carlstein, T., (1977). Innovation, time allocation and time-space packing, *Rapporter och Notiser* No.35, Department of Geography, University of Lund, and in Carlstein, T., Parkes, D.N., and Thrift, N.J. (Eds.) (1978), *Timing Space and Spacing Time,* Volume II, *Human Activity and Time Geography,* London, Edward Arnold, 146-61.

Carlstein, T., Parkes, D.N., and Thrift, N.J. (Eds.) (1978). *Timing Space and Spacing Time, Vol.1, Making Sense of Time,* London, Edward Arnold.

Carlstein, T., Parkes, D.N., and Thrift, N.J. (Eds.) (1978). *Timing Space and Spacing Time, Vol.III, Time and Regional Dynamics,* London, Edward Arnold.

Carlstein, T. and Thrift, N.J. (1978). Towards a time-space structured approach to society and the environment, in Carlstein, T., Parkes, D.N., and Thrift, N.J. (Eds.), *Vol.II Human Activity and Time Geography,* London, Edward Arnold, 225-63.

Carpenter, J. and Cazamian, P. (1977). *Night Work: Its Effects on the Health and Welfare of the Worker,* Geneva, International Labour Office.

Carter, H. (1965). *The Towns of Wales,* Cardiff, University of Wales Press.

Carter, H. (1972). *The Study of Urban Geography,* London, Edward Arnold.

Carter, H. and Davies, W.K.D. (Eds.) (1970). *Urban Essays: Studies in the Geography of Wales,* London, Longman, also, Towns of Wales, Carter.

Cattell, R.B. and Coulter, M.A. (1966). Principles of behavioural taxonomy and the mathematical basis of the taxonomy computer program. *The British Journal of Mathematical and Statistical Psychology,* **19,** 237-269.

Chapin, F. Stuart, Jnr. (1971). Free time activities and quality of urban life, *Journal of American Institute of Planners,* **37,** 411-17.

Chapin, F.S. Jnr. (1978). Human time allocation in the city, in, Carlstein, T., Parkes, D.N., and Thrift, N.J. (Eds.), *Vol.II, Human Activity and Time Geography,* London, Edward Arnold, 13-26.

Chapin, F.S., and Hightower, H.C. (1966). *Household Activity Systems — a Pilot Investigation,* Center for Urban and Regional Studies, University of North Carolina, Chapel Hill.

Chapin, F.S., Jnr., and Logan, T.H. (1968). Patterns of time and space use, in Perloff, H.S. (Ed.), *The Quality of the Urban Environment,* Baltimore, The Johns Hopkins Press, 305-332.

Chapman, G.P. (1978). The folklore of the perceived environment in Bihar, (forthcoming).

Chatfield, C. (1975). *The Analysis of Time Series: Theory and Practice,* London, Chapman and Hall.

Chatfield, C. (1977). Recent developments in time-series analysis, *Journal of the Royal Statistical Society A,* **140,** 492-510.

Chombart de Lauwe, P.H. (1960). *Famille et Habitation,* Vol.2, Centre National de la Recherche Scientifique, Paris (cited by Anderson, 1971 (355)).

Clark, D. (1974). Technology, diffusion, and time-space convergence: the example of the STD telephone, *Area,* **6,** 181-4.

Clemence, G.M. (1968). Time measurement for scientific use, in Fraser, J.T. (Ed.) (1968), 401-414.

Cleugh, M.F., (1937). *Time and its Importance in Modern Thought,* London, Methuen, cited in Benjamin (1968).

Cliff, A.D. (1977). Quantitative methods: time series methods for modelling and forecasting, *Progress in Human Geography,* **1,** 492-502.

Cliff, A.D., Haggett, P., Ord, J.K., and Bassett, K., and Davies, R. (1975). *Elements of Spatial Structure: a Quantitative Approach,* Cambridge, Cambridge University Press.

Cliff, A.D., Haggett, P., Martin, R.L. and Ord, J.K. (1979). *Spatial Diffusion,* Cambridge, Cambridge University Press.

Cohen, J. (1967). *Psychological Time in Health and Disease,* Springfield Ill., Charles and Thomas.

Cohen, J. (1968). *Subjective Time,* in Fraser, J.T. (Ed.) 257-278.

Colquhoun, W.P. (Ed.) (1971). *Biological Rhythms and Human Performance,* London, Academic Press.

Commonwealth of Australia (1975). *Australians' Use of Time,* Canberra, Cities Commission.

Converse, P.E. (1968). Time Budgets, in, *International Encyclopaedia of the Social Sciences,* **42,** New York, Collier Macmillan.

Converse, P.E. (1972). Country differences in the use of time, in, Szalai, A. (Ed.), 145-177.

Conzen, M.R.G. (1960). Alnwick, Northumberland: a study in town-plan analysis, *Transactions Institute of British Geographers,* **27,** 1-81.

Conzen, M.R.G. (1966). Historical townscapes in Britain: a problem in applied geography, in House, J.W. (Ed.) *Northern geographical essays in honour of G.H.J. Daysh,* Newcastle Upon Tyne, 56-60.

Cope, R. (1929). Calendars of the Indians North of Mexico, *University of California Publications in Archaeology and Ethnology,* **16,** 131-53.

Cox, D.R. and Lewis, P.A.W. (1966). *The Statistical Analysis of Series of Events,* London, Methuen.

Cowan, H.J. (1958). *Time and its Measurements,* Cleveland, The World Publishing Company.

Cullen, I.G. (1976). Human geography, regional science, and the study of individual behaviour, *Environment and Planning A,* 397-409.

Cullen, I.G. (1978). The treatment of time in the explanation of spatial behaviour, in, Carlstein, T., Parkes, D.N., and Thrift, N.J. (Eds.) *Vol.II, Human Activity and Time Geography,* London, Edward Arnold, 27-38.

Cullen, I.G., Godson, V., and Major, S. (1972). The structure of activity patterns, in, Wilson, A.G. (Ed.), *Patterns and processes in urban and regional systems,* London Papers in Regional Science, **3,** 281-96.

Cullen, I.G., and Godson, V. (1972). *Networks of Urban Activities, Vol.II: The Structure of Activity Patterns,* Joint Unit of Planning Research, University College, London.

Cullen, I.G., and Godson, V. (1975). The structure of activity patterns, *Progress in Planning,* **4,** 1-96, London, Pergamon.

Cullen, I.G., and Philps, E. (1975). *Diary techniques and the problems of urban life,* Final Report to the Social Science Research Council.

Curry, L. (1971). Applicability of space-time moving-average forecasting. in Chisholm, M.D.I., Frey, A.E., and Haggett, P. (Eds.), *Regional Forecasting,* London, Butterworth, 11-24.

Curry, L., and Bannister, G. (1974). Forecasting township populations of Ontario from space-time covariances, in Bourne, L. (Ed.), *Urban Futures for Central Canada,* Toronto, Toronto University Press, 34-59.

Dagens, Nyheter, (Swedish Newspaper) Torsdagen den 15 Juli 1976. A popularization of the time-geographic model.

Davidoff, L. (1973). *The Best Circles,* London, Croom Helm.

Davis, J.C. (1973). *Statistics and Data Analysis in Geology,* New York, Wiley International Edition.

Daly, D.J. (1969). Business cycles in Canada: their postwar persistence, in, Bronfenbrenner, M. (Ed.). *Is the Business Cycle Obsolete?* New York, Wiley.

Daniels, P.W. (1975). *Office Location: An Urban and Regional Study,* London, Bell.

Davies, P.C.W. (1977). *Space and Time in the Modern Universe,* Cambridge, Cambridge University Press.

Day, R. (1976). The theory of long waves; Kondratieff, Trotsky, Mandel.

496

New Left Review, No. 99, 67-82.

de Grazia, S. (1962). *Of Time, Work, and Leisure,* New York, Twentieth Century Fund.

de Serpa, A.C. (1971). A theory of the economics of time, *Economic Journal,* **81,** 828-846.

Dewey, E.R. and Mandino, O. (1973). *Cycles: the Mysterious Forces that Trigger Events,* New York, Manor Books.

Diamant, A. (1970). The temporal dimensions in models of administration and organization, in, Waldo, D. (Ed.). *Temporal Dimensions of Development Administration,* Durham, Duke University Press, 90-134.

Dickens, C. (1977). *Hard Times,* London, Pan (first published 1854).

Dickinson, R.E. (1964). *City and Region,* London, Routledge, and Kegan Paul.

Doob, L.W. (1971). *Patterning of Time,* New Haven, Yale University Press.

Doob, L.W. (1978). Time: cultural and social anthropological aspects, in, Carlstein, T., Parkes, D.N., and Thrift, N.J. (Eds.), *Vol.I Making Sense of Time,* London, Edward Arnold, 56-65.

Doxiadis, C. (1968). *Ekistics: an Introduction to the Science of Human Settlements,* New York, Oxford University Press.

Dresch, J. (1939). Les genres de vie de montagne dans le Massif du Toubkal, *Revue Géogr. Maroc,* (cited by Ullman, 1974).

Durkheim, E. (1965: 1915). *The Elementary Forms of Religious Life,* New York, The Free Press.

Edholm, O.G. (1970). Ergonomics Research Society, Society's lecture 1970, *Ergonomics,* **13,** 625-643.

Ellegard, A., Hägerstrand, T., and Lenntorp, B. (1975). Activity organization and the generation of daily travel: two further alternatives. *Rapporter och Notiser,* **23,** Department of Geography, University of Lund, and *Economic Geography,* **53,** 126-152 (1977).

Elliott, D.H., Harvey, A.S., and Procos, D. (1973). An overview of the Halifax time-budget study, *Paper to Conference Working Group, Regional and Urban Studies Centre:* Institute of Public Affairs, Dalhousie University, Halifax, Canada.

Elliott, D.H., and Clark, J. (1976). The spatial context of urban activities: some theoretical, methodological and policy considerations. Paper presented at *Symposium on Time Budgets and Social Activities,* IIASA, Austria.

Engel-Frisch, G. (1948). Some neglected temporal aspects of human ecology, *Social Forces,* **43,** 47.

Engels, F. (1892). *The Condition of the Working Class in England,* London, Panther Books 1969 edition.

Evans, A.E. (1972). On the theory of the valuation and allocation of time, *Scottish Journal of Political Economy,* **19,** 236-248.

Evans-Pritchard, E.E. (1940). *The Nuer: a Description of the Modes of Livelihood and Political Institutions of a Nilotic People,* London, Clarendon Press.

Evans-Pritchard, E.E. (1951). *Social Anthropology,* London, Cohen and West.

Ewing, G.E., and Wolfe, R. (1977). Surface feature interpolation on two dimensional time-space maps, *Environment and Planning,* **A,** 9, 429-38.

Firey, W. (1945). Sentiment and symbolism as ecological variables, *American Sociological Review,* **10,** 140-148.

Fischer, R. (1967). The biological fabric of time, *Annals of New York Academy of Sciences,* **138,** 440-488.

Fischer, R. (1968). *Biological time,* in, Fraser, J.T. (Ed.) 357-384.

Foote, N.N. and Mayersohn, R. (1959). Allocations of time among family activities, *Paper to 4th World Congress of Sociology,* Stresa, (cited in Chapin, 1974).

Ford, L.R. (1978). Continuity and change in historic cities: Bath, Chester and Norwich, *Geographical Review,* **68,** 253-73.

Forer, P. (1974). Space through time: a case study with New Zealand airlines, in Cripps, E. (Ed.) *Space-Time Concepts in Regional Science,* London, Pion, 22-45.

Forer, P. (1975). Relative space and regional imbalance: domestic airlines in New Zealand's geometrodynamics, in, *Proceedings of the IGU Conference, New Zealand Geographical Society,* Hamilton, New Zealand Geographical Society, 53-62.

Forer, P. (1978). Time-space and area in the city of the plains, in Carlstein, T., Parkes, D.N., and Thrift, N.J. (Eds.). *Vol.I, Making Sense of Time,* London, Edward Arnold, 99-118.

Fortes, M. (1970). Time and social structure and other essays, *LSE Monographs on Social Anthropology,* **40,** London, Athlone Press.

Fraisse, P. (1964). *The Psychology of Time,* London, Eyre, and Spottiswood.

Frank, L.W. (1939). Time perspectives, *Journal of Social Philosophy,* **4,** 293-312.

Fraser, J.T. (Ed.) (1968). *The Voices of Time: A Co-operative Survey of Man's Views of Time as Expressed by the Sciences and by the Humanities,* London, Allen Lane. New York, Braziller (1966).

Fraser, J.T. (1975). *Of Time, Passion, and Knowledge,* New York, Braziller.

Fraser, J.T. (1977). Personal communication.

Fraser, J.T. (1978). *Time as Conflict: A Scientific and Humanistic Study,* Basle, Birkhauser Verlag.

Fraser, J.T., Haber, F.C., and Müller, G.H. (Eds.) (1972). *The Study of Time,* New York, Springer-Verlag.

Fraser, J.T., and Lawrence, N. (Eds.) (1975). *The Study of Time II,* New York, Springer-Verlag.

Freeman, S.A. (1964). Time perspectives as a function of socio-economic group and age, *Unpublished Ph.D. dissertation,* University of Carolina.

Galbraith, J.K. (1958). *The affluent society.* Harmondsworth, Middlesex, Penguin.

Ghez, G.R. and Becker, G.S. (1975). *The Allocation of Time and Goods over the Life Cycle,* New York, Columbia University Press.

Granger, C.W.J. and Elliott, C.M. (1967). A fresh look at wheat prices and markets in the eighteenth century, *Economic History Review,* **20,** 257-65.

Gauquelin, M. (1973). *The Cosmic Clocks,* London, Paladin.

Gershuny, J.I. (1978). After post-industrial society?, London, Macmillan.

Gerson, E.M. and Gerson, M.S. (1976). The social framework of place perspectives, in, Moore, G.T. and Golledge, R.G. (Eds.)

Gioscia, V. (1972). On social time, in, Yaker, H., Osmond, H., and Cheek, F., (Eds.) *The Future of Time,* London, Hogarth Press, 73-141.

Golledge, R.G. (1978). Time and stress in urban environments, in, Carlstein, T., Parkes, D.N., and Thrift, N.J. (Eds.). *Volume I, Making Sense of Time,* London, Edward Arnold, 76-98.

Golledge, R.G. and Rushton, G. (1973). Multidimensional scaling: review and geographical applications, *Commission on College Geography, Technical Paper* No.10.

Gonseth, F. (1971). From the measurement of time to the method of research, in, Zeman, J. (Ed.). *Time in Science and Philosophy,* New York, Elsevier.

Goodman, P.S. (1967). An empirical examination of Elliott Jaques' concept of time span, *Human Relations,* **20,** 158-170.

Goody, J. (1968). Time: 2. Social organization, *International Encyclo-*

pedia of the Social Sciences, Vol.16, 30-42, New York, Macmillan.

Gottlieb, M. (1976). *Long Swings in Urban Development,* New York, National Bureau for Economic Research.

Goudsmit, S.A. and Claiborne, R. (1970). *Time,* Time-Life International (Nederland) N.V.

Granger, C.W.J. (1969). Spatial data and time series analysis, in, Scott, A.J. (Ed.) *London Papers in Regional Science* Vol.I, London, Pion, 1-24.

Green, H. (1972). Temporal attitudes in four negro subcultures, in, Fraser, J.T., Haber, F.C., and Müller, G.H. (Eds.)

Gregory, D. (1978). *Ideology, Science and Human Geography,* London, Hutchinson.

Greig-Smith, P. (1952). The use of random and contagious quadrats in the study of the structure of plant communities, *Annals of Botany,* (new series), **16**, 293-316.

Grubbstrom, R.W. (1973). *Economic Decisions in Spce and Time: Theoretical and Experimental Inquiries into the Cause of Economic Motion,* Gothenburg, Gothenburg Studies in Business Administration.

Grubbstrom, R.W. (1978): Tensor formulation of economic motion, in, Carlstein, T., Parkes, D.N. and Thrift, N.J. (Eds.), *Vol.III, Time and Regional Dynamics,* London, Edward Arnold, 96-107.

Gurvitch, G. (1953). Hyper-empirisme dialectique, *Cahiers Internationaux de Sociologie,* **15**, cited by Korenbaum (1964).

Gurvitch, G. (1964). *The Spectrum of Social Time,* Dordrecht, D. Reidell Publishing.

Hägerstrand, T. (1952). The propagation of innovation waves, *Lund Studies in Geography, B, Human Geography,* **4**, 3-19.

Hägerstrand, T. (1970). What about people in regional science? *Papers of the Regional Science Association,* **24**, 7-21.

Hägerstrand, T. (1972). Tartorsgrupper som Regionsamhallen, in Regioner att Leva I: in *Rapport fran ERU,* Stockholm: Allmanna Forlagert, 141-173.

Hägerstrand, T. (1973). The domain of human geography, in, Chorley, R.J. (Ed.) *Directions in Geography,* London, Methuen, 67-87.

Hägerstrand, T. (1974). Ecology under one perspective, in *Ecological Problems of the Circumpolar Area,* 271-276.

Hägerstrand, T. (1975a). Survival and arena: on the life-history of individuals in relation to their geographical environment, Clark University: *The Monadnock,* **49**, 9-29, also in Carlstein, T., Parkes, D.N., and Thrift, N.J. (Eds.) *Volume II, Human Activity and Time Geography,* London, Edward Arnold, 122-45.

Hägerstrand, T. (1975b). Space, time and human conditions, in Karlqvist, A., Lundqvist, L., and Snickars, F. (Eds.), *Dynamic Allocation of Urban Space,* Farnborough: Saxon House, 3-12.

Haggett, P. (1971). Leads and lags in interregional systems: a study of cyclic fluctuations in the south-west economy, in, Chisholm, M., and Manners, G. (Eds.) *Spatial Policy Problems of the British Economy,* Cambridge, Cambridge University Press, 69-95.

Haggett, P., Cliff, A.D. and Frey, A. (1977). *Locational Analysis in Human Geography* (second edition). Volume I. Locational Models. Volume 2. Locational Methods, London, Edward Arnold.

Halberg, F. (1969). Chronobiology, *Annual Review of Physiology,* **31**, 675-725.

Halberg, F. (1973). Laboratory techniques and rhythmometry in biologic aspects of Circadian rhythms, in, J.N. Mills (Ed.) *Biological Aspects of Circadian Rhythms,* London, Plenum Press.

Halberg, F. (1975). Chronobiology in 1975, *Chronobiologia,* **3**, 31-53.

Halberg, F., Tong, Y.L. and Johnson, E.A. (1967). Circadian system phase: an aspect of temporal morphology: procedures and illustrative examples, in, von Mayersbach, H. (Ed.) *The Cellular Aspects of Biorhythms,* Berlin-Heidelberg, New York, Springer, 20-48.

Halberg, F. and Katinas, G. (1973). Chronobiologic glossary of the International Society for the Study of Biologic Rhythms, *International Journal of Chronobiology,* **1**, 31-63.

Haldane, N.B.S. (1963). Biologic possibilities in the next ten thousand years, in, G. Wolstenholme (Ed.), *Man and his Future,* London, Churchill.

Hall, E.T. (1959). *The Silent Language,* New York, Doubleday.

Hall, E.T. (1966). *The Hidden Dimension,* London, Bodley Head. New York, Doubleday.

Hallowell, A.L. (1937). Temporal orientation in western civilization and in a preliterate society, *American Anthropology,* **39**, 654 ff.

Hammer, P.G., Jnr., and Chapin, F.S., Jnr. (1972). Human time allocation: a case study of Washington, DC., *A Technical Monograph, Center for Urban and Regional Studies,* University of North Carolina at Chapel Hill.

Hanham, R.W. (1976). Factorial ecology in space and time: an alternative method, *Environment and Planning, A,* **8**, 389-841.

Harrison, J. and Sarre, P. (1975). Personal construct theory in the measurement of environmental images, *Environment and Behaviour,* **7**, 3-59.

Harrod, J. (1958). *Towards an Economic Dynamics,* London, Macmillan.

Harvey, A.S. and Clark, S. (1975). Descriptive analysis of Halifax time budget data. Technical Memorandum. *Regional and Urban Studies Center: Institute of Public Affairs,* Dalhousie University, Canada.

Harvey, A.S. (1978). Discretionary time activities in context, *Occasional Paper No.3,* Institute of Public Affairs, Dalhousie University, Canada.

Harvey, D. (1969). *Explanation in Geography,* London, Edward Arnold.

Harvey, D. (1978). The urban process under capitalism: a framework for analysis, *International Journal of Urban and Regional Research,* **2**, 101-31.

Hastrup, F. (1970). Danish 'Vangelag', *Kulturgeografi,* **114**, 96-100.

Hawley, A. (1950). *Human Ecology: a Theory of Community Structure,* New York, Ronald Press.

Heidegger, M. (1958). An ontological consideration of place, in, *The Question of Being,* New York, Twayne Publishers, (cited by Relph, 1976).

Heirich, M. (1964). The use of time in the study of social change, *Sociological Review,* **29**, 386-97.

Helson, H. and King, S.M. (1931). The tau effect — an example of psychological relativity, *Journal of Experimental Psychology,* **14**, 202-217.

Hepple, L. (1975). Spectral techniques and the study of interregional economic cycles, in, Peel, R.F., Chisholm, M. and Haggett, P. (Eds.) *Processes in Physical and Human Geography, Bristol Essays,* London, Heinemann, 392-408.

Hepple, L. (1978). The econometric specification and estimation of spatio-temporal models, in, Carlstein, T., Parkes, D.N., and Thrift, N.J. (Eds.), *Timing Space and Spacing Time Vol.III, Time and Regional Dynamics,* London, Edward Arnold, 66-80.

Herbert, D.T. (1972). *Urban Geography: a Social Perspective,* New York, Praeger.

Herbert, D.T. and Johnston, R.J. (Eds.) (1976). *Social Areas in Cities,* Volume I, Spatial processes and form, and Volume II, Spatial perspectives on problems and policies, London, John Wiley.

Hillier, W. and Leaman, A. (1977). Space syntax, *Environment and Planning B,* **3**, 147-85.

Hirsch, F. (1977). *Social limits to growth,* London, Routledge and Kegan Paul.

Hoagland, H. (1933). The physiological control of judgements of duration: Evidence for a chemical clock, *Journal of Genetic Psychology,* **9,** 267-287.

Hoagland, H. (1968). Some biochemical considerations of time, in Fraser, J.T. (Ed.), 312-329.

Hobson, R. and Mann, S.H. (1975). A social indicator based on time allocation, *Social Indicators Research,* **1,** 439-57.

Horton, J. (1967). Time and cool people, *Transactions,* University of Washington, **4,** 5-20.

Howse, D.M. (1979). *Greenwich Time and the Discovery of Longitude,* Oxford, Oxford University Press.

Hoyt, H. (1970). *One Hundred Years of Land Values in Chicago: the Relationship of the Growth of Chicago to the Rise in its Land Values, 1830-1933,* New York, Arno Press, (first published 1933).

Huff, and Lutz (1974). Evaluating change in Human Geography. *Economic Geography,* **68,** 253-73.

Hughes, A.O. (1970). Spectral Analysis, *Computer Applications in the Natural and Social Sciences,* No.8, Nottingham, Nottingham University.

Hugill, P.J. (1975). Social conduct in the Golden Mile, *Annals Association of American Geographers,* Vol.65, 214-228.

Huntington, E. (1907). *The Pulse of Asia,* Boston, Houghton Mifflin.

Huntington, E. (1926). *The Pulse of Progress,* New York, Chas. Scribner's.

Huntington, E. (1927). The quantitative phases of human geography, *Scientific Monthly,* **25.**

Huntington, E. (1938). *Season of Birth: its Relation to Human Abilities,* New York, John Wiley.

Huntington, E. (1945). *The Mainsprngs of Civilization,* New York, John Wiley, and reprint, Mentor Books, 1959.

Ilchman, W.F. (1970). New time in old clocks: productivity, development and comparative public administration, in, Waldo, D. (Ed.) *Temporal Dimensions of Development Administration,* Durham, Duke University Press.

Isard, W. (1942). A neglected cycle: the transport building cycle, *Review of Economic Statistics,* **24,** 149-58.

Isard, W. (1943). Transport development and business cycles, *Quarterly Journal of Economics,* **57,** 90-112.

Isard, W. (1971). On notions and models of time, *Papers of the Regional Science Association,* **25,** 7-31.

Janelle, D.G. (1968). Central place development in a time-space framework, *Professional Geographer,* **20,** 5-10.

Janelle, D.G. (1969). Spatial reorganization: a model and concept, *Annals Association of American Geographers,* **58,** 348-64.

Janelle, D.G. (1976). Stagecoach operations in Maine, 1826-1829, *Proceedings, New England-St. Lawrence Valley Geographical Society,* **6,** 15-48.

Jaques, E. (1964). *Time Span Handbook,* London, Heinemann.

Johns, E. (1965). *British Townscapes,* London, Edward Arnold.

Jones, B. (1977). Leisure in Newcastle — a space-time approach, unpublished B.A. Hons. thesis, Department of Geography, Newcastle University, NSW.

Jones, P. (1978). Innovation life-span: the urban tramway, *Area,* **10,** 247-9.

Jones, P.M. (1978). Urban transport and land use planning: a unified approach, Paper to Institute of British Geographers Conference, University of Hull (Mimeo).

Jones, P.M. (1979). 'HATS': a technique for investigating household decisions, *Environment and Planning A,* **11,** 59-70.

Journal of the Royal Statistical Society (1977) Series A, **144,** 411-572.

Kastenbaum, R. (1964). The structure and function of time perspectives, *Journal of Psychological Researches,* **109,** 9-25.

Kelly, G.A. (1955). *The Psychology of Personal Constructs,* New York, Norton (Vol.1 and 2).

Kelly, G.A. (1963). *A Theory of Personality,* New York, Norton.

Kendall, M.B. and Sibley, R.F. (1970). Social class differences in time orientation: artifact? *Journal of Social Psychology,* **82,** 187-191.

King, A.D. (1976). Social anthropological approaches to the study of architecture, Lecture given to the Architektur-Soziologisches Kolloqium, Technische Universitat, Hannover, 23 November (mimeo).

King, L.J., Casetti, E. and Jeffrey, D. (1969). Economic impulses in a regional system of cities, *Regional Studies,* **3,** 213-18.

King, L.J. and Clark, (1978). Government policy and regional development, *Progress in Human Geography,* **21,** 1-16.

Kolaja, J. (1969). *Social System and Time and Space,* Pittsburgh, Pa., Duquesne University Press.

Kondratieff, N.D. (1978). The long waves in economic life, *Lloyds Bank Review,* No.129, 41-60. (English version first published in *Review of Economic Statistics,* 1935).

Korenbaum, M. (1964). Translators preface to Gurvitch, G., *The Spectrum of Social Time.*

Korzybsky, A. (1921). *Manhood of Humanity,* The International Non-Aristotelian Library Institute of General Semantics, Distributions, Lakeville, Conn. 2nd edition, Second printing 1968.

Kranz, P. (1970). What do people do all day? *Behavioural Science,* **15,** 286-293.

Kripke, D.F. (1974). Ultradian rhythms in sleep and wakefulness, *Advances in Sleep Research,* **1,** Spectrum Publications Inc.

Kruskal, J.B. (1964). Multidimensional scaling by optimizing goodness of fit to a nonmetric hypothesis, *Psychometrika,* **29,** 1-27. *And* Nonmetric multidimensional scaling: a numerical method, *Psychometrika,* **29,** 115-130.

Kummell, F. (1968). Time as succession and the problem of duration, in, Fraser, J.T. (Ed.), 31-55.

Kutter, E. (1973). Extension of human constructs and related problems. *Urban Studies,* **42,** 23-38.

Leach, E.R. (1964). *Political Systems of Highland Burma: a Study of Kachin Social Structure,* Boston, Beacon Press (Cited by Kolaja, 1969).

Lee, R.B. (1969). Kung bushmen subsistence: an input-output analysis, in Vayda, A.P. (Ed.) *Environment and Cultural Behaviour: Ecological Studies in Cultural Anthropology,* New York, Natural History Press, 47-79.

Lee, G.S. (1945). *George Herbert Mead: Philosopher of the Social Individual,* New York, King's Crown Press. (Cited in Chapin 1974).

Le Goff, J. (1978). *Pour un Autre Moyen Age: Temps, Travail et Culture en Occident,* Paris, Gallimard.

Lenntorp, B. (1976a). *Paths in Space-Time Environments: a Time-Geographic Study of Movement Possibilities of Individuals,* Meddelanden, Fran Lunds Universitets, Geografiska Insitution.

Lenntorp, B. (1976b). A time-space structured study of the travel possibilities of the public transport passenger, *Rapporter och Notiser 24,* Department of Geography, University of Lund.

Lenntorp, B. (1978). A time-geographic simulation model of individual activity programmes, in Carlstein, T., Parkes, D.N., and Thrift, N.J., (Eds.) *Vol.II, Human Activity and Time Geography.* London, Edward Arnold, 162-180.

Le Shan, L.L. (1952). Time orientation and social class, *Journal of Abnormal Psychology,* **47,** 589-92.

Levi-Strauss, C. (1963). *Structural Anthropology,* New York, Basic Books.

Levy, L.H. and Dugan, R.D. (1966). A factorial study of personal constructs, *Journal of Consulting Psychology,* **20**, (1), 53-57.

Lewin, K. (1935). *A Dynamic Theory of Personality,* New York, McGraw Hill (Cited in Frank 1939).

Lewin, K. (1936). *Principles of Topological Psychology,* New York, McGraw Hill (Cited in Frank 1939).

Lewin, K. (1964). *Field Theory in Social Sciences: Selected Theoretical Papers,* Cartwright, D. (Ed.), New York, Harper Torchbooks.

Lewis, J.P. (1965). *Building Cycles and Britain's Growth,* London, Macmillan.

Lewis, L.T. and Alford, J.J. (1975). The influence of season on assault, *Professional Geographer,* **27**, 214-217.

Linder, S.B. (1970). *The Harried Leisure Class,* New York, Columbia University Press.

Lingoes, J.C. (1965-1966). An IBM-7090 program for Guttman-Lingoes smallest space analysis. Vol.10, 1965, 183-184, p.487. Vol.11, 1966, 75-76.

Lucas, J.R. (1973). *A Treatise on Time and Space,* London, Methuen.

Luce, G.G. (1972). *Body Time: the Natural Rhythms of the Body,* London, Maurice Temple Smith.
See also (1970) *Biological Rhythms in Psychiatry and Medicine,* Maryland, National Institute of Mental Health.

Lundberg, G., Komarovski, M. and McInerny, (1934). *Leisure: a Suburban Study,* New York, Columbia University Press.

Lundeen, R. (1972). The semantic differential technique and personal construct theory in image measurement, Geography Department, *York University, Discussion Paper Series,* No.5.

Lynch, K. (1960). *The Image of the City,* Cambridge, Mass., MIT Press.

Lynch, K. (1972). *What Time is this Place?,* Cambridge, Mass., MIT Press.

Lynch, K. (1976). *Managing the Sense of a Region,* Cambridge, Mass., MIT Press.

McCormick, T.C. (1939). Quantitative analysis and comparison of living cultures, *American Sociological Review,* **4**, 463-474.

Mandel, E. (1975). *Late Capitalism,* London, New Left Books.

Mandel, E. (1978). *The Second Slump: A Marxist Analysis of Recession in the Seventies,* London, New Left Books.

Marshall, A. (1898). *The Principles of Economics,* London, Macmillan.

Mårtensson, S. (1975). Time use and social organisation, *Rapporter och Notiser,* **17**, Department of Geography, University of Lund.

Mårtensson, S. (1978). Time allocation and daily living conditions: comparing regions, in, Carlstein, T., Parkes, D.N., and Thrift, N.J. (Eds.) *Volume II, Human Activity and Time Geography,* London, Edward Arnold, 181-97.

Martin, G.J. (1973). *Ellsworth Huntington; his Life and Thought,* Hamden, Connecticut, Shoe String Press.

Martin, R.L. (1978). Kalman filter modelling of time-varying processes in urban and regional analysis, in, Martin, R.L., Thrift, N.J. and Bennett, R.J. (Eds.) *Towards the Dynamic Analysis of Spatial Systems,* London, Pion, 104-26.

Martin, R.L. and Oeppen, J.E. (1975). The identification of regional forecasting models using space-time. *Transactions, Institute of British Geographers,* **66**, 95-118.

Martindale, D. (1960). *The Nature and Types of Sociological Theory,* Boston, Mass., Houghton Mifflin. (Cited by Korenbaum, 1964).

Maslow, A.H. (1970). *Motivation and Personality,* (second edition), New York, Harper and Row.

Matthews, R.C.O. (1969). Postwar business cycles in the United Kingdom, in Bronfenbrenner, M. (Ed.) *Is the Business Cycle Obsolete?* New York, John Wiley, 99-135.

Matore, G. (1966). Existential space, *Landscape,* **15,** 5-6.

Matzner, E. (1974). Approaches to a theory of urban interventionism, in: H. Swain, R.D. Mackinnon (Eds.). *Issues in the management of urban systems,* Papers and Proceedings from an IIASA Conference on National Settlement Systems and Strategies, Schloss, Laxenburg, Austria.

May, J.A. (1970). Kant's concept of geography, *Research Publication No.4, Department of Geography, University of Toronto.*

Mbiti, J. (1968). The African concept of time, *African Theological Journal,* **1,** 8-20.

Mead, W.R. (1958). The seasonal round: a study of adjustment on Finland's pioneer fringe, *Tidschrift voor Economische en Sociale Geografie,* **49,** 157-62.

Meerloo, J.A.M. (1960). The time sense in psychiatry, in Fraser, J.T. (Ed.) 235-252.

Meier, R.L. (1959). Human time allocation as a basis for social accounts, *Journal American Institute of Planners,* 1959(a) Vol.25, 1959, 27-33.

Meier, R.L. (1959). Measuring social and cultural change in urban regions, *Journal of American Institute of Planners,* 1959(b), Vol.2, No.4, 180-190.

Meier, R.L. (1976). A stable urban ecosystem, *Science,* **192,** 962-968.

Melbin, M. (1976). City rhythms, Paper to *The International Society for the Study of Time,* Alpbach, Austria, (mimeo).

Melbin, M. (1977). *Time Territorality,* Department of Sociology, University of Boston. (Mimeo).

Melbin, M. (1978a). The colonisation of time, in, Carlstein, T., Parkes, D.N. and Thrift, N.J. (Eds.) *Vol.2, Human Activity and Time Geography,* London, Edward Arnold, 100-113.

Melbin, M. (1978b). Night as frontier, *American Sociological Review,* **43,** 3-22.

Mayerhoff, H. (1960). *Time in Literature,* Berkeley, University of California Press.

Michelson, W. (1977). *Environmental Choice, Human Behaviour and Residential Satisfaction,* New York, Oxford University press.

Michelson, W. and Reed, P. (1975). The time budget, in Michelson, W. (Ed.) *Behavioural Research Methods in Environmental Design,* Pennsyl., Halsted Press.

Michon, J.A. (1967). Magnitude scaling of short durations with closely spaced stimuli, *Psychonomic Science,* **9,** 359-60. (Cited in Doob 1971).

Michon, J.A. (1972). Processing of temporal information and the cognitive theory of time experience, in, Fraser, J.T., Haber, F.C., and Muller, G.H. (Eds.), 242-258.

Mikesell, M. (1968). Tribal markets in Morocco, *Geographical Review,* **48,** 494-511.

Mitchell, D. (1975). The location and organization of community health facilities: a space-time approach in Newcastle N.S.W., unpublished B.A. Hons. thesis, Department of Geography, Newcastle University, N.S.W.

Mitchell, W.C. (1927). *Business Cycles: the Problem and its Setting,* London, Pitman.

Moore, G.T. and Golledge, R.G. (Eds.) (1976). *Environmental Knowing: Theories, Research and Methods,* Stroudsburg, Penn., Dowden, Hutchinson and Ross, Inc.

Moore, W.E. (1963). *Man, Time, and Society,* New York, John Wiley.

Mountfield, P.R. (1977). The place of time in economic geography. *Geography,* **62,** 268-85.

Mountford, G.P. (1960). Records of the American-Australian Scientific Expedition to Arnhem Land, Vol.2. *Anthropology and Nutrition,* Parkville, Melbourne University Press.

Mumford, L. (1934). *Technics and Civilization,* New York, Harcourt, Brace and World.

Murdie, R.A. (1969). Factorial ecology of metropolitan Toronto, 1951-61, *Department of Geography Research Papers No.116, University of Chicago.*

Nadel, S.F. (1958). *The Theory of Social Structure,* Glencoe, Ill., The Free Press. (Cited in Kolaja, 1969, 39).

Nakanashi, N. (1966). *A report on the "How do people spend their time survey",* Tokyo, NHK Public Opinion Research Institute.

Neisser, U. (1967). *Cognitive Psychology,* New York, Appleton-Century-Crofts. (Cited in Pocock and Hudson, 1978).

Neuberg, V.E. (1977). *Popular Literature: A History and Guide,* Harmondsworth, Middlesex, Penguin.

Newton, P.W. (1972). Residential mobility in Newcastle, N.S.W., *Unpublished M.A. Thesis,* Department of Geography, Newcastle University, N.S.W.

Niedercorn, J.H., and Hearle, E.F.R. (1964). Recent land use trends in 48 large American cities, *Land Economics,* **40**, 105-110.

Norberg-Schulz, C. (1971). *Social Order and the Theory of Class,* New York, Harcourt, Brace and World.

Nordman, C. (1925). *The Tyranny of Time,* London, Fisher Unwin (Cited in Korenbaum, 1964).

Nystuen, J.D. (1963). Identification of some fundamental spatial concepts, in, Berry, B.J.L., and Marble, D.F. (Eds.). *Spatial Analysis,* Englewood Cliffs, Prentice-Hall, 35-41.

Oatley, K. and Goodwin, B.C. (1971). The explanation and investigation of biological rhythms, in, Colquhoun, W.P. (Ed.), 1-38.

O'Neill, W.M. (1975). *Time and the Calendars,* Sydney, Sydney University Press.

Orme, J.E. (1969). *Time, Experience and Behaviour,* London, Illiffe Books.

Orme, J.E. (1978). Time: psychological aspects, in, Carlstein, T., Parkes, D.N., and Thrift, N.J. (Eds.) *Vol.1, Making Sense of Time,* London, Edward Arnold, 66-75.

Ornstein, R.E. (1969). *On the Experience of Time,* Harmondsworth, Middlesex, Penguin.

Ortiz, A. (1972). Ritual, drama and the Pueblo view of the world, in, Schwarz, D.W. (Ed.), *New Perspectives on the Pueblos,* Albuquerque, University of New Mexico Press, 135-299.

Ottensman, J.R. (1972). Systems of urban activities and time: an interpretative review of the literature. *Urban Studies Research Paper,* Centre for Urban and Regional Studies, University of North Carolina, Chapel Hill.

Palm, R. and Pred, A. (1974). A time-geographic perspective on problems of inequality for women, *Working Paper No.26, Institute of Urban and Regional Development, University of California, Berkeley.* Reprinted in Palm, R., and Lanegren, D. (1978). *An Introduction to Human Geography,* second edition.

Parkes, D.N. (1968). Scale analysis of urban structures: Newcastle, N.S.W. *Unpublished M.A. Thesis,* Department of Geography, University of Newcastle, N.S.W.

Parkes, D.N. (1971). A classical social area analysis and some comparisons: Newcastle N.S.W., *Australian Geographer,* **11**, 555-578.

Parkes, D.N. (1971). Urban clocks? Section 21, Geographical Sciences, *Australia and New Zealand Association for the Advancement of Science,* University of Queensland, Brisbane.

Parkes, D.N. (1973). Formal factors in the social geography of an Australian industrial city: Newcastle, N.S.W. Australian *Geographical Studies,* **11,** 171-200.

Parkes, D.N. (1973). Timing the city: a theme for urban environmental planning, *Royal Australian Planning Institute Journal,* **12,** 130-135.

Parkes, D.N. (1974). Themes on time in urban social space: an Australian study. *Seminar series No.26, Department of Geography, University of Newcastle upon Tyne.*

Parkes, D.N. (1975). The Newcastle urban area 1971: a factorial ecology, *Research Papers in Geography, No.3, University of Newcastle, N.S.W.*

Parkes, D.N. (1977). T-P Graphs, space-time and an experimental city: an extension of factorial ecologies, *Geographical Analysis,* **8,** 277-284.

Parkes, D.N. and Thrift, N.J. (1975). Timing space and spacing time, *Environment and Planning A,* **7,** 651-70.

Parkes, D.N. and Taylor, P.J. (1976). Refutation of conjectures and other comments: a reply to Brown and Boddy, *Environment and Planning, A,* **8,** 839-41.

Parkes, D.N. and Thrift, N.J. (1977). Time spacemakers and entrainment, Paper to Anglo-Swedish Symposium in Time-Geography. Annual Conference of the Institute of British Geographers, University of Newcastle upon Tyne, (forthcoming in *Transactions of the Institute of British Geographers* N.S.4. 1979).

Parkes, D.N. and Thrift, N.J. (1978). Putting time in its place, in, Carlstein, T., Parkes, D.N., and Thrift, N.J. (Eds.), *Vol.I Making Sense of Time,* London, Edward Arnold, 119-29.

Parkes, D.N. and Wallis, W.D. (1976). Digraphs and activity sequences, in, Keats, J.A. and Wallis, W.D. (Eds.) *Spatial and temporal models of behaviour,* Proceedings of an International Conference on Mathematical Psychology, Newcastle University, N.S.W., 130-47.

Parkes, D.N. and Wallis, W.D. (1978). Graph theory and the study of activity structure, in, Carlstein, T., Parkes, D.N. and Thrift, N.J. (Eds.) *Vol.II, Human Activity and Time Geography,* London, Edward Arnold, 75-99.

Paulsson, G. (1959). *The Study of Cities: Notes about the Hermeneutics of Urban Space,* Kobenhaven, Munksgaard.

Phythian-Adams, C. (1972). Ceremony and the citizen: the communal year at Coventry, 1450-1550, in, Clark, P. and Slack, P. (Eds.) *Crisis and Order in English Towns, 1500-1700,* London, Routledge and Kegan Paul, 57-85.

Piaget, J. (1968). Time Perception in Children, in, Fraser, J.T. (Ed.) 202-216.

Pieron, H. (1923). Les problemes psychophysiologiques de la perception du temps, *Année Psychologiques* **24,** 1-25. (Cited in Cohen, 1967).

Pittendrigh, C.S. (1960). *Cold Spring Harbor Symposium on Quantitative Biology,* **25,** 159-184.

Pocock, D. (1964). The anthropology of time reckoning, *Contributions to Indian Sociology,* **7,** 18-29.

Pocock, D. and Hudson, R. (1978). *Images of the Urban Environment,* London, Macmillan.

Pred, A. (1973). Urbanisation, domestic planning problems and Swedish geographic research, in, Board, C. et al. (Eds.) *Progress in Geography,* **5,** London, Edward Arnold, 1-77.

Pred, A. (1977). *City Systems in Advanced Economies: Past Growth, Present Processes and Future Development Options,* London, Hutchinson.

Pred, A. (1977). The choreography of existence: comments on Hägerstrand's time-geography and its usefulness, *Economic Geography,* **53,** 207-221.

Pred, A. (1978). The impact of technological and institutional innovations of life content: some time-geographic observations. *Geographical Analysis,* **10,** 345-72.

Prince, H. (1978a). A future for the city's past, *Geographical Magazine,* 529-32.

Prince, H.C. (1978b). Time and historical geography in Carlstein, T., Parkes, D.N. and Thrift, N.J. (Eds.) Volume 1: *Making Sense of Time,* London, Edward Arnold, 17-37.

Radcliffe-Brown, A. (1952). *Structure and Function in Primitive Society,* London, Cohen and West.

Rapoport, A. (1977). *Urban Aspects of Urban Form: Towards a Man-Environment Approach to Urban Form and Design,* Oxford, Pergamon.

Reinberg, A., Gervais, P., Halberg, F., Gaultier, M., Roynette, N., Abulker, C., and Dupont, J. (1973). Rythmes circadiens et circannuels de mortalité des adultes dans un hôpital parisien et en France, *Nouvelle Presse Medicale,* **2,** 289-294 (Cited in Reinberg, A., 1974).

Reinberg, A. (1974). Aspects of circannual rhythms in man, in Pengelley, E.T. (Ed.), *Circannual Clocks: Annual Biological Rhythms,* New York, Academic Press, 423-505.

Relph, E. (1976). *Place and Placelessness,* London, Pion.

Rezsohazy, R. (1972). Methodological aspects of a study about the social notion of time, in Szalai, A. (Ed.), 449-60.

Richardson, H.W. and Aldcroft, D. (1968). *Building in the British Economy Between the Wars,* London, George Allen and Unwin.

Rickert, J.E. (1967). House facades of the northeastern United States: a tool of geographical analysis, *Annals Association of American Geographers,* **57,** 211-38.

Robb, J. (1977). Activity patterns in time and space for Scone, N.S.W. Unpublished B.A. Hons. thesis, Department of Geography, Newcastle University, N.S.W.

Robinson, J.P., Converse, P.E., Szalai, A. (1972). Everyday life in twelve countries, in, Szalai, A. et al. (Eds.). *The Use of Time,* The Hague, Mouton, 113-144.

Robson, B.T. (1973). *Urban Growth: An Approach,* Cambridge, Cambridge University Press.

Ropke, W. (1936). *Crises and Cycles,* London, William Hodge.

Rose, C. (1977). Reflections on the notion of time incorporated in Hägerstrand's time-geographic model of society, *Tidjschrift voor Economische en Sociale Geografie,* **68,** 43-50.

Rostow, W. (1978). *The World Economy. Problems and Prospects,* New York, Macmillan.

Roth, J.A. (1963). *Timetables,* Indianapolis, Bobbs-Merrill.

Ruiz, R.A., Reivich, R.S. and Krauss, H.H. (1967). Tests of temporal perspectives — do they measure the same construct? *Psychological Reports,* **21,** 849-852.

Rummell, R.J. (1970). The dimensionality of nations project, *Research Report No.40, Department of Political Science, University of Hawaii.*

Sahlins, M. (1972). *Stone Age Economics,* London, Tavistock.

Salter, C.L. and Lloyd, W.J. (1977). *Landscape in literature.* Washington, D.C., Association of American Geographers.

Sant, M. (1973). The Geography of Business Cycles, *Geographical Papers No.5, London School of Economics and Political Science.*

Schlessinger, A. (1933). *The Rise of the City: 1878-1895.* New York, Macmillan.

Schumpeter, J.A. (1939). *Business Cycles: a Theoretical, Historical and Statistical Analysis of the Capitalist Process,* New York, McGraw-Hill (2 vols.)

Schutz, A. (1962). *Collected Papers, Volumes I and II,* The Hague, Martinus Nijhoff.

Schutz, A. and Luckmann, T. (1973). *The Structures of the Life-World,* Evanston, Northwestern University Press.

Schwartz, B. (1978). The social ecology of time barriers, *Social Forces,* **56,** 1203-20.

Seamon, D. (1978). Letter to the editor. *Environment and Planning A,* **10,** 729-30.

Seamon, D. (1979). *A Geography of the Life World: Movement, Rest and Encounter,* London, Croom Helm.

Sechrest, L.B. (1963). The psychology of personal constructs: G.A. Kelly, in, Wepman, J.M. and Heine, R.W. (Eds.) *Concepts of Personality,* Chicago, Aldine, 206-233.

Seeley, J.R., Sim, R.A. and Loosley, E.W. (1963). *Crestwood Heights,* New York, John Wiley.

Shackle, G.L.S. (1972). *Epistemics and Economics: a Critique of Economic Doctrines,* Cambridge, Cambridge University Press.

Shackle, G.L.S. (1978). Time, choice and history, in, Carlstein, T., Parkes, D.N. and Thrift, N.J. (Eds.) *Vol.1, Making Sense of Time,* London, Edward Arnold, 47-55.

Shapcott, M. and Steadman, P. (1978). Rhythms of urban activity, in Carlstein, T., Parkes, D.N. and Thrift, N.J. (Eds.) Vol.2, *Human Activity and Time Geography,* London, Edward Arnold.

Shapcott, M. and Wilson, C. (1976). Correlations among time uses, in *Transactions of the Martin Centre for Architectural and Urban Studies,* **1,** 73-94.

Sherover, C.M. (Ed.). *The human experience of time,* New York, New York University Press.

Shevky, E. and Bell, W. (1955). *Social Area Analysis: Theory, Illustrative Application and Computational Procedures,* Stanford, Stanford University Press.

Shevky, E. and Williams, M. (1949). *The Social Areas of Los Angeles,* Los Angeles, University of California Press.

Skinner, B.F. (1971). *Beyond Freedom and Dignity,* New York, Alfred A. Knopf.

Smith, D.M. (1977). *Human Geography: a Welfare Approach,* London, Edward Arnold.

Smith, N.K. (1918). *Immanuel Kant's Critique of Pure Reason,* London, Macmillan, 2nd edition 1921.

Smolensky, M., Halberg, F. and Sargent, F. (1972). Chronobiology of the life sequence, in, Ito, S., Ogata, K. and Yoshimura, H., (Eds.) *Advances in Climatic Physiology,* Tokyo, Igaku Shoin.

Sollberger, A. (1965). *Biological Rhythm Research,* New York, Elsevier Publishing Company.

Sorokin, P.A. (1943). *Sociocultural Causality, Space and Time,* Durham, Duke University Press.

Sorokin, P.A. and Berger, C.Q. (1939). (a) *Time Budgets of Human Behaviour,* Harvard Sociological Studies Vol.2, Cambridge, Mass., Harvard University Press.

Sorokin, P. and Merton, R.K. (1937). Social time: a methodological and functional analysis, *American Journal of Sociology,* **42,** 615-29.

Soule, G. (1955). *Time for Living,* New York, Viking Press.

Stephens, J.D. (1976). Daily activity sequences and time-space constraints, in Holly, B.P. (Ed.) *Time-Space Budgets and Urban Research: a Symposium,* Department of Geography, Kent State University, Discussion Paper 1, 21-69.

Stone, P.J., Dunphy, D.C., Smith, M.S., Ogilvie, D.M. *et al.* (1966), *The General Inquirer,* Cambridge, Mass., MIT Press.

Stone, P.J. (1972). Models of everyday time use, in Szalai, A. et al. (Ed.) 179-89.

Strachan, A.J. (1978). The Sunday market in Scotland: a case study of Ingliston, *Scottish Geographical Magazine, 94,* 48-58.

Strughold, H. (1971). *Your Body Clock: Its Significance for the Jet Traveller,* London, Angus & Robertson (UK.)

Szalai, A. (1964). Differential work and leisure time budgets, *The New Hungarian Quarterly, 5,* 105-119.

Szalai, A. (1966). Trends in comparative time budget research, *The American Behavioural Scientist, 9,* 3-8.
See also, Differential evaluation of time-budgets for comparative purposes, in Merrett, R.L. and Rokkan, S. (Eds.) *Comparing Nations, the Use of Quantitative Data in Cross-National Research,* New Haven, Conn., Yale University Press, 239-258.

Szalai, A. *et al.* (Eds.) 1972, *Use of Time,* Hague, Mouton.

Taylor, P.J. and Parkes, D.N. (1975). A Kantian view of the city: a factor ecological experiment in space and time, *Environment and Planning, A, 7,* 671-688.

The Economist, (1977, October 22). Time colonization and the structure of the British Parliament, 1977., also 'Early to bed, its a flop'.

Theodorson, G.A. (Ed.) (1961). *Studies in Human Ecology,* Evanston, Ill, Row Peterson.

Thomas B. (1954), *Migration and Economic Growth,* Cambridge, Cambridge University Press, and second edition (1974).

Thomas, B. (1972). *Migration and Urban Development, a Reappraisal of British and American Long Cycles,* London, Methuen.

Thomas, D. (1977). A time-geographic approach to the evaluation of the impact of 'Summer-Time' on the biographies of farmers, Unpublished undergraduate exercise, Department of Geography, Newcastle University, Australia.

Thomas, K. (1971). *Religion and the Decline of Magic: Studies in Popular Beliefs in Sixteenth and Seventeenth Century* England, London, Weidenfeld, and Nicholson.

Thompson, E.P. (1967). Time, work-discipline and industrial capitalism, *Past and Present, 38,* 56-97. Reprinted in, Flinn, M.W., and Smout, T.C. (1974) (Eds.). *Essays in Social History,* Oxford, Clarendon Press, 39-77.

Thor, D.H. (1962). Diurnal variability in time estimation, *Perception and Motor Skills, 15,* 451-454.

Thornes, J.B. and Brunsden, D. (1977). *Geomorphology and Time,* London, Methuen.

Thrift, N.J. (1977a). Time and theory in human geography Part I *Progress in Human Geography, 1,* 65-101. and Part II, *Progress in Human Geography, 1,* 413-457.

Thrift, N.J. (1977b). The diffusion of Greenwich Mean Time: an essay in social and economic history. *School of Geography, University of Leeds, Working Paper 192.*

Thrift, N.J. (1977c). An introduction to time-geography. *Concepts and Techniques in Modern Geography, 13.* Norwich: Geo-Abstracts.

Thrift, N.J. (1978). Letter to the editor *Environment and Planning A, 10,* 347-9.

Timms, D.W.G. (1971). *The Urban Mosaic: Towards a Theory of Residential Structure,* Cambridge, Cambridge University Press.

Timms, D.W.G. (1976). Social bases to social areas, in, Herbert, D.T. and Johnston, R.J. (Eds.) *Vol.1, Spatial Processes and Form,* London, John Wiley, 19-40.

Tinkler, K.J. (1977). An introduction to graph theory. *Concepts and Techniques in Modern Geography* **14,** Norwich: Geo-Abstracts.

Tivers, J. (1977). Constraints on spatial activity patterns: women with young children, *Occasional Paper No.6,* Department of Geography, King's College, University of London.

Tomlinson, J., Bullock, N., Dickens, P., Steadman, P. and Taylor, E. (1973). A model of students' daily activity patterns. *Environment and Planning,* **5,** 231-66.

Tranter, P. and Parkes, D.N. (1977). Time and images in urban space, *Area,* **11,** 115-120.

Tranter, P. (1978). Urban social pathologies in space and time, (Ph.D. thesis in progress) Department of Geography, Newcastle University, Australia.

Tuan, Yi-Fu, (1973). Space and place: humanistic perspective. *Progress in Geography,* **6,** 211-52.

Tuan, Yi-Fu, (1974). *Topophilia: A Study of Environmental Perception, Attitudes and Values,* Englewood Cliffs, Prentice-Hall.

Tuan, Yi-Fu, (1975). Images and mental maps, *Annals Association of American Geographers,* Vol.65, 205-213.

Tuan, Yi-Fu, (1975). Notions of place, Working paper to a workshop at Department of Human Geography, The Australian National University, Canberra, Australia, Aug 21-23, 1975.

Tuan, Yi-Fu, (1977). *Space and Place: the Perspective of Experience,* London, Edward Arnold.

Tuan, Yi-Fu, (1978). Space, time, place: a humanistic frame, in Carlstein, T., Parkes, D.N. and Thrift, N.J. (Eds.) *Vol.I, Making Sense of Time,* London, Edward Arnold, 7-16.

Tuan, Yi-Fu, (1978). The City: its distance from Nature. *Professional Geographer,* **58,** 157-164.

Turnbull, C.M. (1961). *The Forest People,* London, Jonathan Cape.

Ullman, E. (1974). Space and/or time: opportunity for substitution and prediction, *Transactions Institute of British Geographers,* **63,** 125-139.

van Passen, C. (1976). Human geography in terms of existential anthropology, *Tidjschrift voor Economische en Sociale Geografie,* **67,** 324-31.

Veldman, D.J. (1967). *Fortran Programming for the Behavioural Sciences,* New York, Holt, Reinhart, and Winston.

von Wiesse, L. (1924). Allegemeine Soziologie (Vol.1), Munchen, von Dunker and Humboldt, (Quoted in Kolaja, 1969).

Vipond, M.J. (1969). Fluctuations in private housebuilding in Great Britain, 1950-1966, *Scottish Journal of Political Economy,* 196-211.

Waldo, D. (Ed.) (1970). *Temporal Dimensions of Development Administration,* Durham, Duke University Press.

Ward, F.A.B. (1961). How time keeping machines became accurate, *The Chartered Mechanical Engineer,* **8,** 604.

Ward, R.R. (Ed.) (1972). *Living Clocks,* London, Collins.

Ware, J. (1977). The use of time budgets in determining individuals allocation of time, *Discussion Paper, Department of Economics, Griffith University, Queensland, Australia* (Mimeo).

Webb, W.B. (1971). Sleep behaviour as a biorhythm, in Colquhoun, W.P. (Ed.) 149-178.

Webber, M.M. (1964). The urban place and the non-place urban realm, in Webber *et al., Explorations into Urban Structure,* Philadelphia, University of Pennsylvania Press, 79-153.

White, D. (1978). What do people do when they're not working? *New Society,* **46,** 84617, 683-5.

White, L.A. (1960). The world of the Keresan Pueblo Indian, in Diamond, S. (Ed.) *Primitive Views of the World: Essays from Culture in History,* New York, Clarendon Press.

510

Whitehand, J.W.R. (1967). Fringe belts: a neglected aspect of urban geography, *Transactions Institute of British Geographers,* **41,** 223-33.

Whitehand, J.W.R. (1967). The settlement morphology of London's cocktail belt, *Tidjschrift voor Economische en Sociale Geografie,* **58,** 20-7.

Whitehand, J.W.R. (1972a). Building cycles and the spatial pattern of economic growth, *Transactions Institute of British Geographers,* **56,** 39-55.

Whitehand, J.W.R. (1972b). Urban rent theory, time series and morphogenesis: an example of eclecticism in geographical research, *Area,* **4,** 215-22.

Whitehand, J.W.R. (1974). The changing nature of the urban fringe: a time perspective, in Johnson, J.H. (Ed.), *Suburban Growth: Geographical Processes at the Edge of the Western City,* London, Longman, 48-9.

Whitehand, J.W.R. (1975). Building activity and intensity of development at the urban fringe: the case of a London suburb in the nineteenth century, *Journal of Historical Geography,* **1,** 211-24.

Whitehand, J.W.R. (1977). The basis for a historico-geographical theory of urban form, *Transactions, Institute of British Geographers,* N.S.2, 400-16.

Whitehead, A.N. (1920). *The Concept of Nature,* Cambridge, Cambridge University Press.

Whitelaw, J.S. (1972). Scale and urban migrant behaviour, *Australian Geographical Studies,* **10,** 101-106.

Whitrow, G.J. (1962). *The Natural Philosophy of Time,* Oxford University Press.

Whitrow, G.J. (1971). *The Natural Philosophy of Time,* London, Nelson.

Whitrow, G.J. (1972). *What is Time?* London, Thames and Hudson.

Whittlesey, D. (1945). The horizon of geography, *Annals Association of American Geographers,* **35,** 1-36.

Whorf, B.L. (1956). *Language, Thought and Reality,* Cambridge, Mass, MIT Press.

Whorf, B.L. (1968). An American Indian model of the universe, in, Gale, R.M. (Ed.) *The Philosophy of Time,* London, Macmillan.

Williams, R. (1973). *The Country and the City,* London, Chatto and Windus.

Wilson, A.G. (1974). *Urban and Regional Models in Geography and Planning,* Chichester, John Wiley.

Wirth, L. (1938). Urbanism as a way of life, *American Journal of Sociology,* **44,** 1-24.

Yaker, H., Osmond, H., Cheek, F. (Eds.) (1972). *The Future of Time,* London, Hogarth Press.

Yeung, Yue-Man (1974). Periodic markets: comments on spatial-temporal relationships, *Professional Geographer,* **26,** 148, 149.

Young, M. and Ziman, J. (1971). Cycles in social behviour, *Nature,* **229,** 91-95.

Young, M. and Willmott, P. (1973). *The Symmetrical Family,* London, Routledge and Kegan Paul.

Zerubavel, E. (1976). Timetables and scheduling: on the sociological organization of time, *Sociological Inquiry,* **46,** 87-94.

Zerubavel, E. (1977). The French republican calendar: a case study in the sociology of time. *American Sociological Review,* **42,** 868-77.

Author Index

The letters ff indicate that the author is mentioned again on one or both of the next two pages.

512

514

Subject Index

The letters ff indicate that the subject is mentioned again on one or both of the next two pages.